Practical Notions on Fish Health and Production

Edited By:

Maria Manuela Castilho Monteiro de Oliveira

Laboratory of Microbiology and Immunology, Avenida da Universidade Técnica
Lisbon, Portugal

Joana Isabel Espírito Santo Robalo

Eco-Ethology Research Unit, ISPA-IU R. Jardim do Tabaco, 34, 1149-041
Lisbon, Portugal

&

Fernando Manuel D'Almeida Bernardo

Faculdade de Medicina Veterinária, Universidade de Lisboa
Avenida da Universidade Técnica
1300-477, Lisboa, Portugal

advertisements or ideas contained in the Work.

Limitation of Liability:

In no event will Bentham Science Publishers, its staff, editors and/or authors, be liable for any damages, including, without limitation, special, incidental and/or consequential damages and/or damages for lost data and/or profits arising out of (whether directly or indirectly) the use or inability to use the Work. The entire liability of Bentham Science Publishers shall be limited to the amount actually paid by you for the Work.

General:

1. Any dispute or claim arising out of or in connection with this License Agreement or the Work (including non-contractual disputes or claims) will be governed by and construed in accordance with the laws of the U.A.E. as applied in the Emirate of Dubai. Each party agrees that the courts of the Emirate of Dubai shall have exclusive jurisdiction to settle any dispute or claim arising out of or in connection with this License Agreement or the Work (including non-contractual disputes or claims).
2. Your rights under this License Agreement will automatically terminate without notice and without the need for a court order if at any point you breach any terms of this License Agreement. In no event will any delay or failure by Bentham Science Publishers in enforcing your compliance with this License Agreement constitute a waiver of any of its rights.
3. You acknowledge that you have read this License Agreement, and agree to be bound by its terms and conditions. To the extent that any other terms and conditions presented on any website of Bentham Science Publishers conflict with, or are inconsistent with, the terms and conditions set out in this License Agreement, you acknowledge that the terms and conditions set out in this License Agreement shall prevail.

Bentham Science Publishers Ltd.
Executive Suite Y - 2
PO Box 7917, Saif Zone
Sharjah, U.A.E.
Email: subscriptions@benthamscience.org

BENTHAM SCIENCE

CONTENTS

Maria João Fraqueza and *Manuel Abreu Dias*

FOREWORD

Fish health, both of farmed and wild productions, is requesting continuous upgrading not only to avoid significant losses but also to facilitate seafood security and safety. The work in this field is complex, as many different factors need to be considered in relation. Preventive fish health work has to be based on knowledge, not limited to the fish, but relating to all aspects of the conditions under which the fish exist, is traded and processed. Although important literature has been accumulated on this area, there is undoubtedly a need to have contemporary and practical knowledge in this respect summarized in one volume.

This difficult work was undertaken by well-known professors from the Faculty of Veterinary Medicine of the University of Lisbon, authorities on the covered subjects. In this practical, but very comprehensive book, the authors envelop three sections encompassing the general biology, sensible concepts on fish health and useful notions on fish production. The first section includes a brief review on particular features of fish anatomy and physiology as well as the embryonic and larval development. The second section comprises of three chapters devoted to aquaculture and diagnosis practices, infectious and parasitic diseases of fish, and anesthesia and surgery in fish. The last section contains relevant information on the fishery production, catch, trade and processing and the official veterinary inspection practices.

The combined issues offered in this book, well written and well arranged, confirming thus the authors´ rich experience, make this publication an excellent modern key work. Certainly it will have a great impact on the enhancement of fish health and production, allowing also a better understanding of fish health implications on the aquatic systems and furthermore contributing for seafood processing, security, quality and safety. This volume will be undoubtedly appreciated as an important source of information for scientists, university students, authorities, fish farmers, fishermen and fish processors.

The book as a whole is of a high standard, and I am convinced that the readers will find it very valuable for their professional and research activities and a significant contribution for understanding the fascinating complexity of the fauna of our planet. Both the authors and the editor are to be congratulated on this publication.

Maria Leonor Nunes
Principal Scientist
Coordinator of Aquaculture and Seafood Valorization
Portuguese Institute of Sea and Atmosphere
Lisbon, Portugal

PREFACE

Industrial fishing activity is limiting natural fish populations and aquaculture seems to be the only sustainable alternative.

Fish stock management in the wild implies the monitoring of several ecological parameters concerning the evolution of fish population, namely mortality, growth, availability of feed, predatory acts and climate changes. One of the most predictable consequences of global warming is ocean water acidification due to the increasing CO_2 concentration in the atmosphere. This achievement may have a significant impact on trophic chains affecting namely synthesis of exoskeleton of marine zooplankton based on $CaCO_3$.

Aquaculture productivity is conditioned by several factors, including feed and health management, species behavior, reproduction requirements and water characteristics in captivity. Inadequate management procedures may have significant impact on fish health and welfare and subsequently lead to severe economic losses for producers and regular market supply.

In fact, these activities are inter-dependent, since for instance the feed management in aquaculture is dependent on fish capture from the wild. All is connected and related, requiring contributes of different levels of interdisciplinary knowledge for an adequate management including contributions of politicians, judicial authorities, biologists, veterinarians, public health entities, producers and consumers.

In terms of aquatic productivity, interaction between veterinarians and biologists is crucial. Practitioners and researchers of both areas must cooperate, share knowledge, and develop complementary research activities.

This text book aims to contribute to an integrated approach concerning practical notions of ichthyology, fish health and production systems. It is an attempt to contribute to fish related research in biology and veterinary sciences.

The book is organized in three sections. The first includes two reviews on general aspects of fish biology and development and the second included three chapters on fish health related subjects. The remaining three chapters are included in the last section, devoted to fish production aiming food chain supply.

We would like to thank all authors who have contributed to this book and all people who somehow helped us to bring it to daylight, including our family, friends, colleagues and students.

Unfortunately, during the drafting process of this book, two of our major contributors passed away, rendering additional challenges to text edition. However, a comprehensive attitude from our publishers allowed us to overcome the moments of discouragement. For that we would like to express our deepest gratitude.

Professor Cristina Vilela (1958-2013) was a Full Professor with Tenure and the Vice-Dean of the Faculty of Veterinary Medicine of the University of Lisbon. Her main research areas were Clinical Microbiology and Mucosal Immunology, but as a devoted scientist she was also interested on several subjects related to Veterinary Sciences, namely on fish infectious diseases. Her dedication to science and teaching was exceptional, and she will be deeply missed.

Professor VítorAlmada (1950-2013) was a Full Professor with Tenure at ISPA-IU in Lisbon. He started his research activity studying the behavior of littoral fishes, but soon started to spread his research to the ecology, genetics and biogeography of both freshwater and marine fishes. He was a devoted researcher and teacher and also an enthusiastic naturalist. Always happy to build bridges between sciences and research topics, he is, and certainly will be, a reference to present and future fish researchers. His work will last in the form of published papers and in the research of his students. We will always miss him.

Both were wonderful scientists with long and productive carriers. But most of all they were dear friends so this book is dedicated to their loving memory.

Manuela Oliveira
Laboratory of Microbiology and Immunology
Avenida da Universidade Técnica
Lisbon, Portugal

Joana I. Robalo
Eco-Ethology Research Unit
ISPA-IU R. Jardim do Tabaco
34, 1149-041, Lisbon
Portugal
&
Fernando Bernardo
Faculdade de Medicina Veterinária
Universidade de Lisboa, Avenida da Universidade Técnica
1300-477, Lisboa
Portugal

List of Contributors

Ana Faria — Eco-Ethology Research Unit, ISPA — Instituto Universitário, R. Jardim do Tabaco 34, 1149-041, Lisboa, Portugal

António Pedro Correia Margarido — Direção Geral de Alimentação e Veterinária, Largo da Academia Nacional de Belas Artes 2, 1249, Lisboa, Portugal

Cláudia Faria — Instituto de Educação da Universidade de Lisboa, Alameda da Universidade, 1649-013, Lisboa, Portugal

Fernando Bernardo — Faculdade de Medicina Veterinária, Universidade de Lisboa, Avenida da Universidade Técnica, 1300-477, Lisboa, Portugal

João Afonso — Centro de Investigação Interdisciplinar em Sanidade Animal, Faculdade de Medicina Veterinária, Universidade de Lisboa, Avenida da Universidade Técnica, 1300-477, Lisboa, Portugal

Manuel Abreu Dias — Alicontrol, Tecnologia e Controlo de Alimentos, Lda., Rua Fernando Vaz, lote 26-B, 1750 108, Lisbon, Portugal

Maria Gabriela Lopes Veloso — Faculdade de Medicina Veterinária/Universidade de Lisboa, Avenida da Universidade Técnica, 1300-477, Lisboa, Portugal

Maria João Fraqueza — CIISA/Faculdade de Medicina Veterinária, Universidade de Lisboa, Avenida da Universidade Técnica, 1300-477, Lisboa, Portugal

Miguel José Sardinha de Oliveira Cardo — Direção Geral de Alimentação e Veterinária, Largo da Academia Nacional de Belas Artes 2, 1249, Lisboa, Portugal
Faculdade de Medicina Veterinária/Universidade de Lisboa, Avenida da Universidade Técnica, 1300-477, Lisboa, Portugal

Nuno Pereira — Oceanário de Lisboa, Esplanada D. Carlos I, 1990-005, Lisboa, Portugal
Faculty of Veterinary Medicine, Universidade Lusófona de Humanidades e Tecnologias, Lisboa, Portugal
ISPA, Instituto Universitário, Ciências Socias, Psicológicas e da Vida, Lisboa, Portugal
IGC, Instituto Gulbenkian de Ciência, Oeiras, Portugal

Paula Cruz e Silva — Direção de Serviço De Alimentação e Veterinária, Avenida do Mar e das Comunidades Madeirenses, n° 23, 2° andar9000-054 Funchal, Portugal

Rita Borges — Centre of Marine Sciences, CCMAR, University of Algarve, Campus de Gambelas, 8005-139, Faro, Portugal
Eco-Ethology Research Unit, ISPA — Instituto Universitário, R. Jardim do Tabaco 34, 1149-041, Lisboa, Portugal

Practical Notions on Fish Health and Production

2

Practical Notions on Fish Health and Production

Editors: Joana Isabel Espírito Santo Robalo, Ana Cristina Gaspar Nunes Lobo Vilela & Maria Manuela Castilho Monteiro de Oliveira

ISBN (eBook):978-1-68108-267-7

ISBN (Print): 978-1-68108-268-4

First published in 2016.

CHAPTER 1

Particular Features of Fish Anatomy and Physiology – A Brief Review

João Afonso[*]

Centro de Investigação Interdisciplinar em Sanidade Animal, Faculdade de Medicina Veterinária, Universidade de Lisboa, Avenida da Universidade Técnica, 1300-477 Lisboa, Portugal

Abstract: Besides basic anatomical and physiological features common to all fishes, due to the aquatic environment they share, there are some very significant differences. Such differences range from major ones common to all species within each main group, namely cyclostomes, chondrichthyans and osteichthyans, to more limited ones between different species within each group. Even the latter can be very significant, conditioning, for instance, the feeding behavior/strategy of a given species or its reproductive ability. Some of the main anatomical and physiological features of the different fishes are briefly reviewed, since a good knowledge of such features is crucial to understand the behavior of each species in the wild and/or to assure the most correct management of its populations either in captivity or in the wild.

Keywords: Anatomy, Chondrichthyans, Cyclostomes, Osteichthyans, Physiology.

INTRODUCTION

Fishes are aquatic, vertebrate animals, with fins as appendages and gills that allow them to breathe by absorbing water oxygen [1]. They are also ectothermic (or "cold-blooded") animals [1 - 3] (although some fishes, such as tunas, can keep their body temperature some degrees above the temperature of the surrounding water, the former still varies as the latter varies) [4].

While all fishes are very well adapted to the aquatic environment where they live,

---

[*] **Address correspondence to João Afonso:** Faculdade de Medicina Veterinária, Universidade de Lisboa, 1300-44 Lisboa, Portugal; E-mail: jafonso@fmv.ulisboa.pt

Manuela Oliveira, Fernando Bernardo, Joana I. Robalo (Eds.)

there are some very significant differences that justify the division of modern fishes into cyclostomes (hagfishes and lampreys), chondrichthyans (or cartilage fishes) and osteichthyans (or bony fishes). Even within any of these groups there is considerable variability between different species, concerning both external and internal features.

EXTERNAL FEATURES

Body Shape

While the general body structure of a fish is designed for ease of movement in water, fish show a large variety of shapes and sizes, depending on their way of life and on their specific habitat [3, 5]:

- Fish living near the surface usually have a long and fusiform shape, adequate to swim quickly for considerable periods of the time or to allow big bursts of speed.
- Mid-water fish are usually laterally compressed, for easy movement through aquatic vegetation and crevices of rocks and reef where they hide from predators or forage (*e.g.*, angelfish).
- Benthic or bottom-dwelling fish are usually flattened from top to bottom, to conform to the bottom where they live (*e.g.*, rays).

Fins

Fish have fins which they use to maintain position, help balance, move, steer and stop.

Some fins are paired, corresponding, in some manner, to the pectoral and pelvic limbs of mammals – these are the pectoral and pelvic fins, supported by the pectoral and pelvic girdles and placed on both sides of the body. Other fins, such as dorsal, caudal (or tail) and anal (or ventral) fins, are unpaired, being placed in a median position.

In most fishes, pectoral fins are placed just behind the operculum and are used essentially to help a fish to turn, climb or dive, or stop, but may have other functions, such as a propelling one, either helping in swimming (*e.g.*, rays) or, for instance in the case of some bottom-dwelling species (*e.g.*, frogfishes), helping the fish to move around over the surface where it lives [2, 3, 6 - 8]. These fins may also present touch receptors or spines [5, 9]. In rays, the pectoral fins are expanded and connected to the sides of the head, being responsible for the flattened body of these fishes [10, 11].

In osteichthyans the pectoral girdle consists of two sets of endochondral (coracoid, scapula and four radials) and dermal bones (post-temporal, supracleitrum, cleithrum and postcleithrum) articulated with the neurocranium [3, 8, 10 - 12].

In chondrichthyans the pectoral girdle consists of a U-shaped coracoscapular cartilage – the paired coracoid parts are fused ventrally while the scapular parts form the extremities which are projected dorsally on each side [8, 10]. In sharks this cartilage is not connected to the axial skeleton [5], but in rays the scapular parts are connected to the vertebrae [8].

Pelvic fins are placed in the ventral region of the body, in any position cranial to the anal fin (eventually even ahead of the pectoral fins, as in the cods), and add stability in swimming [2, 3]. In some fishes, they are also used to reduce their speed. Some fishes present modified pelvic fins, such as thread-like appendices with a tactile function (*e.g.*, gouramis) or a disc-like sucker structure (*e.g.*, gobies), for instance. Male chondrichthyans have modified pelvic fins, called claspers, which are used as intromittent organs for internal fertilization [13, 14].

The pelvic girdle is much less developed than the pectoral girdle, namely in osteichthyans, and not connected to the vertebral column [11].

In osteichthyans the pelvic girdle consists of two endochondral bones called basipterygia [12], which may be separated or fused, while in chondrichthyans the pelvic girdle is just a cartilaginous bar, the ischiopubic bar, placed transversely on the caudoventral part of the trunk [4, 11].

Cyclostomes have no pectoral or pelvic girdles, nor paired fins [3, 7, 12, 15]. In fact, hagfishes do not have any true fins, but just a rudimentary caudal fin consisting of a skin fold that runs around the caudal end of the body and extends forward, dorsally and ventrally [9].

The dorsal fin is placed on the dorsal surface or border (depending on the body shape) of the fish body and can be divided in two or three parts. It adds stability in swimming and may help the fish to turn or stop suddenly [2, 3]. Some fishes (*e.g.*, eels) present the dorsal fin in continuity with the caudal and anal fins [8]. Differently from actinopterygians, sarcopterygians have two dorsal fins with separated bases. Lampreys also have two separated dorsal fins [3, 12].

The caudal fin is attached to the constricted caudal part of the body, called caudal peduncle [3, 12]. It is the main fin used for propulsion [2].

Depending on the relation between the caudal fin and the vertebral column (hypural join), there are different types of caudal fins [3, 8, 11, 12]:

- Heterocercal – the vertebral column extends into the superior lobe of the tail, which is longer than the inferior lobe (*e.g.*, in sharks).
- Diphycercal – the vertebral column extends horizontally until the tip of the tail, which presents symmetrical superior and inferior lobes (*e.g.*, in lungfishes, lampreys and coelacanth).
- Homocercal – the vertebral column extends slightly into the superior lobe of the tail, but externally the caudal fin looks symmetrical or almost symmetrical; this is the case in most ray-finned fish.

In rays the caudal fin can be small (placed at end of the tail) or absent [5, 11].

The anal fin is placed ventrally, between the anus and the caudal fin, and adds stability in swimming [2, 3].

Within the osteichthyans, the sarcopterygians (*e.g.*, coelacanths and lungfishes) present fleshy lobed fins, while the actinopterygians (or ray-finned fishes) present fins that consist essentially of skin covered rays.

The sarcopterygians' lobed fins are supported by an endoskeleton of bones articulated in succession along each fin, being attached to the body just by one bone [10]. The muscles that make the fin move are connected to that endoskeleton, being responsible for the lobed aspect of the fin [12].

In osteichthyans a fin can present either spiny rays (or simply "spines") or soft rays, or both types. In the latter case, the spines are always placed cranially to the soft rays and the two types can be in perfect continuity or give the idea that there are two different fins, one just behind the other [12]. In contrast to spines, rays are segmented and can be branched. Sharp spines, sometimes associated with venom glands [16], can be used as defense weapons (*e.g.*, catfish leading spines of dorsal and pectoral fins). Each ray or spine of a dorsal or anal fin is attached to a bone named pterygiophore [8, 12]. In fact there are two or three of these in succession – proximal, middle, and distal pterygiophores, the distal being the one that articulates with the ray or spine [3, 8]. The proximal pterygiophores are the longest and closest to the vertebrae and, eventually, may be fused to them, but in general are just embedded in the myomeres, between the neural spines (in the case of the dorsal fin) or the haemal spines (in the case of the anal fin) of successive vertebrae [3, 17].

Chondrichthyans have firm dorsal fins, with soft and unsegmented rays supported by basal cartilages resting on consecutive neural spines [8]. Some sharks have dorsal fin spines [12].

Some fishes (*e.g.*, salmon) present a little and fleshy fin, without rays, placed behind the dorsal fin – it is called adipose fin and has no clear function [3, 8, 17].

Body Covering

Most fishes have their skin covered with scales, to protect the body, and all fishes produce mucus that covers their bodies. This mucus provides additional protection against bacteria and fungi, and enables the fish to swim faster [2, 5, 17]. Particularly well known for their ability to produce large amounts of mucus when they are somehow disturbed are the hagfishes, which have mucus pores in two ventrolateral series along their body [3, 17, 18].

Fish scales are essentially made of connective tissue coated with calcium, but there are different types.

Chondrichthyans have placoid scales [7, 11, 12], which have a quadrangular base with a caudally projected spine, and are quite similar to teeth (around a soft center, with nerves and blood vessels, there is a dentine layer coated with a layer of an enamel-like matter commonly called vitrodentine), reason why they are commonly called dermal denticles [3, 5, 10, 16, 17]. These scales do not overlap and, unlike other scales, they do not grow with the body – as the animal grows and some space starts to appear between the existing scales, a new scale grows to fill it [5]. Besides protecting the body, placoid scales improve the animal's hydrodynamic ability [4].

Some osteichthyans have ganoid scales (typical of chondrosteans and holosteans), but most of them have scales that are either ctenoid (typical of spiny-rayed fish, *e.g.*, perch) or cycloid (typical of soft-rayed fish, *e.g.*, salmon) [10, 11, 17].

Ganoid scales are bony as placoid and cosmoid scales, but diamond shaped and coated with a bright enamel-like matter commonly called ganoin. They cover the body of the animal hardly overlapping each other [10, 11].

Cycloid scales have a discoid shape, with a smooth border, and are quite thin and flexible. They cover the body of the animal showing an imbricated pattern, in which the caudal border of each scale stays free [11].

Ctenoid scales are quite similar to cycloid scales but have a rough caudal border. They are imbricated, as the cycloid scales [11].

Some fishes have very small scales (*e.g.*, eels) and there are fishes that have no scales at all (*e.g.*, torpedo ray) [11]. Others have, instead of true scales, bony plates called scutes. For instance, some catfishes have a naked body, while others (armored catfishes) have imbricated scutes [3, 17].

Body Coloring and Patterns

Fishes present a variety of body colors and patterns that may be used as camouflage, to escape from predators or to catch prey by surprise, as warning of a poisonous nature, or as a mean to attract mates [4, 6].

There are lots of patterns, such as strips from head to tail, bars from top to bottom or spots, for instance, and, while some fishes are very colorful to blend with a colorful environment, a relatively common color in fish such as red is in fact very discrete in deep waters, where it appears gray [4].

Quite often a fish presents patterns and colors adapted to their habitat – benthic fish tend to be brown, to match the bottom surface where they live, but can be darker on the dorsal surface than on the ventral surface, so that it can blend with the lighter water above it, when seen from below. Some special patterns are used not to camouflage but to deceive other fishes about the real body shape or, as in the case of eyespots, about the right extremity of the body [4].

Fish color is essentially due to modified dermal cells called chromatophores [3, 4, 7, 19], usually differentiated in function of the pigments they contain – melanophores (black/brown pigments), erythrophores (red pigments), xantho-phores (yellow pigments), iridophores (iridescent pigments), leucophores (white pigments) and cyanophores (blue pigments) [3, 4, 7, 20]. In some fishes color can change, for instance in males *versus* females, in species that can change sex, or in juveniles *versus* adults. There can also be quick and temporary changes (*e.g.*, due to sudden danger). These changes seem to be under hormonal and/or neural control of the chromatophores [3, 7, 17].

There are some fishes that look luminescent, due to their association with luminescent bacteria, but there are also fishes that are luminescent by themselves, due to the presence, in their skin, of cells (photophores) containing pigments (luciferins) that emit light when they are oxidized [7, 14, 16, 17]. In midwater and deep-sea fishes, this bioluminescence can have the same role of different color patterns, in camouflage, deceiving other fishes or attracting mates [17].

Mouth

While fish may be herbivorous, carnivorous, omnivorous or detritivorous, the mouth's shape is a good indicator of a fish eating habits – for instance, carnivorous fishes tend to have a large mouth, while omnivorous fishes tend to have a small mouth. Also the mouth position is usually related to feeding habits [3]:

• A terminal mouth (both jaws with similar length, meeting each other at the tip of the head) is typical of fish that feed in mid water, generally on other fish.

- A superior or upturned mouth (the lower jaw is longer than the upper jaw, extending beyond the upper jaw) is typical of surface feeding fish (*e.g.*, herring), being useful for feeding on insects or floating prey.
- A subterminal or inferior mouth (the upper jaw is longer than the lower jaw, extending beyond the lower jaw) is typical of bottom feeding fish (*e.g.*, catfish); predatory fish such as sharks, that rip their prey, also have inferior mouths.

Cyclostomes have a jawless mouth:

- Lampreys' mouth is circular and filled with concentric rows of keratinized tooth-like structures (when attached to a fish, a lamprey uses these "teeth" and similar ones present in a protruding tongue to rasp the skin and flesh of the prey) [3, 12, 14 - 16].
- Hagfishes have a lateral-biting mouth with a single keratinized tooth-like structure placed dorsally and two rows of similar keratinized structures on each side of the extremity of an eversible tongue-like structure (commonly called "rasping tongue" it is used to grasp and conduct the food to the pharynx) [2, 18].

Most fishes replace their teeth continually, as they are lost or worn out. In chondrichthyans, the new teeth are organized in parallel rows placed behind the row(s) of functional teeth (usually only the front row and, eventually the second) [12, 13].

Besides other locations such as on body surfaces and/or appendices, taste receptors are also present in the mouth [2, 7, 17, 21, 22].

Barbels

Some fishes (*e.g.*, carps) have whisker-like appendices extending from the head, near the mouth. These appendices, called barbels, are both taste and tactile organs and, in bottom feeding species (*e.g.*, catfish) help in finding food in waters with little visibility [1 - 4, 17, 18].

Nostrils

Most fishes have two small apertures looking like nostrils on each side, above the mouth and the snout [23]. In chondrichthyans they are placed ventrally to the snout. In cyclostomes there is just one of these holes, in a median position [12].

Instead of leading to the throat, as they do in mammals, in most fishes, the nostrils open up into blind sacs lined with sensory pads [1, 23, 24]. In fact, fishes do not breathe through these nostrils, but, through them, they can detect or "smell" chemicals that may signal food or danger, for instance [23]. While these chemicals are diluted in the water, they are different from the chemicals detected by taste buds [7, 18, 21].

In lampreys and hagfishes, the median nostril is the external opening of the nasohypophysial duct. In lampreys this is a blind duct that, after the connection to the olfactory sacs continues to end near the hypophysis [5, 7, 8, 15, 18]. In hagfishes, when the animal breathes, water is taken in through the nostril to pass, *via* the nasopharyngeal duct, through the olfactory organ to the pharynx; then ventilates the gills and is expelled to the exterior through the gill openings [5, 7, 8, 10, 18].

In order to have a good sense of smell, fish must move water quickly in and out through the nostrils – some fishes pump water through their olfactory tract by a muscular movement and others *via* cilia, while others (*e.g.*, smaller species of mackerel) must keep swimming to get water passing through their nostrils [23]. This sense is very important for salmons to find their home spawning stream and it is also particularly well developed in eels, catfishes and cave fishes [1, 7, 11, 23].

Eyes

In most fishes the eyes are placed one on each side of the head, making it possible to see to the right and to the left simultaneously. However, there are some special adaptations, such as the location of both eyes on the same side of the head (*e.g.*, in adult flatfish, which are most of the time lying on one side at the bottom of the sea) or the location of the eyes on the extremity of short appendices projecting from the head, for instance in certain deep-sea fish [3, 5, 7].

In general, a fish's eye is well developed and quite similar to the eye of any other vertebrate. Hagfishes, however, have just rudimentary eyes, without cornea, lens or extrinsic eye muscles, and covered by skin [5, 7, 11, 18]. There are also other fishes that live in lightless waters and are blind – it is the case of the blind cavefish, which has non-functional eyes as an embryo that are lost as the animal develops [5, 11].

Despite the general similarity with the eyes of other vertebrates, a fish's eye presents some particularities. For instance, the lens of a fish eye has a fixed spherical shape (elasmobranch lens excluded), unlike that of a terrestrial vertebrate, and tends to protrude through the pupil. On the other hand, the cornea shows little refractive power [1, 7, 17, 21].

In most fishes, pupils maintain a fixed diameter, but most elasmobranchs have pupils that change in size, depending on light intensity [1, 4].

Fishes have no lachrymal glands, and most of them have no eyelids or a somehow similar structure [6, 11]. The exceptions are some sharks, which have a nictitating membrane, and some osteichthyans, which have an adipose eyelid [9, 13].

In most teleosts there is an annular ligament, connecting the external border of the iris to the cornea, a gland placed in the choroid layer of the eye, near the optic nerve, named choroid gland, and a falciform process, placed ventrally. The choroid gland seems to help in the supply of oxygen to the retina [5, 7, 11, 17]. The falciform process is connected to the lens by the retractor lentis muscle, which pulls the lens caudally, from its relaxed position, changing from long focal length to short focal length [11, 21]. In chondrichthyans there is no falciform process – in the relaxed position of the lens, the eye is focused on close objects and, to focus on distant objects, the protactor lentis muscle pulls the lens cranially [11, 21]. In lampreys, where the lens is simply pressed by the vitreous body against the cornea, without any anterior connections, focus is attained flattening the cornea through the action of extrinsic muscles caudally [21].

In most fishes the sclera is fibrous but, to better protect and maintain the shape of the eye, in some fishes it incorporates skeletal elements. So, many teleosts present scleral ossicles and in elasmobranchs the sclera presents embedded cartilage which, in some species, surrounds most of the eye [25].

The retina of a fish's eye has both rods and cones, even in chondrichthyans, which are quite often referred as lacking cones [13]. Still, rods are present in a much large proportion than cones, especially in chondrichthyans. Given that rods simply detect variations in light intensity and color detection depends on the presence of cones, it is not yet clear how chondrichthyans analyze colors [5]. In general, fishes living in deep waters tend to have larger eyes and relatively less cones in the retina [7, 10, 17].

There are many references to the excellent vision that sharks reveal in poor light conditions, due to the presence of a tapetum lucidum behind the retina [5, 13]. In fact, most chondrichthyans' eyes have such a structure, made of guanine crystals, which reflects back to the retina light that was not detected the first time, increasing light sensibility, but that structure is also present in the eyes of lampreys and many osteichthyans [13, 15, 21].

In lampreys, immediately behind the nostril, there is a transparent spot, covering the pineal gland. This is quite often called pineal eye or third eye since, having an endocrine function, it is involved in the detection of light [1, 11, 14]. However it does not play any role in the identification of visual images [12, 13].

Gill Openings, Operculum and Spiracles

On each side of the body, lampreys have seven gill openings, hagfishes have one to sixteen and most chondrichthyans have five (in most modern species) to seven gill openings [3, 7, 12]. Osteichthyans have just one gill opening on each side, since the gills are covered by a flexible bony plate (the operculum) – water gets in through the mouth, passes through the gills and is expelled from beneath the operculum [3, 12, 16]. Most fish have gill slits, but in cyclostomes gill openings are circular [12]. In rays, given their flattened body, their gill openings are placed ventrally.

Rays and bottom-dwelling sharks present an opening called spiracle on each side of the head (dorsally in rays), just behind the eye [3, 13]. These openings communicate with the gills and help in breathing – instead of taking in respiratory water through the mouth, the animal does it through the spiracles when it is lying on the ocean bottom or buried in the sand [2, 24].

Ampullae of Lorenzini

Around a shark's mouth and nostrils there are small pores and vesicles, called ampullae of Lorenzini [3]. They allow the animal to detect weak electromagnetic fields produced by other fishes, as well as changes in water temperature (translated into electrical information in these ampullae), helping it to locate prey [2, 10, 13, 14].

Lateral Line Organ

The lateral line organ consists of a series of small sensory patches, called neuromasts, running along both sides of a fish, from head to tail. The neuromasts are placed either on the skin surface or in water filled skin ducts that open to the outside through a line of skin pores [2, 10, 12, 13, 26].

Each neuromast presents hair cells innervated by lateral line nerves and is sensitive to the smallest pressure changes in the surrounding water – it is a kind of indirect or distant touch sense [10]. Thanks to the lateral line organ the fish can register movement and the direction of its source [2, 13, 26]. The lateral line organ also helps fish to interpret sounds [10].

Claspers

Claspers are grooved copulatory organs placed along the inner side of the male shark or ray's pelvic fin, near the cloaca. Each male has two claspers and, during the copula, the erect claspers are bent forward, allowing the male to deposit his sperm into the female's cloaca *via* grooves that lie in the upper side of the claspers [10, 13, 14].

Anus and Cloaca

In most fishes, the external openings of the urinary and reproductive tracts are separate from that of the digestive tract [27]. These openings are placed just ahead of the anal fin, the anus being the most anterior. However, lungfishes, elasmobranchs, sarcopterygians and cyclostomes present a common external opening (cloaca) of the digestive, urinary and reproductive tracts [13, 18, 27].

SKELETON

The skeleton of the fish can be made of bone or exclusively of cartilage (chondrichthyans or cartilage fish). Besides having a cartilaginous skull, most chondrichthyans (namely the elasmobranchs) differ from osteichthyans in that their upper jaw is not affixed to the skull [13]. Some osteichthyans (*e.g.*, carps) also have an additional set of jaws in the pharynx [2, 6, 27], while chondrichthyans do not have pharyngeal jaws [13]. Cyclostomes are jawless.

In embryos of all chordates, the longitudinal support of the body is mainly provided by a flexible rod-shaped structure called notochord. This structure persists during the whole life of the animal in cyclostomes [3], for instance, but in most fish, as vertebrae develop, it is incorporated in the vertebral column and gives origin to the nucleus pulposus of each intervertebral disc [14]. Lampreys have cartilaginous rudiments of vertebral arches, but no vertebral bodies at all [12, 15]. Although there is some discussion about the presence or absence of some vertebra-like elements in hagfishes, it is generally considered that hagfishes have no vertebrae at all, having lost them during evolution [28].

Teleosts have a vertebral column made of well-ossified vertebrae, extending from the skull to the tail. Each vertebra presents a central part or centrum (with a constricted remaining of the notochord), which is amphicoelous in most fish, and a neural arch, dorsal to the centrum [3, 4, 14, 17]. There is also a haemal arch, ventral to the centrum in caudal vertebrae [3].

The notochord persists as the main structure of the axial skeleton of sarcopterygians and lungfishes. Besides ossified neural arches, some lungfishes may have rudimentary centra, while in sarcopterygians, each vertebra presents, instead of a centrum, other two bony structures, the pleurocentrum (dorsal to the notochord) and the intercentrum (ventral to the notochord), both enclosed in cartilage [3, 14].

Although primitive chondrosteans had a bony skeleton, sturgeons and paddlefish lost some of the actinopterygian characteristics of their ancestors. Amongst these are an almost completely cartilaginous skeleton and the absence of true vertebral centra, persisting the notochord as the main structure of axial support [3, 14].

In chondrichthyans the cartilaginous vertebrae are partly calcified and, between the concavities of the centra of successive vertebrae, there are spherical remainings of the notochord [1, 24].

Amongst all living fish groups, only lampreys and chimaeras do not have dorsal ribs, but only osteichthyans have ventral ribs. Dorsal ribs are placed at the intersection of the myosepta with the horizontal skeletogenous septum, while ventral ribs are placed at the intersection of the myosepta with the ventrolateral septa just outside the celomic cavity [3]. All fish ribs are connected to the vertebrae by one extremity but the other extremity is free, since there is no sternum [3, 5, 7, 17].

Dorsal ribs separate epaxial and hypaxial muscles and so they are also called intermuscular ribs, but there are many osteichthyans which present other intermuscular bones in the myosepta (*e.g.*, carps) [4, 5, 7, 17].

SKELETAL MUSCLES

Skeletal muscles of jawed fishes are divided into three main groups – head muscles, trunk muscles and paired fin muscles.

The median skeletogenous septum splits the trunk muscles in left and right muscles, in turn divided by the horizontal skeletogenous septum in two major lateral muscular masses – epaxial muscles and hypaxial muscles [3, 4, 5, 12, 17]. In the superficial depression between these two masses and along the lateral line, there is, in elasmobranchs and many osteichthyans, a third and darker muscular mass, sometimes called red muscle, that represents only a small proportion of the total mass of trunk muscles and is particularly noticeable in the tail. In species with high activity levels (*e.g.*, tunas), there is a much higher proportion of red muscle than in other fishes [2, 5]. All these muscles are divided in successive W-shaped segments (myomeres), separated by vertical myosepta of connective tissue extended from the axial skeleton to the skin [3, 4, 11, 12, 17, 24].

Contracting the myomeres of the trunk and tail muscles in sequence, and alternating the side of contraction of the successive myomeres, a fish manages to create an S-shaped movement, and this is its main propelling mechanism [11 -

13].

Being highly vascularized, red muscle is rich in haemoglobin (thus the color) and, so, is prepared for aerobic activity, adequate to steady swimming [3, 4, 10, 14, 18]. In contrast, white muscle is essentially used for short bursts of activity, since it presents thicker fibers and very little haemoglobin – white muscle activity is mainly anaerobic and, used at its maximum potential, glycogen is quickly converted to lactic acid, causing fatigue [2, 4, 10, 14, 17, 18].

Fish muscle can also be pink, which is an intermediate type between red and white muscles in physiological properties [7, 10, 17, 18, 29].

GILLS AND LUNGS

The gills constitute the main breathing apparatus of any fish. They consist of bony or cartilaginous archs, from which soft and paired filaments radiate caudally and bony or cartilaginous rakers point cranially and inward [1 - 3, 5, 13, 17]. In most osteichthyans, beneath the operculum, there are five gill slits and four bony gill arches, each holding a set of gill filaments [3].

On each soft gill filament there is a large number of small projections with very thin membranes concealing an intricate system of blood capillaries and it is here that the gaseous exchanges between the fish blood and the environing water take place [2, 3, 5, 7, 13, 17]. This structure is also involved in osmoregulation – given the different salt concentration in the fish body and in the environing water, it is necessary to control the gain or loss of water and salt, knowing that freshwater fish are hyperosmotic and saltwater are hypoosmotic [2, 10, 13, 17]. Most nitrogenous waste products carried by the blood are eliminated through the gills [2, 10, 24].

The gill rakers help to filter solid substances out of the respiratory tract [2, 3, 14, 17, 27]. Gill rakers long, thin and in large number are used in filter feeding, while large but short (and less numerous) gill rakers help to prevent food from escaping through the gill openings, in fish that eat large preys [2, 3, 7, 13, 17, 27].

To breath, most fishes ingest water through the mouth and then, while oral valves close, they expel this water through the gill openings [7, 8]. Cyclostomes have a set of muscular folds, the velum, which helps to move the water through the mouth (in lampreys) or the nostril (in hagfishes) into the pharynx [2, 10, 12, 15]. In adult lampreys, however, the velum helps essentially in suction, since it isolates the respiratory tract from the esophagus when a lamprey is attached to a prey, sucking blood [12, 15]. In this situation the gill chamber becomes a blind pouch and, to breath, the animal needs to move water in and out of the gill chamber, through the contraction of muscles associated to the gill openings [10, 12].

Lungfishes have reduced gills and a modified swim bladder, linked to the esophagus by a pneumatic duct [2, 11]. In the Lepidosiren of South America and the Protopterus of Africa this modified swim bladder or lung is double and ventral to the gut, while in the Neoceratodus of Australia it is single and dorsal to the gut [8, 11].

SWIM BLADDER

Most osteichthyans have a swim bladder (or "gas bladder"). This is a gas-filled cavitary organ found only in osteichthyans (placed at the top of the celomic cavity) and used to maintain the neutral buoyancy of a fish – controlling the amount of gas in the swim bladder at different depths, the fish can ascend or descend without having to spend energy in swimming [1, 2, 5, 7, 17, 24].

Many fishes present swim bladders adapted to their living habits. It is the case of some fish groups (*e.g.*, lungfishes) with a swim bladder that remains linked to the esophagus – such swim bladder is called physostomous, in opposition to a physoclistous swim bladder, which is completely closed [1, 3, 17].

In physostomous fish, the amount of gas in the swim bladder is controlled by gulping and burping through the pneumatic duct. In physoclistous fish, it is controlled involving the anterior region of the swim bladder, containing the gas gland and the rete mirabile, and, usually (in acanthopterygians and gadiformes), the posterior region of the swim bladder, called oval [5, 8, 11, 14, 24]:

- To inflate a swim bladder, the rete mirabile and the gas gland create a combined gas pressure superior to the pressure inside the swim bladder.
- To deflate a swim bladder, gases from within the bladder diffuse out to the blood

when the circular muscles that surround the capillary network of the oval relax and the radial muscles contract.

Some fishes (*e.g.*, eels and herrings) can use the swim bladder to produce sounds, using special muscles associated with the swim bladder [5, 9, 11, 17, 24].

Some fishes show improved earing ability due to special structures that transmit swim bladder vibrations to the inner ear. The most specialized case is that of ostariophysan fishes (*e.g.*, catfishes and carps), where a set of four bones (Weberian ossicles) links the swim bladder to the inner ear [2, 3, 5, 8, 17, 29].

INNER EAR

In fish the ear is reduced to the inner ear, which consists of a membranous labyrinth that comprises a pouch system and semicircular canals containing endolymph and otoliths [2, 3, 8, 12, 14, 24].

The number of pouches varies from one pouch, in cyclostomes, up to three, in most fishes. The semicircular canals loop from the pouch system (both ends of each canal opening into the vestibule or cavity of the pouch, one of the ends of each canal being dilated) and are in number of one (in hagfishes), two (in lampreys) or three (in osteichthyans and chondrichthyans) [8, 18].

The otoliths are calcium carbonate structures, sometimes called ear stones or fish ear bones, placed in the pouches [16, 17, 29]. Given that the otoliths show higher density than the endolymph and the fish body, their movement, as inertial masses, in response to any vibration passing through the body, stimulates the surrounding hair cells, enabling the fish to detect and analyze sounds [1, 3, 14, 17, 24].

Chondrichthyans and lampreys are often mentioned as not having otoliths. The fact is that, instead of the three otoliths (one in each pouch) found in teleosts, in chondrichthyans the otoliths are in larger number, but small and disseminated, while in lampreys there are some structures analogous to the otoliths of the teleosts, but smaller, called statoliths [5, 8, 9, 24, 30].

The inner ear plays also a role in the maintenance of balance – while the vestibule contains hair cells that detect linear acceleration, the semicircular canals contain in their dilated end hair cells that detect any angular acceleration of the head [8, 10, 12].

ELECTRIC ORGAN

Some fishes (electrogenic fishes) can generate an electric field in modified muscle or nerve cells (electrocytes) organized in a special structure called electric organ [2, 5, 6, 31, 32]. When the electrocytes, stimulated by the brain and linked in series and in parallel, produce a simultaneous electric discharge, they create a strong electric field, up to 600 volts. The most well know electric fishes, capable of producing such strong electric fields, used to stun prey or for protection, are the electric rays or torpedo rays, the electric eel and the electric catfishes [8, 13, 22, 31]. Other electric fishes (*e.g.*, elephantfishes) only produce weak electric fields (in general less than 1 volt), used to sense the surrounding area (electrolocation) and/or for communication (electrocommunication) [22, 31]. This implies that these fishes can also sense electric fields, being simultaneously electrogenic and electroreceptive. There are, however, electroreceptive fishes that are not electrogenic (*e.g.*, sharks) [8, 22, 31, 32].

In most cases the electric organ is found in the tail [32], but there are other locations – in electric rays, for instance, it consists of two masses, one on each side of the head, in the area of the pectoral muscles [22].

DIGESTIVE TRACT

Fishes have a very distensible esophagus – in practice, a fish can swallow anything that passes through its mouth, although prey with erected fin spines can become stuck [2, 8, 24].

While most fishes have a very short esophagus, in some groups which have no stomach, such as lampreys, hagfishes and even some teleosts (*e.g.*, parrotfishes), the esophagus is directly connected to the intestine [2, 15, 18].

In fishes that have a stomach, this organ can vary considerably in size and shape, depending on the species, and can show several adaptations [2]. One of the most remarkable is the distensibility of the stomach or of an anterior gastric evagination, allowing some fishes such as the blowfishes to inflate themselves with water or air, assuming a much bigger and globe-shaped volume [8].

Many osteichthyans present finger-like blind projections of the intestine near the pylorus, the pyloric ceca, which may be very variable in number, depending on the species, from one to thousands [2, 10, 12, 14, 24, 27].

Excluding teleosts (which more commonly have elongated and coiled intestines) and hagfishes, fishes with a short intestine and a small stomach or no stomach at all present a spiral valve connected to the rectum, instead of a small intestine – externally the intestine is quite straight, but the mucosa is folded in a spiral manner, greatly increasing the internal surface [5, 13, 15, 24].

The liver is a relatively large organ, with variable shape, depending on the body shape of the different fishes, in general with a gall bladder. Besides performing various other functions the liver stores fats and carbohydrates [2, 5, 17, 24].

In lampreys, lungfishes and teleosts the pancreas presents a diffuse structure, and one can find pancreatic tissue incorporated in the liver as well as dispersed in the omentum and the peritoneal folds attached to the intestine [2, 7, 14, 17, 18].

URINARY TRACT

In most fishes the kidneys are elongated, reddish dark and paired organs, placed just under the vertebral column (and over the swim bladder, when there is one) [5, 7, 11].

Each kidney excretes the urine through a duct that may be fused to the one from the other kidney caudally. In some fishes the caudal end of these ducts is slightly enlarged, forming the urogenital sinus (in lampreys and elasmobranchs) or the urinary bladder (in many osteichthyans) [2, 5, 7, 8, 12, 13].

Besides other functions, namely in hematopoiesis and filtering waste substances from the blood, the kidneys are responsible for osmoregulation – in freshwater fishes they reabsorb salts and excrete large volumes of urine, which is very diluted, counteracting the diffusion of large amounts of water into the body; in saltwater fishes it is the opposite, with the kidneys excreting a very concentrated urine [2, 5, 10, 13].

CARDIOVASCULAR SYSTEM

The circulatory system of a fish consists of a single circuit – blood pumped by the heart to the ventral aorta goes to the afferent branchial arteries and then to the gills, where it is oxygenated; then it follows through the efferent branchial arteries to the carotid arteries (for the head region) and the dorsal aorta (for the rest of the body), to exchange gases, nutrients and wastes (which are transported to the kidneys and the liver to be eliminated) with the body tissues; finally the blood returns to the heart, through the common cardinal veins and the hepatic veins [5, 7, 8, 11, 12].

The heart of most fishes presents four compartments in sequence, namely the sinus venosus, the atrium, the ventricle and the bulbus arteriosus (in teleosts) or the conus arteriosus (in chondrichthyans), although in adult fishes the heart is bent in a sigmoid manner [2, 5, 7, 8, 10, 24]. Sinoatrial valves isolate the sinus venosus from the thin-walled atrium, which, in turn, is isolated from the thick-walled ventricle by atrioventricular valves (the atrium and ventricle being the two main chambers of the heart) – the blood passes from the atrium to the ventricle when the atrium contracts and the ventricle relaxes; when the ventricle contracts, the blood passes to the fourth compartment that constitutes the base of the ventral aorta [2, 5, 7, 8, 10]. The bulbus arteriosus is an elastic, valved and enlarged structure, while the conus arteriosus is a contractile muscular structure, with several valves [4, 5, 8, 10, 17].

Most commonly it is considered that in the heart of cyclostomes only the sinus venosus, the atrium and the ventricle [8, 10, 12, 21] are present. However, some authors mention the presence of a bulbus arteriosus in the heart of lampreys [10, 15].

In lungfishes there is a partial division in two of the atrium and the ventricle that continues into the ventral aorta. The Australian lungfish presents two pulmonary arteries (each originated on the fourth aortic arch of each side), while the African and the South American lungfishes present one pulmonary artery originated on the dorsal aorta. In all lungfishes the pulmonary vein opens to the left side of the sinus venosus [8, 11]. Commonly, the conus arteriosus is considered typical of chondrichthyans but, despite the partial segmentation in two, there are several references to the existence of a conus arteriosus in the heart of lungfishes [5, 7, 10, 11].

REPRODUCTIVE SYSTEM

Most adult fishes are either male (with testes) or female (with ovaries), although most often they look alike externally. However, many fishes are hermaphrodites. Some of these (*e.g.*, salmon) are simultaneous or synchronous hermaphrodites (with ovaries and testes at the same time), while others are sequential or serial hermaphrodites, changing sex during their lifetime. In the latter case, both testes and ovaries may be present, but only one type of gonad is active at any moment, and the change can be from male to female (protandry), as in anemonefishes, or from female to male (protogyny), as in most wrasses [2, 5 - 7].

Both male and female gonads (or testes and ovaries, respectively) are paired organs (although they may be fused in a variable degree) lying longitudinally in the celomic cavity, suspended from the dorsal wall by the mesenteries. When there is a swim bladder, the gonads are placed under or, in the case of the testes, laterally to it [8].

In elasmobranchs and some osteichthyans the testes are connected to the anterior part of the kidneys – the sperm duct is in fact a modified mesonephric duct, caudally enlarged in a seminal vesicle, before ending in the urogenital sinus, which presents a pair of sperm sacs [3, 12]. In lungfishes and teleosts the testes are completely separated from the kidneys, but there are variations concerning the sperm duct, which in trouts and other salmonids are absent [10].

In most teleosts there is a peritoneal fold encapsulating the ovary and in continuity with the oviduct (cystovarian condition), which conducts the oocytes to the exterior. In cyclostomes, chondrichthyans, sturgeons, lungfishes, the bowfin and some teleosts (*e.g.*, salmonids) the oocytes are released to the celomic cavity and then to the exterior (gymnovarian condition) [5, 7, 15]. Cyclostomes and some teleosts (*e.g.*, trouts and other salmonids) don't have oviducts [10, 15].

Chondrichthyan males use their paired claspers on the pelvic fins for internal fertilization [13], but in lampreys and most osteichthyans fertilization is external [5, 12, 13, 24]. In case of internal fertilization, some fishes are oviparous, while others are ovoviviparous or even (more rare) viviparous [2, 3, 5, 7, 13]. Oviparous females present a specialized segment in the anterior part of the oviduct, the shell gland, which adds a protein layer, produces the egg shell and can store sperm [3, 5, 7, 13, 24]. Ovoviviparous or viviparous females present a uterus, which corresponds to a posterior enlargement of the oviduct, where embryonic development takes place [5, 7, 13].

BRAIN

Fishes have a relatively small brain. In general, the elasmobranchs have relatively larger brains than the teleosts, but the latter show a large variability in this regard and, comparatively to their body dimension, elephantfishes have the largest brains among fishes [5].

A fish brain can be divided in forebrain (or telencephalon, with the two cerebral hemispheres plus the two olfactory lobes at front), midbrain (or mesencephalon, with the two optic lobes) and hindbrain (or metencephalon, with the cerebelum), more the diencephalon, connecting the forebrain with the midbrain, and the myelencephalon (or medulla oblongata) as the most caudal part of the brain, in continuity with the spinal cord [1, 5, 12].

The telencephalon in fishes processes essentially olfactory information, coming from the nostril(s), and the olfactory lobes are particularly well developed in cyclostomes and elasmobranchs, but also in some osteichthyans (*e.g.*, catfishes) that depend largely on their sense of smell, for instance for hunting [5, 8, 18]. The cerebral hemispheres are more developed in osteichthyans and chondrichthyans than in cyclostomes, showing a large variability in size between different species [8, 10].

The mesencephalon or midbrain of fishes is relatively large and quite often the most developed part of the brain [8, 10]. In osteichthyans the optic lobes are particularly well developed in those species depending largely on their sight for feeding (*e.g.*, trouts) [1, 5].

The cerebellum, mainly involved in the maintenance of muscular tonus and balance, is more developed in osteichthyans and chondrichthyans than in cyclostomes [8, 10, 17]. In most gnathostome fishes it makes the largest part of the brain, although there is a large variability – being relatively small in some fishes (*e.g.*, dogfishes), it is particularly prominent in elephantfishes (apparently in relation to their electroreception ability) [5, 8, 10].

CONFLICT OF INTEREST

The author confirms that author has no conflict of interest to declare for this publication.

ACKNOWLEDGEMENTS

Declared none.

REFERENCES

[1]   Springer JT, Holley D. An introduction to zoology: investigating the animal world. Boston: Jones & Bartlett Learning 2013.

[2]   Evans DH, Claiborn JB, Eds. The physiology of fishes. 3rd ed., Boca Raton, FL: CRC Press 2006.

[3]   Wake MH, Ed. Hyman's comparative vertebrate anatomy. 3rd ed., Chicago: University of Chicago Press 1992.

[4]   Moyle PB, Cech JJ Jr. Fishes: An Introduction to Ichthyology. 3rd ed., Englewood Cliffs, N.J.: Prentice-Hall 1995.

[5]   Helfman GS, Collette BB, Facey DE, Bowen BW. The Diversity of Fishes: Biology, Evolution, and Ecology. Chichester, UK; Hoboken, NJ: Blackwell 2009.

[6]   Dawes J, Campbell A, Friel JP. Exploring the World of Aquatic Life. New York: Chelsea House Publishers 2009.

[7]   Kapoor BG, Khanna B. Ichthyology Handbook. New Delhi: Narosa Publishing House 2004. [http://dx.doi.org/10.1007/978-3-662-07844-0]

[8]   Lagler KF, Bardach JE, Miller RR, Passino DR. Ichthyology. New York: John Wiley & Sons, Inc. 1977.

[9]   Nelson JS. Fishes of the World. John Wiley & Sons, Inc. 2006.

[10]   Kardong K. Vertebrates: Comparative anatomy, function, evolution. Boston: McGraw-Hill 2008.

[11]   Kotpal RL. Modern Text Book Of Zoology: Vertebrates. Rastogi Publications 2009.

[12]   De Iuliis G, Pulerà D. The Dissection of Vertebrates: A Laboratory Manual. Amsterdam, Boston: Elsevier/Academic Press 2007.

[13]   Carrier JC, Musick JA, Heithaus MR. Biology of sharks and their relatives. 2nd ed., Boca Raton, FL: CRC Press 2012.
[http://dx.doi.org/10.1201/b11867]

[14]   Walker WF Jr, Liem KF. Functional Anatomy of the Vertebrates – An Evolutionary Perspective. 2nd ed., Saunders College Publishing 1994.

[15]   Hardisty MW, Potter IC. The Biology of Lampreys. New York: Academic Press 1981; 3: p. 469.

[16]   Romer AS, Parsons TS. The Vertebrate Body. Philadelphia, PA: Holt-Saunders International 1977; pp. 437-42.

[17]   Patino R, Redding JM. Handbook of Experimental Animals – The Laboratory Fish. San Diego, CA: Academic Press 2000; pp. 489-500.

[18]   Jørgensen JM. The Biology of Hagfishes. Chapman & Hall Ltd Thompson Science, 2-6 Boundary Row, London SE1 8Hn, UK. 1998.
[http://dx.doi.org/10.1007/978-94-011-5834-3]

[19]   Smith DC. Color changes in the isolated scale iridocytes of the squirrel fish, holocentrus ascensionis (osbeck). Proc Natl Acad Sci USA 1933; 19(10): 885-92.
[http://dx.doi.org/10.1073/pnas.19.10.885] [PMID: 16587805]

[20]   Menter DG, Obika M, Tchen TT, Taylor JD. Leucophores and iridophores of Fundulus heteroclitus: Biophysical and ultrastructural properties. J Morphol 1979; 160: 103-19.
[http://dx.doi.org/10.1002/jmor.1051600107]

[21]   McKenzie DJ, Farrel AP, Brauner CJ. Primitive Fishes. In: Farrel AP, Brauner CJ, Eds. Fish Physiology. New York: Elsevier, Inc. 2007; 26: p. 560.

[22]   Moller P. Electric Fishes: History and Behavior. London: Chapman and Hall 1995.

[23]   Glas D. Can Fish Smell? , 2004 [2013 Jan 21]; Available from: http://indianapublicmedia.org/amomentofscience/can-fish-smell/.

[24]   Hildebrand M, Goslow GE Jr. Analysis of Vertebrate Structure. 5th ed., New York: John Wiley & Sons 2001.

[25]   Pilgrim BL, Franz-Odendaal TA. A comparative study of the ocular skeleton of fossil and modern chondrichthyans. J Anat 2009; 214(6): 848-58.
[http://dx.doi.org/10.1111/j.1469-7580.2009.01077.x] [PMID: 19538630]

[26]   Coombs S, Braun CB, Donovan B. The orienting response of Lake Michigan mottled sculpin is mediated by canal neuromasts. J Exp Biol 2001; 204(Pt 2): 337-48.
[PMID: 11136619]

[27]   Stevens CE, Hume ID. Comparative Physiology of the Vertebrate Digestive System. 2nd., Cambridge University Press 1995.

[28]   Ota KG, Fujimoto S, Oisi Y, Kuratani S. Identification of vertebra-like elements and their possible differentiation from sclerotomes in the hagfish. Nat Commun 2011; 2: 373.
[http://dx.doi.org/10.1038/ncomms1355] [PMID: 21712821]

[29]   Lucas M, Baras E. Migration of Freshwater Fishes. Oxford; Malden, MA: Blackwell Science 2001.

[http://dx.doi.org/10.1002/9780470999653]

[30]   Renaud CB. Lampreys of the world An annotated and illustrated catalogue of lamprey species known to date FAO Species Catalogue for Fishery Purposes, No 5. Rome: FAO 2011.

[31]   Albert JS, Crampton WG. Electroreception and electrogenesis. In: Evans DH, Claiborne JB, Eds. The physiology of fishes. 3rd ed. Boca Raton, FL: CRC Press 2005; pp. 431-72.

[32]   Nelson ME. Electric fish. Curr Biol 2011; 21(14): R528-9.
        [http://dx.doi.org/10.1016/j.cub.2011.03.045] [PMID: 21783026]

CHAPTER 2

Embryonic and Larval Development

Rita Borges[1,2,*], Ana Faria[2] and Cláudia Faria[3]

[1] *CCMAR, CIMAR-Laboratório Associado, University of Algarve, Campus de Gambelas, 8005-139 Faro, Portugal*

[2] *MARE-Marine and Environmental Sciences Centre, ISPA-Instituto Universitário, Rua Jardim do Tabaco, 34, 1149-041 Lisboa, Portugal*

[3] *Instituto de Educação da Universidade de Lisboa Alameda da Universidade, 1649-013, Lisboa, Portugal*

Abstract: Fishes have a great diversity of reproductive strategies and associated traits, as well as complex life-cycles with multiple stages. This chapter briefly introduces different, commonly found, reproductive strategies and focus mainly on the early stages of fish life cycles, in particular eggs and larvae. Most fishes are iteroparous, spawning several times during their life, but some species, semelparous, adopt an extreme strategy, spawning only once in their life cycle.

There is a great diversity of fish egg types and adaptations, and they can be classified as either pelagic or demersal, depending on where they occur in the environment. The embryonic development depends on the species, and also on abiotic factors such as temperature, oxygen and salinity. The ontogenetic development that starts during the embryonic stage continues during the larval life. During this phase, fundamental structural and functional changes will occur in a short period, increasing the larval abilities to interact with the environment. Depending on the species reproductive strategy and life history, newly hatched larvae can vary from very small and poorly developed, to larvae that hatch larger with developed sensorial and functional capabilities; some larvae can even resemble the adults at hatching.

Keywords: Early ontogeny, Embryonic development, Fish eggs, Fish larvae, Larval development.

---

* **Address correspondence to Rita Borges:** MARE-Marine and Environmental Sciences Centre, ISPA-Instituto Universitário, Rua Jardim do Tabaco, 34, 1149-041 Lisboa, Portugal; E-mail: rborges@ispa.pt

Manuela Oliveira, Fernando Bernardo, Joana I. Robalo (Eds.)

INTRODUCTION

Reproduction can vary according to the breeding system, gender role, spawning habitat, spawning season, fecundity, among other features [1]. Most fishes are iteroparous, spawning several times during their life [2]. For these species, there are seasonal cycles of reproduction that are usually controlled by endogenous rhythms and synchronized in response to physical variables such as temperature and photoperiod [3]. The spawning strategy in each seasonal cycle can also vary: some species spawn only once in each breeding season (total spawners), while others release eggs in several batches over the spawning season (batch spawners). Some species adopt however an extreme strategy spawning only once in their life cycle (semelparous breeders) and investing all energy in a single, massive reproductive episode, followed by death [2].

Developing embryos can depend on maternal or yolk provisioning [2]. Viviparous species, also known as live-bearers, have internal fertilization and embryos develop within the female, giving birth to free-living larvae or juveniles. The vast majority of fish species are, nonetheless, oviparous [2], producing eggs that are spawned and made fertile afterwards. In this case, an egg membrane is present and the embryo is entirely nourished by the yolk.

Terminology

After spawning and fertilization, distinct developmental stages characterized by dominant physiological processes will succeed during fish life cycles. Several terminologies for intervals of fish development have been proposed, and there is not a single, widely accepted classification system given the variety of ways in which fishes develop. The terminology used in the present chapter for the early stages of fish life-cycles is based on the widely used system of [4] (Fig. **1**) as follows:

- "Egg stage": from fertilization to hatching;
- "Larval stage": from hatching to attainment of complete fin rays development and beginning of squamation (arrangement of scales on the skin). An important event during this stage is the flexion of the tip of the notochord that accompanies the hypochordal development of the caudal fin;

- "Juvenile stage": besides fully developed fin rays, there is a loss of larval characters and the completion of squamation. The individual already resembles the adult. The juvenile stage lasts until fish enters adult population or attain sexual maturity.

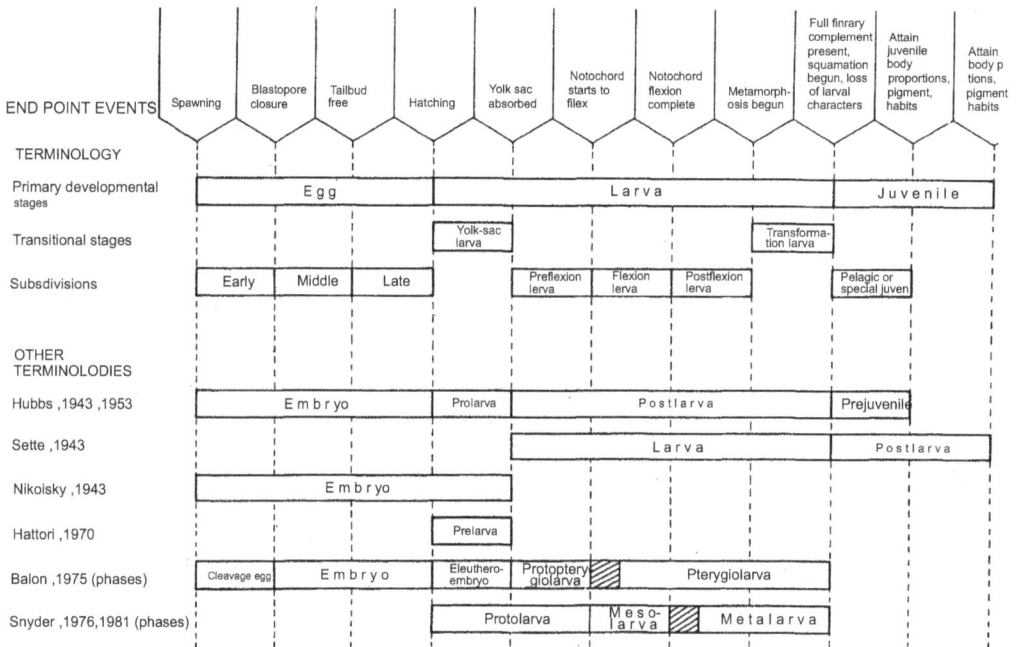

Fig. (1). Terminology of early life history stages [Adapted from [4], permission granted by ASIH].

Transitional stages can sometimes be recognized, such as a "yolk sac larval stage" (development stage beginning with hatching and ending with exhaustion of yolk reserves and characterized by the presence of a yolk sac); and a "transformation stage" (usually coincident with the beginning of metamorphosis, which is only considered complete when the fish assumes the juvenile features).

FROM FERTILIZATION TO HATCHING: THE EGG

There is a great variety of fish egg types and adaptations, with a large diversity in morphological and physiological traits. Fish eggs can be classified as pelagic or demersal, depending on where they occur in the environment. Pelagic spawning is the most common spawning mode in the marine environment, regardless of adult habitat (demersal or pelagic), species distribution (coastal or oceanic), latitude (tropical to temperate), and taxonomy [5]. Out of about the 12,000 existing marine teleost species, ca 9,000 (75%) produce pelagic (buoyant) eggs [5] that are spawned, fertilized, and develop in the water column as part of the plankton.

Most offshore marine teleostean fishes spawn pelagic eggs that are fertilized externally and float individually with the plankton in the water column (egg buoyancy can be variable depending on the species or developmental stage). These species must be able to produce large numbers of small eggs for successful recruitment, given the large dispersal that eggs will suffer with currents into areas far beyond the optimal conditions for survival, and facing high mortality before hatching. These forms are thus expected to have a wider geographic distribution.

On the contrary, many coastal marine fishes living in nearshore turbulent environments lay demersal eggs. These eggs may sink, or be anchored to some kind of substrate, avoiding being swept offshore by currents into areas less favourable to their development. Eggs can be laid attached to substrata such as stones, shells, weeds or rock or in a specially constructed nest, singly or in layers or clumps, that are frequently guarded by one or both parents [4 - 6]. Common expressions of parental care by these species are nest building, territoriality, nest cleaning (removal of dead eggs and other decaying matter), fanning (for aeration of eggs), or egg guarding (predator removal) [7]. In most cases the male guards a nest where females lay eggs and where fertilization takes place. Other types of care are mouth brooding and carrying the eggs, externally, in skin pouches or internally (as in viviparous species).

Like inshore marine species, nearly all freshwater fishes deposit eggs that sink or are sticky and become attached to various substrates [8]. The main evolutionary forces acting on this are the fact that upper layers of freshwater environments do not provide rich food resources, as it happens in the marine environment, and that fast moving rivers and streams can remove nearly all pelagic eggs and larvae from a local population, preventing recruitment.

Pelagic fish eggs can be very different from demersal eggs in several aspects. For instance, incubation time, which is also temperature dependent and related to egg size, is generally longer in demersal eggs [5, 6]. Pelagic eggs are usually smaller (with 1mm mode, ranging from 0.5 to 5.5 mm in diameter) and transparent, having a very high water content (>90% of their wet weight), which makes them approximately neutrally buoyant [9].

On the other hand, demersal eggs are larger (than the 1mm mode of pelagic eggs), reflecting a higher concentration of yolk, often coloured, and have a much thicker and more resilient egg envelope than pelagic eggs. Therefore, demersal spawners produce fewer but larger eggs [5, 6].

Within a single species there is little variation in egg characters (size, number and size of oil globules, chorion surface, yolk, pigmentation, and morphology of the developing embryo).

The majority of pelagic eggs are spherical (Fig. **2**). Nonetheless, species in some groups produce eggs with ellipsoidal chorions (*e.g.* Engraulidae). In addition, demersal eggs tend to be spherical (*e.g.* Blenniidae), flattened (*e.g.* Gobiesocidae) or ellipsoid (*e.g.* Gobiidae).

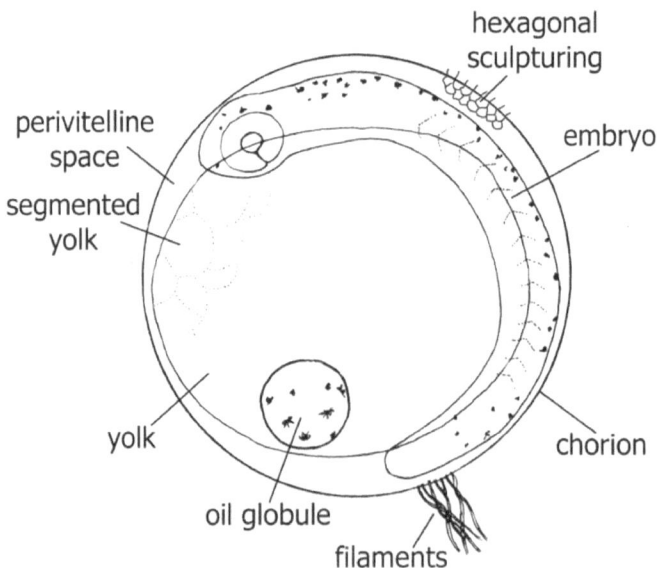

Fig. (2). Egg characters [Adapted from [10]].

There is an outer egg membrane, the chorion, consisting of a double layer permeated by fine pores. The egg membrane is slightly thickened at the animal pole, to form the micropyle, where the sperm enters the oocyte membrane [4]. The structure of the micropyle opening can be variable among species and can be used in egg taxonomy [4]. The chorion can be smooth or ornamented with spines and filaments (*e.g.,* Belonidae, Atherinidae), hexagonal or polygonal networks of different sizes (*e.g.,* Callinonymidae, Macrouridae) or a single protuberance or swelling (*e.g.,* Centrachantidae) [4, 6, 10].

Inside the membrane of the fertilized egg, there is a spherical yolk mass surrounded by a thin protoplasmic layer. In most species the yolk almost completely fills the available space leaving only a marginal area free, known as the perivitelline space, but in some groups this space can be considerably large (*e.g.,* Clupeidae, Anguiliforms). The yolk is the main component of eggs. It provides energy and materials for the developing embryos. Along with external factors, the amount and quality of yolk are decisive for the successful embryonic and post-embryonic development of many fish species as well as for successful recruitment into the adult population [11].

In many species the yolk itself is homogeneous in appearance, but in some species it may be partially or wholly divided into small segmental masses which can be a useful taxonomical character. Eggs of some species have one or more spherical globules of oil that can be important for egg buoyance and embryo nourishment [4]. The presence or absence of oil globules, their size, number and position inside the yolk, are also important taxonomic characters. In about 60% of the pelagic species for which egg descriptions exist, a single oil globule is present, and 15% have multiple oil globules [4, 6, 10].

Patterns of Embryonic Development

Immediately after spawning, fertilization occurs, *i.e.* the sperm penetrates the chorion of the egg and the process leading to sperm-egg fusion starts. After sperm penetration, a plug forms at the base of the micropyle to prevent polyspermy. Upon fertilization, the perivitelline space forms between the inner edge of the chorion and the egg membrane. This space protects and lubricates the egg, and helps with osmotic regulation [5, 6].

Inside the egg, the embryo develops from a single cell into a complex organism. This period can be divided into three stages: i) early stage embryo (from fertilization to blastopore closure); ii) middle stage embryo (from blastopore closure to the time the tail begins to curve laterally away from the embryonic axis) and iii) late stage embryo (from the moment the tail is curved away from the embryonic axis to hatching) [5, 6].

The early development of fish eggs generally exhibits a meroblastic pattern of cleavage, in which the cells form only at the animal pole of the egg, and cleavage does not go through the entire yolk. Depending on the species and temperature, the early cell divisions can take place in less than an hour to several hours. By the time the cell mass is halfway around the yolk, the differentiation of the embryo can be seen as a thickened line on top of the yolk. As the embryonic shield forms there is a thickening of the caudal margin of the blastoderm cap. The outer layer of the cells will originate the skin, and the inner layer the gut and mesoderm. The cells continue to divide around the yolk until only a small circle of yolk is not covered (the blastopore), and the embryo begins to take shape. Along the neural keel the neural tube and notochord are forming. At this point, myomeres gradually develop on either side of the notochord, the three main portions of the brain can be seen, and the optic vesicles appear as the blastopore closes [5, 6].

During the middle stage of development, internal organs (*e.g.* the liver and the gut) can be seen and the heart forms and starts to beat when the embryo circles about halfway around the yolk. Auditory organs and the eye lenses appear during this period. The embryo pigmentation also often first appears at this stage. Myomeres continue to be added both anteriorly and posteriorly, and the tail bud margin is defined [5, 6].

The late stage of development begins when the tail bud lifts off the yolk. During this period, pectoral fin buds and the otic capsules appear, the body lengthens, the various organs become better defined, and pigmentation increases. The embryo begins to move within the egg by this time [5, 6].

Hatching can occur at various stages of development, depending on the species.

Some fishes hatch when the embryo reaches a full circle around the yolk, while others hatch before or after this. Before hatching, special gland cells *e.g.* on the anterior part of most fish embryos [12] produce proteolytic enzymes that dissolve or weaken the egg envelope in order to release the embryo. Hatching is not a fixed threshold, but it is triggered by environmental cues such as low oxygen tension, light intensity or the release of hatching enzymes from adjacent eggs [13].

The rate of development of fish eggs and the length of time from fertilization until hatching depends on the species considered, and on factors such as temperature, oxygen and salinity. Fish species differ greatly in the optimal temperature for embryo development [11]. Nevertheless, within the usual range of temperature in which the eggs of a species develop, there is a linear inverse relationship between developmental time and temperature [5]. Reduced oxygen content and lowered salinity retard embryonic development for many species [5]. [14] also found a significant positive relationship between incubation period and the size of eggs of pelagic fish species [11]. Very small ova usually with little and low density yolk require the external acquisition of additional nutrients and in those cases the larval period becomes a natural prolongation of the life-history for getting food supply [15]. On the contrary, the maternal deposition of more and richer yolk into larger eggs enables the provision of enough nutrients to produce larger juveniles. Directly developing fishes start with large demersal eggs provided with an adequate volume of high density yolk and so require no or little external nutrients to develop into the definitive phenotype [13].

FROM HATCHING TO METAMORPHOSIS: THE LARVAL PHASE

The ontogenetic development that starts during the embryonic stage continues during the larval stage. Additionally, it is during this period that fish biomass starts to increase [9]. During the larval stage, fundamental structural and functional changes will occur in a short period, increasing the larval abilities to interact with the environment. Except in very few situations where direct development occurs, larvae hatching from both pelagic and demersal eggs will develop in the pelagic environment. The duration of this pelagic larval phase (PLD) is variable across species and environmental conditions. Given the small size, incomplete development and vulnerability to predation, mortality during this pelagic period can be extremely high (>95%, [9]).

Body transparency and some transient larval features that characterize many early larvae are often viewed as adaptations to the pelagic existence. Examples of such features are elongated body parts, head and/or fin spines, enlargement of soft rays, extended guts or stalked eyes [4]. Furthermore, the general body form, the gut shape and position of the anus, the form and size of the yolk relative to body size, or the number and positioning of oil globules inside the yolk sac, can be useful features in larval taxonomy at this stage.

Depending on the species reproductive strategy and life history, newly hatched larvae can vary between very small and poorly developed (altricial development, Fig. (3A)), and larvae that hatch larger (Fig. 3B) with developed sensorial and functional capabilities. Some larvae can even resemble the adults at hatching (precocial species) [9].

A

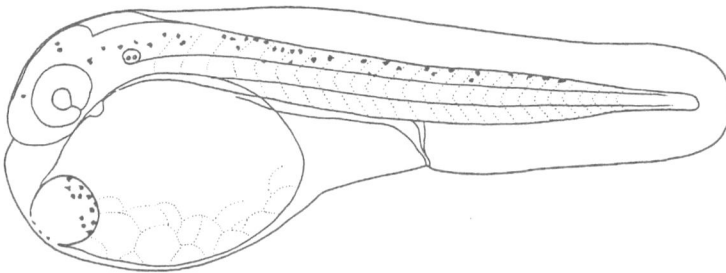

B

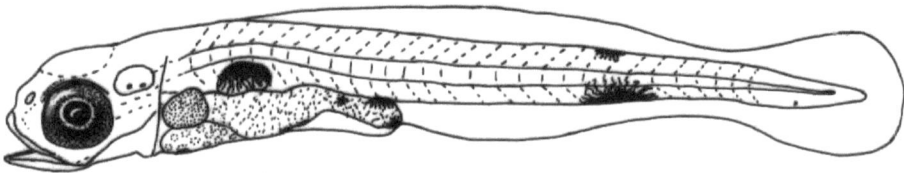

Fig. (3). Newly hatched larvae of **A**) *Trachurus trachurus,* with 2.2 mm, hatched from a pelagic egg [Adapted from [31]] and **B**) *Gobius paganellus,* with 3.9 mm, hatched from a demersal egg [Adapted from [32]].

However, the newly hatched larvae of most species are poorly developed and have a large yolk sac in a ventral-anterior position (Fig. **3A**). The yolk sac can vary in shape and size among species, and have one or several oil globules that are also metabolized after the yolk is absorbed. At this stage, the mouth and anus can still be closed; the intestine is not functional, and nourishment is ensured mainly from yolk absorption.

Usually most myomeres are formed upon hatching (although some can still differentiate during larval development in the caudal region) and their number approximately corresponds to the number of vertebrae that will later develop.

The transparent larval body facilitates myomer counting, which can also be useful in larval identification. For most species, the internal pigmentation is not very intensive at first. Nevertheless, at this earlier stage, significant changes in pigmentation patterns can also occur, as a result of melanophore migration between body parts [4] which will form the general larval pigmentation pattern, also variable among taxonomical groups. Although live larvae can have other cromatophores, the brown or black pigmentation (melanin) from melanophores is the most useful pigmentation for identification purposes, as the other pigments tend to disappear with fixatives and preservatives. The number, type and positioning of pigments can be very useful in this regard [6, 16]. Pigmentation tends to increase in the head and dorsal part of the body [9]. When approaching metamorphosis, internal pigments can develop close to the vertebrae; later dermal pigmentation appears and extends to both sizes of the body [9].

Growth is usually fast during the yolk phase, slowing down near yolk resorption and accelerating again with exogenous feeding [5]. Larval growth is allometric [9], *i.e.* body proportions change during development, until isometric growth is achieved in the juvenile stage [9].

As ontogeny proceeds, the development of the respiratory and digestive systems will improve the metabolism. In many species, branchial arches only develop during the larval phase and can be absent at hatching; kidneys also need to develop before getting functional. This means that respiration and osmoregulation in the larval stage can be different from the adults. Early in development, these functions can be achieved through the skin and through gas exchanges in the subdermal space [5]; this space exists in marine larvae and is filled with a gelatinous liquid of low specific gravity, which also helps larval buoyancy in the pelagic environment. This subdermal space is, nonetheless, absent in freshwater larvae [5].

While depending on yolk for nourishment, the digestive system of yolk-sac larvae develops while the mouth and anus open and jaws develop. The gut, initially straight or coiled and only fixed by mesenteries and integument within the subdermal space, also increases in complexity and differentiates into sections (foregut, midgut and hindgut), while muscles develop around the intestine [5]. Teeth appear usually later in larval ontogeny.

A mixed feeding period, in which larvae start feeding while the yolk is not yet depleted, may occur [9]. The transition to exogenous feeding is a critical period and increased mortality can occur at this stage due to starvation. When starvation is severe, an irreversible "point of no return" may be reached. The ability to resist starvation varies from species to species and depends on factors such as temperature (at higher temperatures shorter periods without food can lead to starvation) or body size and the degree of development (as the risk of starvation decreases as larvae have more developed sensorial and swimming capabilities and improved metabolism) [9]. The type of prey items will also depend on the mouth size and burst swimming performance [5, 9]. Small larvae can feed on protozoans, phytoplankton, or copepod eggs but the most common preys are planktonic copepod larvae. The size and type of preys also change along larval development [5] as the mouth gape size rapidly increases during this period. Sometimes even larval cannibalism can occur [9].

The little developed yolk-sac larvae lack differentiated fins and ossified structures; locomotion is achieved by a median finfold (a continuous membrane from the mid-dorsal body surface contouring the body until the anus, where the dorsal, caudal and anal fins will later develop) and sometimes a preanal finfold anterior to the anus; pectoral and pelvic fins are usually absent [9]. Furthermore, the notochord ends in a straight tip, the urostyle, which at this stage is still aligned with the body axis [9].

Fin development during the larval stage is of extreme importance in fish ontogeny. It will allow a significant increase of locomotor capabilities; when reaching the juvenile stage the adult complete fin ray number will be fully differentiated. Although the timing of fin development can also vary among species (and within the same species depending on environmental factors), the sequence of appearance usually follows the same pattern [9]: pectoral fin buds appear early in development, usually still lacking fin rays. The first rays to start differentiating will develop in the caudal fin region. This occurs simultaneously with the flexion of the notochord, a fundamental ontogenetic transition. During flexion, the urostyle bends upward and hypural bones first start to differentiate; caudal rays start to develop in that region of the median finfold [5, 9]. The urostyle flexion marks an important transition and larvae are often classified as (see [5, 16]):

i. Preflexion larva- from hatching to the beginning of notochord flexion.
ii. Flexion larva- from the beginning of the notochord flexion to the hypural bones assuming a vertical position.
iii. Postflexion larva- from formation of the caudal fin to attainment of full development of fin rays.

This classification is a good proxy to the functional stage in the ontogenetic development that is sometimes more useful for comparisons rather than using body lengths, extremely variable among species.

Rays will next differentiate in the caudal and dorsal fins, while the median finfold is reabsorbed between the fins; pectoral fin rays also develop from the initial buds and usually the pelvic fin buds are the latest to differentiate rays [9]. Fin development will influence the larval swimming performance, their ability to capture prey and avoid predators, and larval movements initially dominated by a viscous environment [17, 18].

Ossification of cartilaginous structures starts early after hatching. Jaws, gill arches and bones of the pectoral girdle are amongst the first structures to ossify [9]; vertebral ossification occurs in the postflexion stage, and further increases larval locomotor capabilities. It is frequently possible to count vertebrae while the body is still transparent, which is also important for taxonomy. Together with increased locomotor capabilities, the gas bladder filling that occurs usually during the flexion stage [5], will also contribute to an enhanced control of larval positioning in the water column. In fact, recent evidence revealed that late stage larvae can be strong swimmers, swimming for long periods against currents, which can influence their ability to disperse or be retained close to shore, and/or to find adequate shore habitats where to settle [19, 20].

The development of sensorial capabilities during the larval stage has also major ecological importance (see Review by [20]). The ability to sense the environment is variable across species and also improves throughout ontogeny. Most likely, larval fish use multiple sensorial channels to orient and detect suitable habitats.

Fish larvae are visual feeders and thus the ability to search, locate and capture prey is dependent on eye development, occurring during the larval period [5]. In early yolk-sac larvae, the eyes are formed and can have large dimensions in relation to body length, but they are not functional in many species, as the retina still lacks pigments [9]; nevertheless, eye development is very fast during the yolk-sac stage and it can be functional even in altricial species when the larvae starts exogenous feeding, *ca.* 2 days after hatching [20]. The first photoreceptors to develop in the retina during ontogeny are usually cones; rods will only increase in number by the end of the larval stage [9, 20]. On the other hand, in many species with demersal eggs, the eyes are already functional at the embryo stage and thus they are able to see at hatching (Fig. **3B**). Vision can also be important to find settlement habitats in the end of the larval stage; recent findings reveal that polarized light can be important for larval fish orientation [20, 21]. However, research is still needed on this field.

As opposite to vision, hearing has the potential to operate over larger distances (tens to hundreds of kilometres) as sound travels well in water with little attenuation [22], and is current-independent [23]. Fish depend on the inner ear and, to a lesser extent, on the mechanosensory lateral line system, to detect sounds. The actual role of the lateral line system on hearing abilities is still poorly understood, although it is known that it can be used to detect vibrations and water flow [24]; furthermore, the inner ear, which contains otoliths (calcium carbonate structures), is responsible for the detection of acoustic signals [24]. Otoliths may already be formed at hatching, or many times even before hatching [25], and fish are known to have good hearing capabilities throughout their larval phase [26, 27].

Other than sound and vision, olfaction can also play a role in orientation, or at least, in habitat detection. Studies on settlement-stage coral reef fish larvae have revealed that some species can detect odours, change behaviour in response to them, and localize their sources, at least over scales of tens of metres [28, 29]. As with hearing, what remains to be determined about olfaction is which substances can be detected, at what stage in development this ability is present and over what spatial scales such orientation can actually operate.

It is now clear that fish larvae of several species can be good swimmers and respond to environmental cues. However, when during ontogeny such behaviours and capabilities develop and what are the exact cues that are used are yet open fields of research.

The above description shows that it is during the larval stage that fish develop fundamental systems to become more competent to interact with the environment. By the end of the larval phase, the transition to the juvenile stage can also be a critical phase where high mortality can occur and influence recruitment variability. This is particularly true in those species with adults living in association with a demersal habitat, in which metamorphosis is usually coincident with larval settlement.

CONCLUDING REMARKS

The above described patterns express the most common ontogenetic transitions in the early phases of fish life cycles. Nevertheless, although frequent, the exact order of ontogenetic transitions is not universal. Many species will hatch in a more advanced developmental stage or may have distinct sequences of development. Marine fish larvae hatching from demersal eggs are usually larger and more developed at hatching, with little or no yolk and ready to start external feeding, having the eyes already pigmented and other developed sensorial and morphological features that allow a more capable perception and interaction with the surrounding environment earlier in development. Larvae of freshwater species also hatch from demersal eggs with a large size, a much higher dry weight (ca 10-fold heavier) and at a more developed stage than marine fish larvae hatching from pelagic eggs [9, 30].

The study of the variety of egg and larval traits and developmental patterns displayed by fishes brings to light some interesting and complex relationships that tie early-life history traits to environmental characteristics, and reproductive strategies. The comprehension of these relationships will provide a better basis for the understanding of fish population dynamics.

CONFLICT OF INTEREST

The authors confirm that they have no conflict of interest to declare for this publication.

ACKNOWLEDGEMENTS

The authors would like to thank E.J. Gonçalves for his comments and English review. This work was supported by a post-doc grant to AF (SFRH/BPD/ 68673/2010), and through the Pluriannual Program (PEst-OE/MAR/UI0331/ 2011), funded by Fundação para a Ciência e a Tecnologia.

REFERENCES

[1]   Wootton RJ. Ecology of Teleost Fishes. 1st. London: Chapman and Hall 1990; p. 404.

[2]   Helfman GS, Collette BB, Facey DE. The Diversity of Fishes. Blackwell Science 1997.

[3]   Bye VJ. The role of environmental factors in the timing of reproductive cycles. In: Potts GW, Wootton RJ, Eds. Fish reproduction: strategies and tactics. London: Academic Press 1984.

[4]   Kendall AW, Ahlstrom EH Jr, Moser HG. Early life history stages of fishes and their characters. In: HG Moser, Ed. Ontogeny and systematics of fishes Special Publication No 1. Lawrence, Kansas: American Society of Ichthyologists and Herpetologists. 1984; pp. 11-25.

[5]   Miller BS, Kendall AW Jr. Early life history of marine fishes. USA: University of California Press 2009.
      [http://dx.doi.org/10.1525/california/9780520249721.001.0001]

[6]   Russell FS. The eggs and planktonic stages of British marine fishes. London: Academic Press 1976.

[7]   Potts GW. Parental behaviour in temperate marine teleosts with special reference to the development of nest structures. In: Potts GW, Wootton RJ, Eds. Fish Reproduction: Strategies and Tactics. London: Academic Press 1984; pp. 223-44.

[8]   Balon EK. Patterns in the evolution of reproductive styles in fishes. In: Potts GW, Wootton RJ, Eds. Fish Reproduction: Strategies and Tactics. London: Academic Press 1984; pp. 35-53.

[9]   Fuiman LA. Special considerations of Fish Eggs and Larvae. In: Werner RG, Fuiman LA, Eds. Fishery Science – The unique contributions of early life stages. Blackwell Science Ltd 2002.

[10]   Ré P, Meneses I. Early stages of marine fishes occurring in the Iberian Peninsula. Lisboa: IPIMAR 2008.

[11]   Kamler E. Early life history of fish. An energetics approach. Fisheries Series 4. London: Chapman & Hall 1992.

[12]   Mabee PM, Cua DS, Barlow SB, Helvik JV. Morphology of the hatching glands in *Betta splendus* (Teleostei: Perciformes). Copeia 1998; 1021-6.
        [http://dx.doi.org/10.2307/1447351]

[13]   Balon EK. Alternative ways to become a juvenile or a definitive phenotype (and on some persisting linguistic offenses). Environ Biol Fishes 1999; 56(1-2): 17-38.
        [http://dx.doi.org/10.1023/A:1007502209082]

[14]   Ware DM. Relation between egg size, growth and natural mortality of larval fish. J Fish Res Board Can 1975; 32: 2503-12.
        [http://dx.doi.org/10.1139/f75-288]

[15]   Balon EK. Types of feeding in the ontogeny of fishes and the life-history model. Environ Biol Fishes 1986; 16: 11-24.
        [http://dx.doi.org/10.1007/BF00005156]

[16]   Leis JM, Carson-Ewart BM. The larvae of Indo-Pacific coastal fishes An Identification guide to marine fish larvae. Brill 2000.

[17]   Batty RS, Blaxter JH. The effect of temperature on the burst swimming performance of fish larvae. J Exp Biol 1992; 170: 187-201.

[18]   Fuiman LA, Batty RS. What a drag it is getting cold: partitioning the physical and physiological effects of temperature on fish swimming. J Exp Biol 1997; 200: 1745-55.
        [PMID: 9319652]

[19]   Leis JM. Are larvae of demersal fishes plankton or nekton? Adv Mar Biol 2006; 51: 59-141.
        [http://dx.doi.org/10.1016/S0065-2881(06)51002-8] [PMID: 16905426]

[20]   Leis JM, Siebeck U, Dixson DL. How Nemo finds home: the neuroecology of marine larval-fish dispersal and population connectivity. Integr Comp Biol 2011; 51: 826-43.
        [http://dx.doi.org/10.1093/icb/icr004] [PMID: 21562025]

[21]   Paris CB, Guigand C, Irisson J-O, *et al.* Orientation With No Frame of Reference (OWNFOR): A novel system to observe and quantify orientation in reef fish larvae. In: Grober-Dunsmore R, Keller B, Eds. Caribbean Connectivity: Implications for marine protected area management, US Department of Commerce, NOAA, National Marine Sanctuary Program, NMSP-08-07. Silver Spring, MD 2008; pp. 52-62.

[22]   Popper AN, Carlson TJ. Application of the use of sound to control fish behavior. Trans Am Fish Soc 1998; 127: 673-707.

[http://dx.doi.org/10.1577/1548-8659(1998)127<0673:AOSAOS>2.0.CO;2]

[23]   Armsworth PR. Modelling the swimming response of late stage larval reef fish to different stimuli. Mar Ecol Prog Ser 2000; 195: 231-47.
[http://dx.doi.org/10.3354/meps195231]

[24]   Hawkins AD. Underwater sound and fish behavior. In: T Pitcher, Ed. The behavior of teleost fishes. Baltimore: The Johns Hopkins University Press; 1986; pp. 114-51.

[25]   Simpson SD, Meekan M, Montgomery J, McCauley R, Jeffs A. Homeward sound. Science 2005; 308(5719): 221.
[http://dx.doi.org/10.1126/science.1107406] [PMID: 15821083]

[26]   Montgomery JC, Jeffs A, Simpson SD, Meekan M, Tindle C. Sound as an orientation cue for the pelagic larvae of reef fishes and decapod crustaceans. Adv Mar Biol 2006; 51: 143-96.
[http://dx.doi.org/10.1016/S0065-2881(06)51003-X] [PMID: 16905427]

[27]   Arvedlund M, Kavanagh K. The senses and environmental cues used by larvae of marine fish and crustacean decapods to find tropical ecosystems. Ecological interactions between tropical coastal ecosystems. Springer 2009.
[http://dx.doi.org/10.1007/978-90-481-2406-0_5]

[28]   Myrberg AA, Fuiman LA. The sensory world of coral reef fishes. In: Sale PF, Ed. Coral reef fishes: dynamics and diversity in a complex ecosystem. San Diego: Academic Press 2002.
[http://dx.doi.org/10.1016/B978-012615185-5/50009-8]

[29]   Leis JM, McCormick MI. The biology, behavior, and ecology of the pelagic, larval stage of coral reef fishes. In: Sale PF, Ed. Coral reef fishes: dynamics and diversity in a complex ecosystem. San Diego: Academic Press 2002; pp. 171-99.

[30]   Houde ED. Differences between marine and freshwater fish larvae: implications for recruitment. ICES J Mar Sci 1994; 51: 91-7.
[http://dx.doi.org/10.1006/jmsc.1994.1008]

[31]   Brownell CL. Stages in the early development of 40 marine fish species with pelagic eggs from the Cape of Good Hope. Ichthyological Bulletin. 1979. J.L.B. Smith Institute of Ichthyology, Rhodes University, 40.

[32]   Borges R, Faria C, Gil MF, Gonçalves EJ, Almada VC. Embryonic and larval development of *Gobius paganellus* (Pisces: *Gobiidae*). J Mar Biol Assoc UK 2003; 83: 1151-6.
[http://dx.doi.org/10.1017/S0025315403008415h]

CHAPTER 3

Introduction to Aquaculture

Paula Cruz e Silva[*]

Direção de Serviço De Alimentação e Veterinária, Avenida do Mar e das Comunidades Madeirenses, n° 23, 2° andar, 9000-054 Funchal, Portugal

Abstract: Farmed fish have been produced for more than four thousand years, but the most significant developments only arise in the last 50 years. Nowadays, fish and crustacean obtained in aquaculture systems represent almost 45% of all the fish products placed in the global market, about 160 million tons. Aquaculture involves human intervention in the life cycle of the cultivated organisms, and requires special techniques applied to housing, reproduction, feeding, fattening, healthcare and package to market distribution. The fish farms location depends on the species, space availability, climatic characteristics and environmental impacts. Most common regimes of exploitation are intensive, especially when they are intended for productions to be placed on the global market. Different production systems have distinct requirements in terms of ecological, reproductive and sanitary management, and their control is mandatory for the economic success of the aquaculture system. Health management is a key issue for the success of animal exploitation system.

Keywords: Aquaculture in ponds and raceways, Diseases diagnosis, Fish growing and fattening, Fish necropsy, Hatchery systems, Open-water aquaculture, Treatment and prophylaxis.

PRACTICAL NOTIONS FOR AQUACULTURE ESTABLISHMENT

One of the primordial factors to establish sustainable fish farm activities is the selection of place for its implantation. Aquaculture structures may be located in land, by the seashore, inshore and offshore (marine aquaculture). Most of the times, the most adequate locals for implantation of fish of farms of fresh, salt or

[*] **Address correspondence to Paula Cruz e Silva:** Direção de Serviço De Alimentação e Veterinária, Avenida do Mar e das Comunidades Madeirenses, n° 23, 2° andar, 9000-054 Funchal, Portugal; E-mail: paula.silva@netmadeira.com

Manuela Oliveira, Fernando Bernardo, Joana I. Robalo (Eds.)

brackish water, compete with agricultural practices or are located in geographic areas belonging to natural reserves. Due to this fact, a clear balance between environmental protection and economic development must be found [1].

Fish cultures implantation local is still dependent on the species to be harvested. Indigenous species should be preferentially chosen, since the physical and chemical composition of the water is more adequate for the development of the species. Market research and consumer preferences must also be taken into consideration in order to evaluate if the species are accepted by the consumer, in order to assure sustainability [2,3].

Another factor that can influence the choice of the implantation site is the material nature of the layout structures (ponds, cement and fibreglass tanks, metal cages, silos, frame nets) and also the regime of exploitation (intensive, semi-intensive, extensive).

In addition, other factors must be considered, including the proximity to the markets for products distribution (freshness), transportation facilities and the region climate, including water and air temperature, wind's regime and rainfall.

Regarding the water characteristics, which are major determinants on the success of a fish farm, several factors must be considered, including physical and chemical characteristics such as temperature, pH, suspended solids, solved gases and mineral content. Among these, it is important to refer the temperature, which affects fish life and growth. An optimum temperature can be established for the development of each fish species, and outside the eugenics temperature range, a decrease on the conversion rate and growth may be observed. Temperature variations also represent a stress factor and can also be responsible for disease development and for the success of spawning and egg hatching at the maternities.

AQUACULTURE IN PONDS AND RACEWAYS

In ponds, the constitution and characteristics of the land must also be taken into consideration, as well as the facilities for water runoff and ponds renewal. For an optimal water renewal, which improves fish handling, the pond must have a minimum inclination of 2% and water channels at the bottom.

In pond aquaculture systems, the soil characteristics are especially relevant, due to the necessity of dike construction to guarantee the detainment of the water column. Soil texture and porosity are the most important physical properties.

Intensive culture systems can also be applied, using raceways and circular tanks. Raceways are long tanks, with a water column of approximately one metre and preferably with a water inlet and outlet at each extreme. In circular tanks, there is usually a central water outlet and a peripheral water inlet. Therefore, the water flows with a vortex effect that promotes walls hygiene.

In raceways the fish stock density may be higher due to the constant water renewal, while the ponds just have one or two water renewals per day. They are grouped in series, by size, corresponding also to different age groups. In this production system, disease prevention is more difficult, as water moves between tanks. In fact, the water that comes out from the one tank must not be used in others to avoid pathogen dissemination.

OPEN-WATER AQUACULTURE

Open-water aquaculture can be performed in the ocean or in fresh water bodies, and may be applied to molluscs and fish. Rearing fish in off-shore cages is a quite promising practice in several European countries, including Portugal and Norway, but is a frequent traditional practice in some Asian countries and in America, the viability of the producing units in offshore systems is dependent on the quality of the marine environment and must take in consideration environmental impacts, namely aquatic animals' migration [4]. This production should not cause conflict with the ecosystem and other economic activities. Due to these facts, it is fundamental to determine which coastal areas are appropriate. When choosing the location of sea water cages there are logistic, oceanographic and environmental factors that must be considered (Table **1**).

Regarding the ecosystem characteristics, several parameters must be considered (Table **2**).

In addition to these factors, there are also official regulatory restrictions and concurrent recreational activities that may take place in the same spatial locations as the cages that must be taken into consideration for the site selection (Table **3**).

Cages structures may assume many configurations, designs and dimensions, which may reach 15000 m³ each. There is a trend in offshore aquaculture towards a larger cage dimension. Most commonly used cages have a circular cross-section. They can be submersible or have rigid-walls, but the majority is floating units that include a framework and a suspended flexible mesh-net. Cages construction material must take in consideration their intrinsic resistance, the combined forces of currents and waves, the need for maintaining cages depth and for surfacing and orientation.

Table 1. Factors to consider in open-water aquaculture.

Logistic	Oceanographic	Environmental
Proximity to the ports (handling operations; disposal facilities) Equipment availability (freezer, accessibility, fuel) Technological empowerment	Salinity Water renewal Sea currents (structures stability, pathogen dissemination including vectors, waste products removal) Period, amplitude and waves direction Amplitude of the tide Turbidity Bathymetry	Water temperature (constant) Trophic condition River entries Eolic factors pH Primary production Aquatic animals' migration routes

Once the cages are installed, it is necessary to monitor and control fish circulation, so they must be organized by fish ages and origin. Smaults should be located on the initial cages, so the organization is adjusted to the direction of water currents, to avoid pathogens dissemination from the older fishes. General management should also avoid the accumulation of unused feed and fish faeces on the bottom of the site, causing oxygen depletion or generation of detrimental gases, which represent one of the major problems in these culture systems.

When the aim of the production is to obtain only one species with uniformed characteristics, the reproductive management is a major factor for a successful aquaculture activity.

HATCHERY SYSTEMS

For seed production, some aquaculture facilities promote the complete production cycle, while others acquire the fries from a third operator.

Table 2. Ecosystem criteria for implantation site selection according to Katacic & Dadic [adapted from 4].

Criteria	Optimal	Acceptable	Not Acceptable
Exposition	Partially exposed	Protected	Exposed
Amplitude of the wave (m)	1 a 3	<1	>3
Bathymetry (m)	>30	15-30	<15
Flow velocity (m.s-1)	0.2-1.0	1.0-1.5	> 2.0
Contamination	Low	Medium	High
Maximum temperature (°C)	22-24	24-27	>27
Minimum temperature (°C)	12	10	<8
Average Salinity (‰)	25-35	15-25	<15
Salinity (fluctuation)	<5	5 a 10	>10
Dissolved oxygen (% saturação)	>100	70 a 100	<70
Slope (%)	≤30	30 a 45	>45
Substrate	Rock, sand or gravel	Sand or mixture with rock	mud
Trophic condition	Oligotrophic	Mesotrophic	Eutrophic
Fouling	Low	Moderate	High
Pedation	Low	Moderate	High
Tide effect	Low	Moderate	High
Faecal coliforms (MPN/100ml)	≤14	14-88	>88
Turbidity(m)	>3	1 a 3	<1
Port distance (km)	<2	2-5	>5

To achieve a complete cycle, it is crucial to have a very accurate reproductive program. The first stage of these programs is unfertilized eggs and sperm extraction. After fertilization, the eggs are placed in trays with a perforated base in a maximum of two layers. These trays usually have a height of 15 cm and an area of 50 cm^2 and can be placed side by side in tanks with running water. This arrangement facilitates eggs observation and subsequent removal of the unfertilized ones, to avoid fungal development. Unfertilized eggs have a whitish appearance and are easy to be identified by trained operators [5].

Besides trays, eggs can also be placed in other containers, such as fiberglass jars or vertical spawning systems. Within these systems, the water entrance must occur by the inferior part and pass through the eggs before overflowing by the surface. A mesocosm can also be used as a hatchery system to obtain higher volumes of eggs, using tanks with a slight water flow [6, 7].

Table 3. Restrictions to be considered in open-water aquaculture.

Regulatory restrictions	Co-spatial activities
Airport servitude Ports and marinas limits Protected areas Undersea telecommunications cables	Recreational areas Sewage from farms and undersea outfalls Landscape interest Beaches Ships routes Extraction areas of marine inert Fishing areas Hotels and tourism

Besides water flow, which should be characterized by a low constant volume of inflow, in a fish maternity there are other aspects that must be controlled, like water temperature and the amount of oxygen.

The first feeding of the fry takes place in the hatchery, using live planktonic culture composed by microalgae or zooplankton, until fries reach the minimal dimension necessary to be fed with commercial formula. To monitor this production step, fries can be placed together with some larvae on a petri dish, to observe if the young fish grab them. At this stage it is essential to give special attention to the temperature, water oxygenation and light cycle, since fries are very young and their immune system is not fully competent.

Afterwards, fries and fingerlings are removed and placed in cement or fibreglass tanks for subsequent development phases (growing and fattening). In these initial phases of fish development, special attention must be given to water circulation.

FISH GROWING AND FATTENING

When fingerlings have the appropriate dimension, they are transferred for outdoor ponds with a low water renewal rate, where the subsequent production steps take place. These ponds could be located in salterns, small lagoons, tanks or river dikes. As a rule, fishes usually remain in the ponds or are transferred to cages, where they are fed and fattened until they reach the sale commercial size. Afterwards, they are harvested, without applying evasive handling operations to guarantee animal welfare protection, using large nets called seines.

In ponds, the conversion and growth rates are lower comparing to the results obtained in cement or fibreglass tanks, where the control of the environment parameters is optimized due to the lower dimensions and the predicted performance of the construction materials. When pond systems are used, it is necessary to consider the abrupt differences of dissolved oxygen level during day/night cycle. If this level falls below a critical point, fish may suddenly suffocate and die.

The most relevant handling operation in raceways, circular tanks and floating cages are good hygienic practices. In cages, a diver is necessary, to avoid dirt accumulation and collapse of the nets, compromising the water flow in the interior of the cage and consequently the deterioration of the biotope (water mechanical agitation is sometimes needed). Nowadays, vacuum cleaners are used, to avoid perturbation of the faeces and unused feed in the bottom.

Other good principle practices include fish handling and separation by size, to avoid competition for feeding and the increase of size and development discrepancies. In addition, the nutritional requirements and size of the feed pellets are distinct. Feed analysis is also extremely important for guaranteeing fish health, so one of the most relevant facilities in aquaculture are laboratories for feeding analysis, with qualified and experienced technicians, and also laboratories for diseases diagnostic.

GENERAL NOTIONS OF DISEASES DIAGNOSIS, TREATMENT AND PROPHYLAXIS

For the diagnosis of fish diseases in aquaculture systems it is advisable to promote the interactions between the producers, pathologists and the competent authorities responsible for establishing the official sanitary programmes. Without this complementary work it is difficult to elaborate an anamnesis, able to conducing the clinical exam and the specific laboratorial methodology, leading to a differential diagnosis indispensable to obtain the final confirmation. The following scheme demonstrates this situation.

The need for diagnosis could overcome contingency situations, due to a sudden emergency related with high morbidity and mortality rates, or being a routine procedure for sanitary certification, aiming the statutory health classification of the production system within its geographic context (compartmentalization).

Fish diseases are usually multifactorial and may involve many aspects of the aquaculture system: nutrition, handling, reproduction, exploitation regimen, biology, pathology and epidemiology.

FIELD EXAMINATION AND ANAMNESIS

Previous clinical history should contain information regarding pisciculture systems (ponds, raceways, cages, tanks); fish stock density and its origin (complete cycle or acquisition from third exploitations); species and age group specific vulnerabilities; epizootiology (quantification and localization of tanks harbouring affected animals); nature and characteristics of the culture systems, including water entrances, shape and construction material; disease and mortality assessment, including evaluation of the acute, pre-acute or chronic character of the disease; symptoms and/or injuries onset; feed management (nature and frequency); handling operations (weighing, screening, cleaning procedures); variations in water quality and quantity; climatic changes; and exogenous factors with implications in fish farm exploitation (biocides uses and effluent discharges).

Complementarily, water analysis may be necessary to better understand disease physiopathology in pisciculture, regarding the following parameters: oxygen and temperature, pH, ammonia, nitrites, salinity and hardness. Each tank or pond must be treated as one unit. The water sample should be kept at room temperature if the analysis is performed within one hour; or if not, then it should be refrigerated.

Observation of the animals' behaviour in the farm environment should also be part of the anamnesis. Therefore, producers and pathologists should be aware of fish swimming patterns (normal, erratic, leaping, scraping behaviour, spinning, side swimming); fish response to external stimulus, like feeding habits; animal distribution in the tanks; respiratory movements; presence of external parasites; body pigmentation (affected animals present a darker colour); presence of ulcers, necrosis or scales flaking; ocular distresses (blindness, exophthalmia, loss of the ocular globe); deformations of the vertebral column and abdominal arching.

Assembling all of the collected information from anamnesis is a major challenge for the veterinarians. This assessment allows them to evaluate the performance of the productive system and to distinguish if the responsible factors for disease development are handling procedures or environmental characteristics. This approach allows scheduling and integrating prophylactic measures.

It is also important to take into consideration previous information concerning previous diseases outbreaks, since they may contribute with important elements to direct specific diagnosis, including most affected age groups, seasonal disease variation and interrelation with the origin of fish stock and feeding.

CLINICAL EXAMINATION

In the fish farm, the first clinical observations should be performed in a subgroup of six to ten animals that present morbidity or typical symptoms. The purpose of this exam is to complement the anamnesis. Whenever possible, the clinical responsible should have an appropriate knowledge of the fish farm, its management and structure, and to obtain complementary elements required for an assertive diagnosis. When the fish farm is not equipped with its own laboratory, necropsies must take place in a distant local, to avoid risk of cross-contamination, and using appropriate asepsis devices and procedures.

The first step includes fish measurement (from the tip of the snout to the end of caudal peduncle) and weighing, and should be followed by a systematic necropsy procedure (Table **4**).

When a skin biopsy is required, fish should not be involved in paper, since this material absorbs the skin mucus and may remove external parasites. Skin biopsy preparations can be observed using a stereoscopic magnifying glass and an optical microscope.

The first anatomical structure that must be observed is the gills, resorting, if necessary to a stereoscopic magnifying glass or an optical microscope. The medial gill arches should be preferentially observed, because they are sheltered from the water dragging action and may present a larger number of parasites and other pathogenic agents. This observation allows detecting the presence of ectoparasites (protozoa and metazoan), the existence and/or the effect of therapeutic schemes (confirmed by the outline of the distal portion of the primaries lamellas), the presence of chemical products in the water (a haemorrhagic appearance) and anaemia signals (gills paleness or discolouration).

Table 4. Systematic schedule for fish necropsy.

EXTERNAL EXAM	INTERNAL EXAM
External habit	Blood observation
Weighing and measurement	Gills observation
Skin observation	Eyes observation
Nostrils observation	Opening the abdominal cavity
Mouth observation	Removal and observation of the heart
	Removal of the intern organs
	Separation and observation of the intern organs
	Removal and observation of the kidney
	Removal and observation of the brain

In case of anaemia suspicion, blood smears and posterior evaluation should be performed. After assessing the abdominal cavity, internal organs should be observed, lesion patterns should be registered and samples should be taken for the histopathology analysis (tissues fixation in formaldehyde with phosphate buffer) or others (bacteriology, mycology, parasitology, toxicology) (Table **5**). For microbiological testing, haematopoietic organs are the preferential targets to be collected (liver, spleen and kidney). Furthermore, heart samples should also be collected. In particular, samples should be kept refrigerated for viral detection, while for parasitology examinations, specimens can also be fixed in a 70% alcohol. Fish freezing should not be considered as a preservation method for laboratory analysis, with the exception for toxicology and virology.

THERAPEUTIC METHODOLOGIES

To prescribe and apply the appropriate therapeutic, the clinician must have in consideration the specific etiological primarily responsible for fish disease or mortality. Therapeutic schedules can be performed as short-term or a long term approaches. The long term treatment concerns to prophylactic management practices, including: not overcome the facilities capacity, and stock density and animal size should be monitored, as it varies according to the animal species, the type of production and the amount of water; handling animals and maintaining water quality according to the target species; applying quarantine periods appropriated to the fish species, age and animals origin, quarantine is fundamental to the control of the introduction of new pathologies; Disinfecting tanks and devices, adequately identified by the tank they belong; Increasing of the host resistance, using nutritional supplements in feed, probiotics and vaccinations; eradicating the potentials intermediary, vectors and predators hosts; providing disinfection and general hygiene devices for the workers and wheel sets for vehicles [8, 9].

Table 5. Systematic schedule for fish necropsy; Legend: *** Optimal, ** Acceptable, * Conditionally accepted, - Inappropriate.

	Live fish	4°C	Frozen	Formaldehyde
Direct exam	***	**	*	*
Toxicology	***	**	***	-
Parasitology	***	**	*	**
Bacteriology	***	**	-	-
Virology	***	***	**	-
Histology	***	-	-	***

The short term therapeutic plan involves the use of drugs for disease combat. Before starting any therapeutic schedule, it is crucial to:

a. Diagnose the specific cause of the disease. The lack of an accurate diagnose can lead to the misuse of ineffective medicines and these also may induce antimicrobial resistances;

b. Enforce a fasting period of 24 h before the start of each treatment, especially when *per osvia* is used. The fasting period increases the animals' appetite, ensuring that the first dose of drugs will be ingested, because the animals will eat all the available feed. It will also contribute to reduce the amount of detritus in the tanks. Therefore, the quality of the water will be favoured.

c. Check the water volume in the tanks and input flow;

d. Register the applied treatments in the fish farm book of therapeutics and enforce the withdrawal period when fish are harvested for market;

e. Study the cost/benefit relationship of treatment administration;

f. Monitor the qualitative parameters of the water;

g. Observe the gills. It is very important in case of bath treatments: Generally drugs affect the gills and, for that reason, it is necessary to assure the better possible conditions;

h. Check the aeration devices so that it may be activated in case that the animals demonstrate shortness of breath;

i. Maintain the availability of the staff necessary to act in case of urgent and quick need;

j. Assure that staff is wearing specific protective clothing;

k. Perform periodic surveillance of small number of fishes (samples) to ensure sure that the animals are not having adverse reactions to used drug.

ROUTES OF ADMINISTRATION

Drugs administration to farmed fish may be applied using bathes or water treatments including dipping, bathing, diving, flushing or in continuous; *per os* or oral administration, incorporating the medicine in the used feed; parental administration, applicable when the number of fish is scarce (brood stock or aquarium fish); and topical administration.

BATHES OR WATER TREATMENTS

Medicines may be applied to the affected fishes using different modalities of immersion, especially adequate for small structures (aquariums, tanks).

Dipping is a route of the medicine administration that may only be applied for a limited period of time, even seconds. Before starting using a mixture of boiled, the volume of the water must be reduced and the aeration system checked. The drugs are diluted in a recipient with the tank water. The water entrances are closed and the drug is added avoiding its direct contact with the fish. Once the time of treatment is finish, the initial volume of water is replaced.

During the treatment, the animals must be carefully observed, so that any alarm signal may be early detected allowing a quick response in order to restore the normal conditions of the tank. The calculation of therapeutic doses is crucial and and the volume of remaining water in tank in which fishes were treated must be taken into consideration.

Bathes may be applied for several hours, it can even take one day. Before the beginning of the treatment procedure, the level of the water is reduced and the aeration system has to be checked. Drugs are diluted in the same way as on the dipping procedure.

The water influx has to be stopped to guaranty the maintenance of the medicine concentration and it should be poured carefully into water drugs avoiding direct contact with the fishes (previous dilutions in suspension water). Before treatment ending, the initial volume of water is replaced. The animals must be monitored continuously as referred for the dipping procedure.

Diving is a methodology usually chosen for the disinfection of eggs and small groups of fishes. It involves the addition of drugs in higher doses to the water contained in recipients usually made from fibreglass, where the eggs and the fishes are immersed for a few seconds.

Flush is a process having permanent water ingress, not closed, however water flow is reduced. The drugs are added on high concentration in the water ingress and, with the water flow, they are dragged into the entire tank. After finishing the treatment, the initial volume of water is replaced as soon as possible.

The quantity of product to apply should be determined according to the following formulas:

Volume to be handled = Caudal (1/second) X Time of treatment

Quantity of product to be used = Dose (ml/m³) X Volume to be handled

The flush is only used on explorations having a high water flow.

The continuum water ingress is controlled in order to have a specific water flow. A portion of the drug is added to the water tank and the remaining part is placed in a recipient that allows a constant dose (example, a plastic barrel with a tap), placed on the tank water ingress. With this the drugs are dragged into the tank.

SYSTEMIC *PER OS*

Most common medicines used in this method are generally antibiotics. It is necessary to access the weight and the total number of animals to be treated (biomass), in order to calculate the amount of product that is needed to be applied. Whenever possible, antibiotic prescription should be based on the results of a test for antibiotic sensitivity (TAS) as a way for better targeting, although this procedure may take excessive time and the mortality requires a quick treatment attack. If this is the case, the decision is based on symptomatology and pisciculture clinical history, for immediate treatment application.

During treatment process, there is a need to reduce the amount of feeding, in order to ensure that all the medicated feed is eaten. Whenever possible and in case of especial fish exemplars (specimen of exhibition or collection, reproducers) manual feeding should be tried, allowing to better realize the animals' reactions.

During prolonged antibiotics' treatments, antimicrobial resistances must be monitored. Due to this fact, an antibiotics turnover is recommended and the interdiction of their use as a prophylactic treatment (good practices of prudent use of antimicrobial drugs are mandatory).

A very relevant practice is the respect of the dosages and the time of treatment that shall not exceed the time prescribed by veterinary.

In those circumstances where it is difficult to obtain medicated feed, there are practical methods that can be used by allowing incorporating the drugs in the feeding stuff. In feed having moderate level of fat, a mixture of the medicine with vegetal oil (1 litre of oil for 25 Kg of feeding) may be tried. After this preparation, and with the help from sprayer or a sprinkler, the mixture is homogeneously added to the feeding.

When the feed has high levels of fat, or it assumes the form of large size pellets, instead of the procedure, other methods should be tried, using a mixture of boiled water with gelatine. When the mixture reaches room temperature, a specific dose of the medicine product is added. This mixture is afterwards added to the feeding with the help of a mixer. The amount of gelatine generally recommended is about 5% of the feeding weight.

Drugs dose to apply can be calculated by finding first the tank biomass (number of fishes X unit weight); then reduce 1 to 1.5% the feeding volume; and the quantity of drugs = (gr of drugs X kg of feeding) X days of treatment.

SYSTEMIC PLUS IMMERSION

Association of these two procedures is used when a septicaemia condition is concomitant with external ulcerative lesions or when simultaneous mixed infections are present (bacteria, parasites and fungi).

The technical procedures are the combination of those ones previously described for each treatment.

PARENTERAL

Parenteral treatments, normally used on fishes are of great economic value, in small number, for example the breeders. Fishes are anesthetized (in small tanks) before the application of the injection, by intraperitoneal or intramuscular routes.

After this operation the fishes are placed in tanks with water without aesthesia for recover.

TOPICAL

Topical treatment is a stressful procedure but that can be applied on fishes having a special economic value (collections, breeders) and it is used in the treatment of wounds, ulcers or to remove external parasites.

After therapeutics, fishes stress should be minimized and avoided for at least the consecutive 48 hours. Treatment efficiency should be monitored through a confirmatory negative result obtained by new laboratorial analysis, the disappearance of symptoms or injuries and stop of mortality and normal swimming behaviour.

CONFLICT OF INTEREST

The author confirms that author has no conflict of interest to declare for this publication.

ACKNOWLEDGEMENTS

Declared none.

REFERENCES

[1]   Goldburg R, Naylor R. Future seascapes, fishing, and fish farming. Front Ecol Environ 2005; 3: 21-8.
[http://dx.doi.org/10.1890/1540-9295(2005)003[0021:FSFAFF]2.0.CO;2]

[2]   Das B, Khan YS, Das P. Environmental impact of aquaculture-sedimentation and nutrient loadings from shrimp culture of the southeast coastal region of the Bay of Bengal. J Environ Sci (China) 2004; 16(3): 466-70.
[PMID: 15272725]

[3]   Wurts WA. Sustainable aquaculture in the twenty-first century. Rev Fish Sci 2000; 8: 141-50.
[http://dx.doi.org/10.1080/10641260091129206]

[4]   Torres C, Andrade C. Processo de decisão de Análise Espacial na Selecção de Áreas óptimas para a Aquacultura Marinha: O Exemplo da Ilha da Madeira. J Integr Coastal Zone Manage 2010; 10: 321-30.

[5]   Lekang IODD. Aquaculture Engineering. Oxford, UK: Blackwell Publishing 2007; p. 340.
[http://dx.doi.org/10.1002/9780470995945]

[6]   Andrade C, Nogueira N, Silva P, Dinis MT, Narciso L. Mesocosm hatcheries using semi-intensive methodologies and species diversification in aquaculture. J Agric Sci Technol 2012; 2: 428-33.

[7]   Andrade C, Abreu A, Branco A, *et al.* Red Porgy, *Pagrus pagrus* L. (Pisces: *Sparidae*) larvae culture and live food density under mesocosm culture conditions. Bol Mus Munic Funchal 2010; 60: 45-56.

[8]   Lindsay GR, Ross B. Anaesthetic and Sedative techniques for Aquatic Animals. Hoboken, New Jersey: Blackwell Publishing 2008; p. 222.

[9]   Milne PH. Fish and shellfish Farming in Coastal Waters Fishing News Books. Oxford, UK: Wiley-Blackwell 1972; p. 208.

CHAPTER 4

Microbial and Parasitic Diseases of Fish

Fernando Bernardo[1,*]

[1] *Faculdade de Medicina Veterinária, Universidade de Lisboa, Avenida da Universidade Técnica, 1300-477 Lisboa, Portugal*

Abstract: Diseases of fish, caused by biological agents (bacteria, fungi, virus or parasites), are better known in farmed or ornamental fish, since the access to the affected fish coming from the natural environment is less probable. In fact, the fish that are infected by a specific pathogenic agent, natural inhabitants of the marine, estuarine or freshwater ecosystems, are rapidly eliminated from its biotopes by other predatory animals, due to its higher vulnerability and susceptibility to the social interactions.

Diffusion of pathogenic bacteria, fungi, virus or parasites in the aquatic environments is more efficient than in the terrestrial ones. Diagnosis and therapeutics of fish disease have specific difficulties and the application of preventive measures is a very complex issue. Some of those diseases have the same epidemiological problems of the infectious diseases of the terrestrial animals: High spread of diffusion, very significant economic losses, restrictions to fish travel or commerce and some (few) have zoonotic impacts.

Keywords: Diagnosis of fish disease, Disease of fish, Health status of fish farms, Icthyopathology, International sanitary certification.

INTRODUCTION

Bruno Hofer (1861-1916) was the first author to describe diseases of fish on a scientific perspective and the creator of a new discipline - Icthyopathology. Since those early times, in the begining of the 20th Century, many evolutions were achieved concerning the study of microbial and parasitic fish diseases. Transmission of pathogenic bacteria, fungi or virus in the aquatic environments

* **Address correspondence to Fernando Bernardo:** Faculdade de Medicina Veterinária, Universidade de Lisboa, Avenida da Universidade Técnica, 1300-477 Lisboa, Portugal; Email: fbernardo@fmv.ulisboa.pt

Manuela Oliveira, Fernando Bernardo, Joana I. Robalo (Eds.)

is certainly more efficient than in the terrestrial one. Water, where fishes leave, is simultaneously its table, food store, bedroom, nursery, bathroom, sanitation plant and grave for most of them. Exception is those that are fished for consumption, whose latest graves are the human dishes. Infectious diseases of fishes arise when susceptible species are exposed to microbial pathogens under adverse ecological conditions. Compared with the terrestrial animals, the volume of research workload carried out to assess aquatic animal health is markedly inferior, including the effect of water sanitation status on infectious fish diseases outbreaks. Although, there is much literature showing the occurrence of infectious diseases in stressed fishes due to inappropriate temperatures (stenohaline fishes), water eutrophication, polluted discharges, metabolic products of sub aquatic living organisms and many chemical or physical residues and contaminants thrown into water, its health impacts in fish are not appropriately assessed. In disgenesic conditions, and in natural ones, fishes are vulnerable to a wide range of parasites and microbial agents which potentially impact with its health status, welfare and lifetime expectations. Some of those infectious diseases have an epizootic evolution, with high rates of mobility and morbidity [1]. In the natural environment, the morbidity and mortality rates triggered by fish diseases, may impact the life of other creatures dependent of them to survive in the given ecosystem (aquatic birds, mammals and invertebrate) and may also affect the subsistence of the local human communities (piscatorial). For those fishes produced in aquaculture systems, the consequences of infectious and parasitic diseases also have very relevant economic and social impacts [2]. The potential wide spread of those diseases to other regions where the specific involved pathogens are not present, they enhance a very serious threat to the indemnity status of a previous free geographic zone. This is why a very particular attention must be given to the rules of fish commerce (local, regional and international). Of special concern are the pathogens carried accidentally by ornamental fish, traded all over the world, especially if local official veterinary services have not sufficient capacity building to diagnose and to control fish diseases in the international markets. Global commerce is a major source of concern for the spread of the infectious and parasitic disease agents. Due to this potential threat, many international agreements have been stated, aiming to avoid such problem. The international organization for animal health, OIE, published and review, regularly, the "Aquatic animal health code", an essential document for the risk management of fish health, at local, national and international levels. Systematic veterinary controls at borders are a crucial tool to avoid spreading of those agents. At local level, farmed fishes for food or ornamental proposes, must be scrupulously scrutinised, monitored, surveyed, with the propose to early detect the presence of any the infectious agents, especially those that may have major sanitary and economic impacts. Veterinary authorities also need to have in place official efficient plans for control, eradication or surveillance. Registration of all the nosologic events is absolutely necessary. Without that, sanitary certification for international trade is not possible. Most relevant infectious and parasitic diseases of fishes are reported in the "Aquatic animal health code" published by OIE, but, in general, the recognition of a specific infectious disease in fish is not an easy task. Major difficulties come from the lack of clinical signs, unless high mortality, stop growing, feed refuse, abnormal movements or colour modifications. Many fish diseases have the same symptoms or clinical signs.

BACTERIAL DISEASES OF FISH

There are many bacterial species living in the aquatic ecosystems (freshwater, estuary and marine) that are able to infect multiple fish species. Most are opportunistic agents attacking whenever the natural defenses of fish organism breakdown. Most frequent bacteria that may be incriminated in fish diseases are: *Aeromonas hydrophila* and *A. salmonicida* sbp. *salmonicida; Edwardsiella tarda* and *E. ictaluri; Vibrio* spp.; *Yersinia ruckeri; Renibacterium salmoninarum; Flavobacterium psychrophilum, F. columnaris* and *F. johnsonae; Piscirickettsia salmonis; Mycobacterium marinum* and *M. fortuitum; Nocardia asteroides* and *N. kampachi; Carnobacterium piscícola;* and *Pseudomonas anguilliseptica.* Some of these bacteria express their virulence inducing morbid situations, expressing or not clinical signals having some specificity, although this is not mandatory [3]. Those bacteria species enhance morbidity in different fish species, yielding a series of specific nosologies.

BACTERIAL HEMORRHAGIC SEPTICAEMIA

Aetiology: *Aeromonas hydrophila* is the opportunistic pathogen responsible for this infection. It is a Gram-negative rod, motile through polar flagella. It is a facultative anaerobe, oxidase-positive, natural inhabitant of aquatic ecosystems; Able to grow in wide range temperatures (from 4 °C to 37 °C).

Susceptible fish species: Most of freshwater fish species (aquaculture, ornamental and natural) are vulnerable to this bacteria, but predisposing factors must be associated (stress, overpopulation, sudden alterations in waters) [4].

Major clinical signs and lesions: Symptoms and lesions are not specific enough. Most frequent events are fins, skin, mouth and muscles hemorrhages, epidermidis and celomic cavity ulceration (similar to *A. salmonicida*), exophthalmos, ascite, swollen kidney and liver [5].

Diagnosis: Clinical and anatomopathological and histological examinations are not conclusive, as unspecific multiple areas of focal necrosis enclosing bacilli can be observed in several organs, including the liver, kidney and heart. Only performing conventional bacterial cultures for isolation and characterization of the organism, using aseptic samples collected from fresh organ biopsies of the affected fish, is conclusive. Even so, it is important to devote some attention to the fact that this bacterium is a common water inhabitant, with a large variety of serotypes with different degrees of virulence or even none [6].

Therapeutics and prevention: *Aeromonas* spp. are natural resistant to some β-lactam antibiotics (penicillin, ampicillin, carbenicillin and ticarcillin) but are, generally, susceptible to broad- spectrum cephalosporins, aminoglycosides, carbapenems, chloramphenicol (not allow in fish for food, only for exhibition or decorative aquarium fish), tetracycline, trimethoprim-sulfamethoxazole and the quinolones. Experimental vaccines have been developed [7]. Control of predisposing factors, is crucial to avoid the deflagration of the disease, namely: water pH, temperature, under nourishment and overpopulation. Sanitary impacts: This bacterial infection can be disseminated by contact with contaminated water or fish. Introduction of new infected fish in the system is the main source. Affected fishes or contaminated fish farms must block the commerce. Eventual public health impact of the affected fish has been discussed.

EDWARDSIELLA SEPTICAEMIAS

Aetiology: *Edwardsiella tarda* and *E. ictaluri* (Enteric septicaemia of catfish) are rod-shaped Gram-negative bacteria, pleomorphic, belonging to the *Enterobacteriaceae* family. They are facultative anaerobe and motile due to peritrichous flagella. They are found in the normal gut microbiota of fresh and marine water fishes. In general, fish faeces are heavily charged with these agents. Susceptible fish species: These bacteria primarily infect channel catfish, but it has also been reported in tilapia, golden shiners, goldfishes, brown bullhead and largemouthbasses. They are also one of the most serious threats to eel culture in East Asia. There are some reports of the zoonotic potential of *E. tarda* (food poisoning). Major clinical signs and lesions: these are analogous to those found in *A. hydrophila* infections, as fish can present skin and muscle haemorrhages and skin ulceration [8]. Gas can be found inside muscular lesions of the newborn cavity, smelling malodorously. Disease affects primarily yearling blue catfish and *Centrarchidae* (largemouth bass). Gross clinical signs of this catfish septicemia are similar to those other systemic bacteria infections. Most typical external lesion is a fistula located in the eyes regions [8]. Diagnosis: Gross morbid signals do not have not a specific pattern, being similar to all acute systemic infections due to Gram-negative bacteria (Vibriosis, infections due to *Aeromonas* or *Pseudomonas*) [10]. Only conventional bacteriological examinations allow differentiation. Routine standard protocols are available for bacteriological culture, isolation and identification of *Edwardsiella* species and for differentiation from other Gram-negative facultative bacteria. Therapeutics and prevention: Antibiotherapy has proved to be effective, based in prudent use protocols and *in vitro* antimicrobial susceptibility tests. Infection dissemination may be limited through reduction of stock fish densities. Experimental vaccines have been tested [11].

Sanitary impacts: Infections induced by Edwarsiella species do not have very significant sanitary impacts, neither in traffic or fish commerce. Also does not enforce international notification. However, since morbidity rates are high and economic losses are significant, aquacultures where these specific infections occurs must avoid commercial exchange when fish have clinical signs.

VIBRIOSIS (VIBRIO SEPTICAEMIA)

Vibrioses occur in fishes in special disbiotic conditions due to some disequilibrium in the ecosystem. In general, these are opportunistic infections emerging when the water is organically or chemically polluted or abnormal temperatures of water, by stressing factors, or when other primary pathogens are active (virus, parasites).

Aetiology: *Vibrio parahaemolyticus, V. alginolyticus, V. anguillarum, V. salmonicida, V. tubiashii, V. carchariae, V. splendidus, V. pelagius* and *V. vulnificus* [12]. These bacteria are Gram-negative rods, slightly incurved, living primarily in a salted water environment (strictly Na dependent). Halophylic *Vibrio* species are aquatic bacteria characteristic of the marine and estuarine environments [13]. A special pathologic entity – "Ulcer Disease of Damselfish or photobacteriosis" is caused by a former designated *"Vibrio damsela"*, then *"Listonella damsela"* and, nowadays, classified as *Photobacterium damselae* subsp. *piscicida*.

Susceptible fish species: Vibriosis may occur in almost all fish species living in a salted water ecosystem and even in some catadromous (eels) and anadromous (*Salmonidae*) fish. Fishes belonging to *Salmonidae, Anguillidae, Ciprinidae* and *Cichlidae* families seem to be more vulnerable. Major clinical signs and lesions: Some authors refer to distinct nosologic entities when a particular species of *Vibrio* is responsible for illness conditions in specific fish species, for instance: "red pest of eels", "haemorrhagic syndrome", "Hitra disease" or "cold-water vibriosis". For more than five centuries ago those diseases were described in the Italian coastal zones as: "red pests", "red plague", "red boil" or "saltwater furunculosis". Infection may have three different evolutions: hyper acute, acute or chronic. Most common signs are found in the acute form of the disease, namely: swimming movements are slower due to prostration and stiffness of the muscles; ulcers and darkened coloration of the skin and fins, changes in the eyes, with distension and cloudiness and periorbital area with swelling; gills gain a purple tonality; gut become filled with gas and mucus material; haemorrhages may be seen in intestines walls, coelomic cavity, spleen and muscle; intestine may be distended with a mucoid and necrotic aspect, and serosa' petechiation; white/grey lesions (necrotic focus) may be found on the intestines, liver and spleen; splenomegaly may be present. Infection of farmed tilapia by *V. parahaemolyticus* in the overwintering period may rise to rapidly fulminating septicaemia, affecting most of the fish (hyper acute) after 48 to 72 hours [3]. In farmed tilapia, vibriosis may also be chronic, with a mortality rate of 10 to 20%. Pathological manifestations in acute form of the disease include internal organs hemorrhages, haemorrhagic ascites and gas accumulation in the intestines. In the chronic forms of disease, splenic necrotic nodules can be observed. Asymptomatic carries have also been detected, with Vibrio present in their blood (unapparent infection, passive carriage).

Diagnosis: Histological observations may reveal focal necrosis points in several organs, including the liver, kidney, spleen and intestines. A more definitive diagnosis is obtained through bacterioscopic and cultural examinations of *Vibrio* that may be isolated from blood, kidney or liver and intestinal contents of fish clinically affected. From those samples *Vibrio* is often obtained in pure culture. Thiosulfate citrate bile salts sucrose agar (TCBS) is the most commonly used culture media for *Vibrio* isolation. Several halophilic *Vibrio* may be isolated from the affected fish. *Vibrio* species are distinguished from one another throughout their biochemical metabolic competences: *Vibrio* are able to use several sugars at 42 °C, including lactose, galactose, cellobiose, sucrose and salicin, and aminoacids, such as lysine and ornitine. Conventionally, arginine dihydrolase and Vogues Proskauer tests are used. Miniatured galleries are consistent devices for biochemical identifications propose [3]. Molecular techniques (ELISA, PCR) for indirect detection of the presence of some *Vibrio* in specific matrixes are also available.

Therapeutics and prevention: Farmed and ornamental fish affected with vibriosis may be treated with antimicrobial drugs (antibiotics and biocides), like ampicillin, nalidixic acid derivatives (quinolones), sulphonamides and trimethoprim. Antisepsis may be used in aquariophylia, using halogenated and organic compounds. Some resistances to antimicrobial drugs appeared in *Vibrio* strains because of the extensive misuse of these chemicals. Like in almost all infectious diseases, prevention is the most effective weapon: control of water characteristics, temperature, pH, organic pollution, fish overpopulation, starvation, parasitic diseases, predator attacks and stressing factors. In special circumstances, when *Vibrio* infection are recurrent or endemic, in especially valuable fish exemplars (reproducers), formalin-killed *Vibrio* vaccines are available, or may be produced (fish schools vaccines), and can be administered *via* intraperitoneal injection, water immersion or oral administration. Sanitary impacts: Although these *Vibrio* infections are not an international notifiable disease, fish vibriosis has become a major cause of aquaculture economic losses, sometimes to the extent of being a limiting factor of fish farm viability. Disease control enhances additional costs to the fish production systems and may compromise economic viability of fish farms. Additionally, some species (*V. parahaemolyticus, V. vulnificus* and *V. damsela*) have been proven to have zoonotic potential, through fish products consumption or direct fish contact and manipulation (fisherman).

BACTERIAL KIDNEY DISEASE (BKD)

Aetiology: *Renibacterium salmoninarum* is a nonmotile Gram-positive diplobacillus member of the *Micrococcaceae* family. Morphologically it seems like *Corynebacteriaceae*. It is an intracellular bacterium that is pathogenic to some young fish of *Salmo* genera.

Susceptible fish species: BKD is a serious disease of *Salmonidae*, being especially vulnerable the Pacific salmon and brook trout. Major clinical signs and lesions: BKD has significant sanitary and economic impacts due to its effect on health of farmed and wild *Salmonidae*. Disease follows an almost subclinical course without evident clinical signs until the fish is ready for market introduction. Salmonidae may show darkening of the skin, exophthalmia, fins and skin haemorrhages, presence of skin ulcers, abscesses, muscular cavitations with fistulas, splenomegaly and ascitis. Kidney and liver can present edema, surface nodules or be covered by an opaque membrane. Clinical manifestations severity is variable. There may be no outward clinical signs, or fish may show signs of lethargy and anaemia.

Diagnosis: On post-mortem examination, signs of necrosis and granulomatous inflammation may be seen on the internal organs, particularly in the kidney. Suspicion of BKD may be presumptively assumed attending to the lesional aspects and tissues histopathological evaluation, which can reveal agglomerations of Gram-positive bacilli inside host cells. It is not possible to achieve a definitive diagnosis for this disease based on clinical manifestations. Instead, laboratorial tests based on specific bacterial cultures (isolation and identification), ELISA, PCR and fluorescent antibody testing, are needed to assess the definitive identification of the causal agent, and preferably more than one test should be used. The etiological bacteria, *Renibacterium salmoninarum,* are Gram-positive diplobacilli, small and non-motile. This is a fastidious bacteria although it grows well on brain heart infusion agar (BHIA) and charcoal agar. Primary cultures of the organism are therefore quite difficult due to concomitant microbiota (contaminants). Complementarily, confirmatory identification may be performed using immunoassays, ELISA and PCR.

Therapeutics and prevention: Farms having infected fish may experiment sporadic outbreaks throughout the year, but, generally, they become more frequent in spring, with the rise in the temperature of waters. Although juvenile fishes are rarely affected, disease can be observed in all ages. Disease is usually chronic, with high mortality rates.

Oral or parenteral antimicrobial drugs should be useful to treat BKD. Intraperitoneal vaccination can also be used to treat fish in an outbreak. Antimicrobial agents used to treat *R. salmoninarum* include almost all effective drugs for Gram-positive bacteria, namely erythromycin, bacitracin, novobiocin, penicillin, cefazolin, cephradine, cephalosporins, gentamicin, clindamycin, lincomycin, oleandomycin, kitasamycin, spiramycin, tiamulin, tetracycline, chlortetracycline and oxytetracycline. Prevention is the most relevant strategic vector to avoid the disease. Husbandry measures such as segregation, culling of the affected nucleus and decontamination should be used to ensure infection is not introduced (sanitary certification of new introduced fish).

Sanitary impacts: Disease is found in fish of different regions (North America, Chile, Europe and Japan), being spread both vertically, through eggs and sperm, and horizontally, *via* direct contact with contaminated water and fish. Most common *via* of transmission is the second one, with the bacteria entering through the skin and then disseminating throughout the fish body. Pacific salmon seems to be the most susceptible specie. BKD had been previously designated as "White Boil Disease", "Corynebacterial Kidney Disease", "Dee Disease" and "Salmonid Kidney Disease". BKD is a notifiable disease under many national and international official acts since 1937. It is a nosological entity included in the List III of the European Community Council Directive 91/67/EEC. In endemic regions, BKD must be monitored once a year in salmon or trout aquacultures, by physically examination of fish to detect clinical signals of the infection and by taking samples from suspect tissues for laboratorial analysis. Some European zones have official approved programs for controlling and eradicate BKD. This program requires the sampling of 30 farming salmonids, tested for BKD every two years. Infected areas are not allowed to trade susceptible fish or fish products to other areas of similar or higher fish health status, with the exception of the fish farms which become approved through an official monitoring plan. Fish farms confirmed that BKD-positives are surveyed to control movement restrictions of its fish and fish products. It must be compulsory to apply a specific eradication plan. Infected farms are not permitted to use or trade fishes for broods stock. Risk controlling strategies applied to BKD control can comprise sanitation, disinfection, stress control, quarantines and depopulation measures.

ENTERIC RED MOUTH DISEASE (ERMD)

Aetiology: ERMD agent is a bacterium belonging to Enterobacteriaceae family, *Yersinia ruckeri*. Like all members of its family, it is a motile rod shaped Gram-negative bacterium, anaerobe facultative, oxidase negative, and it can use glucose, ornithine and lysine. There are several serotypes and the antigenic mosaic discrimination system can be performed by serotyping or molecular typing. The serotype related to ERMD is the O1a, the "Hagerman strain", being specially virulent. For virulence expression, the siderophore ruckerbactin (iron absorption), specific haemolytic and proteolytic activities, and a higher resistance to immune defenses' skills, were proved to be relevant.

Susceptible fish species: ERMD can affect several fish species from all ages, but *Salmonidae* and *Morronidae* seem to have special vulnerabilities. Major clinical signs and lesions: Disease evolves as sepsis, enhancing low mortality rates in the beginning that rises in later stages of the disease evolution. Diseased fish can present loss of appetite, body darkening, abdominal ascites, exophthalmia and haemorrhages, that can be found in the fins, vent, gill, mouth, tongue, eye, liver, pancreas, ceca, swim bladder, muscles, spleen and intestines [14].

Diagnosis: Affected or suspected fish are those showing generalized haemorrhagic septicaemia; bacterial dissemination to the more vascularized tissues (gills, liver, kidney, spleen); necrosis of hematopoietic tissue (kidney and spleen). ERMD diagnosis confirmation requires laboratorial analyses. For that propose, biological samples (tissues or organs with lesions) must be aseptically collected and tested with conventional bacteriological methods or molecular examinations (PCR ELISA, agglutination test and immunofluorescence antibody technique). Yersinia ruckeri is a variable motility bacterium of the Enterobacteriaceae family that grows at 25° C for 48 hours, being sorbitol and gelatinase-positive. Taxonomy is not definitively set. Six different serotypes of two distinct biotypes have been described.

Therapeutics and prevention: Vaccines to prevent ERMD had been difficult to develop, especially due to the antigenic mosaic variation. However, one company had developed one vaccine that must be applied carefully. Administration temperature must be between 15 and 18 °C to obtain the maximum protection. Probiotics has also proved to have health prevention benefits [14]. Treatments using sulphamethazine for five days followed by oxytetracycline administration during three days, sulphonamides and oxolinic acid. *Y. ruckeri* is susceptible to several antimicrobial classes, but acquired resistance has already been described in some strains. Secondary antimicrobial resistances had been also reported in ERMD treatments with sulphamerazine and oxytetracycline. Prudent use policies must be always in perspective, including withdrawal times in fish intended for consumption. Sanitary impacts: ERMD has been found all around the world (USA, Canada, Australia, Europe, and South Africa among others). *Y. ruckeri* is generally spread through direct contact with infected fish or their faeces. Outbreaks usually begin with low mortalities but progressively it becomes very high. ERMD outbreak severity depends on the strain and presence of stress determinants. Due to its economic impacts, ERMD is a one of fish diseases that must be internationally monitored (although it is not a notifiable fish disease to OIE). For adequate sanitary protection, infected zones would avoid to access the international fish market (restriction of circulation).

FURUNCULOSIS

Aetiology: Furunculosis aetiological agent is *Aeromonas salmonicida* subsp. *salmonicida,* one of the first bacterial fish pathogens to be described. This bacteria is a short Gram-negative, very motile, pleomorphic curved rod, with a worldwide distribution. Its biology is very complex and many aspects of its survival remain obscure. Some authors refer its condition of obligatory pathogen, other state that some specific ecological adaptations allow them to live in the farm water environment. Actually, it is assumed that this bacterium can maintain its potential virulence in freshwater conditions for about 9 months and in salted water conditions for up to 10 days without needing a host.

Susceptible fish species: Furunculosis is a very common infection in fish. Bacterium affects primarily Salmonidae but a very wide range of fish families is susceptible to the agent and can become affected (*Angillidae, Cyprinidae, Anoplopomidae, Serranidae, Gadidae, Pleuronectidae*). Most frequent sites of entry of the pathogen into fish body are the gills, the mouth mucosa, anus and/or the injured skin (horizontal transmission). Major clinical signs and lesions: Infected fish may show one or more of the following signs: lethargy, swimming just below the water surface, respiratory distress and random jumping attempts from water before the onset of an outbreak and, at the end, refuse of feeding with subsequent high mortality rate. Initial clinical expressions of the hyper acute form of the disease are the same of all other septicaemia, with haemorrhages in the muscles and in the body surface (mouth, skin and in the bases of paired fin). The main lesion is an ulcerative dermatitis designated by furuncle, which evolves from the internal inflammation. In the chronic form of this disease, furuncles can spread to adjacent muscles. Sometimes darkening of body colour and exophthalmus (pop eye) are observed. In the septicaemia form of the disease, swelling of the kidneys, splenomegaly, and ascites is observed; also focal necrosis of the liver; abundant mucus in gastric contents; congestion of intestine walls; blood and sloughed epithelial cells and fusion of gill lamellae. Infection may also be present without expression of clinical signs (unapparent) – is the subclinical form of the disease. While the clinical form of furunculosis occurs in both juvenile and adult salmon, peracute infections may be expressed only by high rates of death in juvenile salmon, without any clinical signs, unless the skin darkening. The most relevant form of the disease is the haemorrhagic septicaemia, with the presence of lateral furuncles (boils), visible on the fish flanks.

Diagnosis: The gross lesions aspect of the clinical forms and morbidity rates could be relevant indicators for the suspicion of the disease. Histologically, tissue necrosis can be observed, with the presence of numerous bacteria and a low number of inflammatory cells. Confirmatory methods include conventional bacterial cultures, serology and molecular biology techniques. The agent can be cultured on tryptone soya agar or brain heart infusion agar at 15-25 °C, but supplementation with blood or serum may be necessary for the isolation of atypical strains. Pre-incubation of clinical materials (tissues lesion macerates) for 24-48 hours in tryptone soya broth followed by a presumptive isolation on Coomassie brilliant blue agar, may allow detecting the pathogen. However, there are not an effective selective medium. Alternative approaches to culturing tests for *A. salmonicida,* include serological and molecular methodologies. *A. salmonicida* isolation from subclinical fish is difficult, and more difficult yet from the aquatic environment, even during epizootics phases of the disease. Serological tests (slide latex agglutination and ELISA) have revealed to be successful on pure and mixed cultures and with pathological material to confirm the presence of the bacterial agent. ELISA enables reliable diagnoses in 30-60 minutes, and appears to be effective for use with asymptomatic carrier fish. Also DNA probes have been used successfully with samples obtained from the clinical materials and from the environmental. PCR technique is the adequate procedure to detect the presence of *A. salmonicida* in effluents, waters, faeces and sediment from a commercial freshwater fish. However, molecular methods such as PCR, as well as serological methods, can originate false-positive results, as they can detect dead bacteria and vaccine strains. And also the significance of bacterial cells that cannot be cultured from fish infection, remains to be solved (viable not cultivable). With the development of molecular detection methods, such as PCR, *A. salmonicida* allows to detect the presence of its genome in the aquatic environment in the absence of colony-forming units. Also free living non virulent cells of this *Aeromonas/* are found the environment.

Therapeutics and prevention: Depending on the clinical form of the disease (acute, chronic) and the fish age, therapy may be tried to reduce economic losses. Tetracyclines, quinolones and suphametoxazol-trimethoprim had been proved efficient, but in recurrent situations, antimicrobial tests must be used to detect secondary resistances. Antimicrobial drugs are administered through fish feed. Regarding the economic impacts of furunculosis into farmed fish (salmon), special emphasis must be given on the development of effective preventive strategies. Risk management measures to evaluate include hygiene measures (such as suitable water, equipment's and eggs disinfection, low population densities), selection of resistant fish strains, enhanced foods, immune-stimulation, antimicrobial therapy, administration of probiotics and vaccines. In the past, some researches confirm the value of immune-stimulatory compounds, like β-1,3-glucans, synthetic peptides and inactivated mycobacterial cells, are able to enhanced disease resistance on salmons [15]. Also, attempts to develop effective vaccines able to prevent the disease have been assayed since 1942, and include inactivated, oil-adjuvanted and injectable anti-furunculosis vaccines for salmonids. Members of the indigenous fish microbiota, such as *Vibrio alginolyticus*, may also be administered as probiotics. Generally, transmission of *A. salmonicida* can result from contact with sick fish or carriers, with contaminated water or fomites, or vertically through contaminated fish eggs. Therefore, eggs should be disinfected before incubation, using iodine compounds. Low professional hygiene behavior may enhance contaminations through human clothing or equipment. It is also possible that aquatic birds may act as vectors of the agent.

Preventive strategies may also fail if other relevant factors impacting with fish exploitation system are not adequately cared. Among them, excessive fish density, traumatic injuries of skin (density of solids suspended in water) and avoidance of carrier fish introduction in farms are relevant in the prophylactic plan. Natural disease resistant fish breed lines should be utilized as a relevant risk management tool.

Sanitary Impacts: *A. salmonicida* sub sp. *salmonicida* is classified as an obligate pathogen for fish. A basic step to reduce its sanitary impacts is the official notification of this communicable fish disease. Most national veterinary authorities have in place a specific inspection program applied to farmed fish and a sanitary classification system for the water production areas. In indemne fish farm, all nucleuses should be sampled at least once a year for laboratorial surveillance. Data generated from this plan are crucial to avoid relocations of diseased or carrier fish. If eggs have to be introduced from an external hatchery source, only inspected and certified furunculosis-free eggs should be used. Only fish stocks coming from a furunculosis certified free farm are allowed to be transacted.

COLUMNARIS DISEASE, COLDWATER DISEASE AND BACTERIAL GILL DISEASES

Aetiology: These include several syndromes due to some Flavobacterium species infections, including: *Flavobaterium columnaris*, responsible by "Columnaris disease". This agent has been previously identified as *Bacillus columnaris, Chondrococcus columnaris, Cytophaga columnaris* and *Flexibacter columnaris; Flavobacerium psychrophilum* (previously known as *Cytophaga psychrophila*), the aetiological agent of the "bacterial cold water disease" complex [16]; and *Flavobacterium branchiophilum*, that causes "bacterial gill diseases". These Gram-negative slender rods belonging to *Flavobacteriaceae* family are found in soil and fresh water in a wide range of ecosystems [17]. Susceptible species: A huge range of hosts can be infected with those bacteria, including young fish from the *Salmonidae, Anguillidae, Cyprinidae, Siluridae* and *Centrarchidae* families. Aquaculture fish and aquarium fish can all be infected although farmed fish are the major concern. It is not recognized as a significant problem in wild fish populations.

Major clinical signs and lesions: Skin and fins of affected fish become darker or lighten in colour and circular yellow-grey opalescent necrotic ulcers or erosions are visible on the skin and/or gills. Lamellae of the brachial filaments may have whitish spots on their tips. Excessive amounts of mucus accumulate often on the gills, head and dorsal regions. In advanced phases of this disease complex, erythematous spots are generalized. Inespecific signals like anorexia and lethargy are common. Mortality rates are also irregular. Gill lesions cause respiratory distress and its associated clinical signs – gulping air, sinking and erratic swimming. In catfish specifically, severe tissue necrosis develops into "saddleback" lesions on the dorsum. *F. psychrophilum* is responsible by a nosologic entity designed "Cold water disease" or "Peduncle disease" affecting fresh water fish. Fish (rainbow trout fry) exhibits darker skin, haemorrhage at the fins' base and anaemia (pale gills with increases of mucus). Muscles haemorrhages are frequently found; periostitis in the cranial and vertebral bones is commonly seen in chronical forms of the disease; abdominal distention; exophthalmos; lethargy; balance loss; skin and caudal fins ulcers and necrosis; cutaneous hyperaemia; increased mucus production; enlarged spleen and liver; multiple focus of necrosis in the spleen, liver and kidney. A chronic meningoencephalitis with abnormal and erratic swimming movements may occur [18].

Diagnosis: Lesional aspects do not have enough specificity. Microscopy of wet tissue mounts from branchia or skin scrapings often reveals haystacks or columns of bacteria. Cells can also form chains, giving the impression of a longer, single prokaryotic cell. Etiological agents (*Flavobaterium* spp.) may be isolated from gills, skin and internal organs (kidneys). These pathogens can be cultured on reduced nutrient agar, like modified "Cytophaga agar" incubated at 16 °C [16]. Culture is often more successful by inhibiting growth of contaminants on the agar using neomycin or polymyxin B supplements. Some of the *Flavobacterium* spp. will also grow at 37 °C which will inhibit other contaminant aquatic bacteria, so it can be used as an inhibitory factor [17]. Colonies are small, 3-4 mm in diameter and growing within 24 h to 16 days, depending of incubation temperatures. Its structure is typically rhizoid, pale yellow and adherent to the agar surface. These typical findings can be used as a definitive diagnosis for columnaris disease, especially where clinical signs are present. Phenetic tests can be carried out using suspicious isolates. Therapeutics and prevention: For ornamental fish affected with "columnaris disease", emersions in biocides solutions may be successful (copper sulphate, potassium permanganate and hydrogen peroxide, applied to adult fish and fry but can it may have some safety problems at high concentrations). Antimicrobial treatments using tetracyclines or quinolones added to feed are used for adults, fingerlings or broodstock farmed fish, proved to be efficient although secondary resistances are emerging. Preventive strategies may also be based on vaccination programs, especially when some circumstances make evident the emergency of an outbreak. A vaccine based on anacultures, formalin killed, against *F. columnare,* is commercially available.

Sanitary Impacts: Columnaris disease is a highly diffusible disease, being frequently related with stress, water temperature from 25 to 32 °C, over-population, wounds, and deficient water quality, especially regarding oxygen and ammonia levels. Columnaris disease can also occur in cooler or warmer waters, but is most prevalent in weather temperatures above 14 °C. Cold water disease and bacterial gill diseases morbidities can approach 100%; mortality rates are highest in younger fish (fry, fingerlings). Skin abrasions, husbandry stressors such as overcrowding, poor nutrition, underlying diseases and warm water all increase the risk and prevalence of infection. Contaminated water is the most frequent *via* of disease transmission.

PISCIRICKETTSIOSIS

Aetiology: Piscirickettsiosis is promoted by *Piscirickettsia salmonis*, a Gram-negative and facultative intracellular bacteria [19]. These rickettsia-like organisms are coccoid or pleomorphic and occurred singly or in groups in cytoplasmic vacuoles, being able to form biofilms [20]. They do not stain by PAS or Ziehl-Neelsen, blue by May-Griinwald- Giemsa. It occurs globally in many fresh and salt water species including some Salmonidae species, but its pathogenesis and environmental persistence mechanisms are not fully understood [20]. Susceptible fish species: Most affected species belong to Salmonideae family, specially the Atlantic and "coho" salmons. It has been found in other fish species of different families either from the fresh as from the salted water environments.

Major clinical signs and lesions: Clinical forms of the disease may have acute or chronic evolutions and may also be unapparent (subclinical). In hyper acute cases, fish may die without showing any other signals.

Fish affected with the acute clinical form show lethargy, darker in colour of skin and swim at the surface; may express unspecific signals like: anorexia, respiratory difficulties, ulcers in the skin and also anaemia due to haemorrhages lesions; internal organs are swollen (kidney, liver and spleen); haemorrhages may occur on the swim bladder and other abdominal organs; frequently, the abdominal cavity is filled with a bloody ascitic fluid. In salmon, the most frequent clinical form is the acute evolution, due to the agent systemic invasion, targeting gills, kidneys, liver, spleen, heart, intestines, ovaries and brain. Multifocal inflammatory and reparative lesions may be seen in the epicardium and in the myocardium (spongy and compact). Minor changes can be observed as small foci of mononuclear inflammatory cells in the endocardium. Additionally, focal necrosis of the white pulp of the spleen may be observed, and an inflammation of the meninges [21]. Fish chronically affected may exhibit ring- shaped, cream-colored lesions on liver (focal lesions) and a granulomatous inflammation in the dermis.

Diagnosis: Preliminary diagnosis may be based on gross lesions observation, especially in fish exhibiting cream colour ring-shaped nodules. Diagnosis must be complementary and confirmed with histopathological examination and isolation of the specific agent and using serological or molecular typing. Histopathological examinations may reveal multifocal necrotic areas in the liver parenchyma with an inflammatory response. Cases with dominance of polymorphonuclear granulocytes and scattered macrophages are recognized in acute and sub acute stages. Perivascular inflammation and necrosis may also be observed; sporadically, thrombi are seen. In chronic disease, inflammatory changes are characterized by macrophage infiltration developing into a focal granulomatous type of lesion: multinuclear giant cell formations may be seen in this granuloma. Kidney may exhibit focal degeneration and necrosis of the hematopoietic tissue and, in severe cases, degeneration and necrosis of the tubular epithelium; sometimes, fibrosis of exocrine pancreatic tissue is consistent with infectious pancreatic necrosis. Intracellular, intravacuolar bacteria-like inclusions were seen by light microscopy in apparently normal cells in the perimeter of inflamed areas and in degenerated tissues. Rickettsia isolation from biological samples requires aseptic collection of the attainted tissues (kidney and liver) and conservation between 4 to 10 °C (shipping on refrigerated container). Freezing fish tissues will not affect the performance of PCR assays. Methods for *P. salmonis* identification using serology, immunohistochemical tests, molecular diagnostics (RFLP, PCR) and pathogen isolation in cell culture in monolayers of CHSE-214 line cells (ATCC CRL 1681) at 17 °C have been described [22]. Multiple primers were tested and RFLP analysis of PCR-products showed no differences between the Norwegian strains and the Chilen type isolate.

Therapeutics and prevention: *P. salmonis* is susceptible to a wide range of antimicrobial drug families including commonly used tetracyclines, but being a facultative intracellular bacteria, the administration of medicated feed may not be sufficient to control this disease, as the *in vivo* antimicrobial concentration may not be high enough.

Rigorous biosafety procedures must be applied in the affected areas. Diseased and dead animals must be removed in a daily base from their cages or tanks. Fish from an infected farm should never be sold for breeding. Prevention is also achieved by routine disinfection of fish eggs by iodophors. However, egg disinfection is perhaps not efficient enough, because as there is a possibility of an intraoval localization of the bacterium.

Sanitary Impacts: Piscirickettsiosis has a worldwide distribution, being already reported in the Atlantic salmon in Ireland, Chile, Norway and Japan, where most significant epizooties has been referred, being responsible for major economic losses. Disease is primarily associated with fish inhabitant of seawater ecosystems, although it has been also refereed in fresh water due to Salmon migrations [23]. Disease transmission occurs rapidly in aquaculture salmon by fecal *via* and vertically through infected gonads [24, 25]. Piscirickettsiosis occurs in late summer and autumn and mortality rates may range from 10% to 90% in the most severe cases. As piscirickettsiosis is listed on OIE aquatic code, being referred as a disease that must be officially notified and internationally declared. Standard diagnostic procedures for the identification of *P. salmonis* are outlined in the OIE Manual of Diagnostic tests for Aquatic Animals .

Histology, immunohistochemistry and conventional PCR methods to be performed are those accorded in OIE. Once declared in a fish farm, fish commerce is not allowed.

MYCOBACTERIOSIS

Mycobaterioses are very common bacterial infections of fresh and sea water fish that had been firstly described in the end of 19th century. Mycobacterial infections are among the most frequent chronic diseases of ornamental or aquarium fish [26], either from temperate or tropical zones, being responsible not only for fish suffering but also by major economic losses [27].

Aetiology: These infections are caused by Gram-positive acid fast rods, namely *Mycobacterium marinum, M. chelonae, M. trivial, M. avium subsp. hominissuis, M. fortuitum* (most commonly) and more rarely by *M. aurum, M. gordonae, M. parafortuitum, M. poriferae* and *M. triplex* [28, 29]. Susceptible fish species: All fish species are susceptible to these bacterial agents. These chronic infections have been found in Mediterranean fish produced in aquaculture (*Dicentrarchus labrax, Sparus aurata*), and from the wild (*Mugil curema, Seriola dumerili*). Mycobacterial infections are also frequently described in sea fish in Japan, causing economic losses [30].

Major clinical signs and lesions: Signs are quite diverse and may include anorexia, emaciation, columnar deformations, exophthalmos, discoloration, nodules in external and internal organs, ulcers in the skin, fins and gills, opercula petechial, abdominal enlargement, liver congestion and necrosis. Fish may also express listlessness, emaciation, abnormal spinal curvature and sunken eyes.

Diagnosis: Lesions found in organs of the affected fish may be subjected to anatomopathological and histopathological examinations, especially the granulomatous lesions of skin and parenchymatous organs (hepatopancreas, spleen, kidneys and intestine). Direct prints and smears performed from the tissue lesions and from fresh faecal materials and wet mounts, Ziehl-Neelsen stained, allow finding very abundant acid alcohol resistant intracytoplasmic mycobacteria. However, these examinations are presumptive and inadequate to obtain a definitive diagnosis. To achieve this, other analytical tests are needed [30]. Conventional bacterial culture for mycobacteria detection without isolate identification cannot lead to a final diagnosis due to the presence of atypical mycobacteria in the water environments (public tanks, aquariums). Tissue samples macerates can be cultured, after centrifugation and previous acidic and alkali treatments (HCl – NaOH) and incubated at 25 °C and 37 °C, for 2 weeks to 2 months [31]. Several culture media have been suggested, based on egg yolk and serum. Isolates are identified according to their morphological appearance, Ziehl-Neelsen staining, biochemical methods and serotyping [30]. Molecular methods are also available, namely PCR with primers for IS901 or IS1245 [31]. Therapeutics and prevention: This infection is not treatable. Affected fish need to be quickly removed and destroyed, avoiding more unnecessary contacts. Oral route is the major probable *via* of mycobacteria infections dissemination, including necrophagia. Other possible infections include injured skin or transovarial transmission of mycobacteria. Early detection is a key point of the prevention strategy, through regular monitoring of fish and dead embryo (neo or post-natal occurrences).

Sanitary impacts: Most mycobacteria isolated from fish are potentially pathogenic for humans (zoonotic potential - atypical myco-bacteriosis). Infected aquarium fish represent a risk factor for humans (tank workers, immunocompromised patients). Once an aquarium or an ornamental tank becomes infected, it is difficult to eliminate the agent except through total depopulation and deep tank disinfection. This disease is not internationally declared, although introduction of infected imported fish could be a relevant source. Health status of imported fish must be always adequately cared.

Besides those bacterial diseases described above there are many other that have been referred to affect fish. Some of them are opportunistic pathogens other are obligatory. That is the case of *Pseudomonas anquilliseptica*, a small Gram-negative flagellated rod, responsible by serious problems in Japanese eels. Disease manifests as a septicaemia with subsequent haemorrhages in fins and ulcers in tails and skin, similar to those produced by *Aeromonas salmonicida*. Infected fish are the main source of new infections. Another relevant fish bacterial pathogen is *Streptococcus iniae*. This is a beta- haemolytic Streptococcus able to cause illness in rainbow trout, hybrid striped bass and tilapia. *Streptococcus iniae* infection may be expressed through an extremely acute septicaemia or a chronic disease affecting the central nervous system, being a major problem in the tilapia industry [34, 35]. Septicemic form is characterised by haemorrhage of the skin and and ulcers. Histopathological analysis can reveal the present of cocci or diplococci in the inflammatory tissue, and granulomatous inflammation can be observed in several organs, including the liver, kidney and brain, in chronic diseased animals.

In meningoencephalitis cases, brain is the organ that must be elected as target for bacterial culture. *Streptococcus iniae* infection is a primarily management problem on aquacultures, due to recirculation of culture system. Overcrowding and poor water characteristics (high levels of nitrates) may be predisposal factors. Fish depopulation, tanks and water decontamination and repopulation are major strategies to eliminate the organism. Vaccination may be also used. Sometimes aquaculture workers become infected, developing hands cellulites and endocarditis [36].

Another referred fish pathogenic bacteria is *Nocardia* sp. It is a Gram-positive bacillus, being a major issue in aquarium fish and occasionally in aquacultures of *Salmonidae*. This disease is generally chronic, and its signals include the presence of granulomas in several locations and organs, including the mouth, jaw, gills, skin and intestines, which ulcerate. The transmission *via* is not fully established , although skin wounds and abrasions could be considered the most common *via* of infection. Also, oral route has been implicated in bacteria dissemination. Pasteurella piscicida outbreaks have been referred in tilapia hybrids and red sea bream from fish farms [37, 38]. Predominant clinical signs include the presence of nodules in the viscera. Bacteria are short bipolar Gram-negative bacilli, non-motile, catalase and oxidase positive, being able to use glucose and nitrates and to produce indole. Histopathological observation reveals very characteristic lesions, including persistent encapsulated granulomas, enclosing and surrounded by abundant bacteria, that can also be observed inside the macrophages.

This agent seems to have a broadly distribution, as it has already been found in both oceanic and estuarine environments, and can affect several fish species, including striped bass (*Morone saxatilis*), white perch (*M. americanus*), grey mullet, sea cultured European sea bass (*Dicentrarchus labrax*) and bass (*Morone* spp.) [39].

Risk factors for disease development include the use of poultry manure as ponds fertilizer [40] and stress factors. To prevent those situations it is need to enforce sanitary barriers.

FUNGAL DISEASES OF FISH

Aquatic fungi are natural inhabitants of water, aquaculture tanks and aquaria. Some fungi may become infectious when the immunocompetence of fish is compromised or if there are some injures in the gills or skin. Most fungal fish infections are secondary infections. The main signal of a mycotic disease is the presence of a whitish, "cottony" envoled lesion in the fish body surface. Excessive use of antibiotic drugs may also predispose to mycotic infections in fish due to biotic imbalance disruption [4]. Once again, there are ecological determinants that play a major role on fish fungal diseases: overpopulation, organic matter excess in waters, low levels of dissolved oxygen, and sudden water temperatures changes. Although quite relevant for fresh water fish farms and aquariophily, the number of nosologic entities caused by fungi in fish is low. Fish mycosis is especially relevant in ornamental and aquarium fish and they have some economic relevance in fish produced in freshwater aquacultures systems. Most biocides used in fish mycosis therapeutics are also effective against bacteria, thus it can provide double protection. Few fish mycosis have legal impacts on international fish trade.

EPIZOOTIC ULCERATIVE SYNDROME (EUS) OR RED SPOT DISEASE

EUS is an ulcerative disease that may strike fish in both salted and fresh water, being found in several regions, including Asia-Pacific, South Africa, North America and Australia. Economic impacts may be very high due to mortality and morbidity rates. This disease had also been referred as Mycotic granulomatosis.

Aetiology: EUS is caused by *Aphanomyces invadans*. This agent belongs to the kigdom Chromalveolata and to class *Oomycetes*. Like other member of *Saprolegniaceae* family, it produces motile spores (zoospores). Those structures play a very relevant role in infection dissemination. Susceptible species: EUS may affect a wide range of fish species (mullet, bream, whiting, pikey bream, African pike, tigerfish, channel catfish, spangled perch, Murray cod, redbreast tilapia, sand whiting, dusky tripletooth goby and catfish). Juvenile and young adults are usually more susceptible, and fries and larvae seem to be resistant to this disease. Common carp, Nile tilapia and European eel (*Anguilla anguilla*) are also resistant to these infections [41]. Disease is more frequent when water temperatures fall as consequence of sudden heavy rainfall in tropical and sub-tropical regions.

Major clinical signs and lesions: Attained fish present cutaneous red spots, which can ulcerate and be covered with necrotic tissue and fungi. These ulcers are the earliest signs to be observed, and can be present in several locations, including in the head. Consecutively, ulcers start to affect deeper tissues or the coelomic cavity, granulomas develop on internal organs and death occurs, due to osmoregulation problems or opportunistic infections. Mortality and morbidity rates may be superior to 50%. More severely diseased fish may survive but cannot be sold in the market, due to their abnormal aspect.

Diagnosis: Septate fungal mycelium can be identified in squash microscopic preparations of skeletal muscle attaint by ulcers. Final diagnosis can be confirmed through histopathology analysis, bacteria isolation or molecular techniques (PCR). Fish with minor clinical signs are recommended for laboratorial examination, and collection of the muscles located near the ulcers is recommended for Oomycete isolation, but cutaneous lesions can also be used. For isolation, agar supplemented with glucose, peptone, penicillin G (100 units/ml) and streptomycin (100 µg/ml) should be used, and incubated 25 °C for atleast 24 hours. The most adequate material for histopathological examination is the muscle from the ulcers edge. In histopathological examination, using haematoxylin-eosin technique, the stained fungus will show typical granulomas and invasive hyphae. There are also available specific techniques for *A. invadans* identification (PCR).

Therapeutics and Prevention: As there are no efficient treatments available, infected fish have to be relocated in a clean tank for recovery. To minimize losses, the water quality of the original tanks has to be evaluated in terms of pH and salinity, by adding lime and salt [42], and routine disinfections should be performed, including not only the water but also eggs and larvae. Malachite green solutions are now forbidden. Some other palliative cares have been evaluated, including supplementation with vitamins C, B1, B2, B6 and B12.

Sanitary Impacts: EUS must be internationally notified [41]. Confirmation of a EUS case in a specific geographical area, blocks the international commerce of susceptible fish from that region. Once fish show gross signs of infection, fish and also the farm and its location region must be classified as suspect and be subjected to fungi isolation and laboratorial confirmation. PCR is the methodology used for targeting surveillance and to declare a free *A. invadans* area after gross signs examination. Biosafety procedures must also be implemented. According to the OIE code, targeted active surveillance must be performed twice a year, including two distinct seasons. Higher morbidity and mortality rates occur during the cold season.

Disease onset requires predisposing factors, including secondary infections and acid waters, leading to skin injury [42]. Many other fish fungal infections have been described, although their report to official authorities is not mandatory by law. Some of these mycosis are relevant in aquaculture (Saprolegniosis, Achlyosis, Branchiomycosis), and others to aquarist (Pytiosis). *Saprolegnia* spp. are also *Myxomycete (Oomycetes)* that can be incriminated in freshwater fish infections, in aquaculture tanks, aquariums or ponds.

SAPROLEGNIOSIS

Saprolegniosis is a frequent infection of Koi carp, common trout and European eels. Fungal spores or their "cotton wool" hyphae can be present on the whole body of the Koi carp, including on the gills, which infections are difficult to detect. *Saprolegnia* spp. and *Achlya* spp. infections can affect multiple body systems and usually occur when the fish are weak, due to injury or trauma. There are consistent evidences suggesting that *Saprolegnia* and *Achlya* can be primary pathogens in weakened and immunocompromised fish and their eggs. These fungi can also develop in poor environmental conditions (substandard water quality, containing decaying organic matter and ammonia; overpopulation; inadequate temperatures). Clinical signals of Saprolegniosis and Achlyosis include light grey, cottony growths on the skin, fins, gills and eyes. If left untreated, it may cause fish death. Treatment of the *Saprolegnia* and *Achlya* infections is accomplished by including biocides in the water (potassium permanganate, iodophorus), while increasing salt levels, combined with good electrolyte and calcium levels. These funguses are more virulent in colder waters. It is important to thoroughly clean the fish tank, aquarium or fish ponds, removing dead infected fish and sanitizing the environment, to prevent these fungal infections. Fish eggs systematic treatment is a key issue concerning the prevention of Saprolegniosis in freshwater aquaculture fish.

BRANCHIOMYCOSIS OR "GILL ROT"

Branchiomycosis or "Gill rot" is another important mycotic disease of freshwater fish having relevant sanitary and economic impacts. It is also promoted by *Oomycete* species, *Branchiomyces demigrans* or *Branchiomyces sanguinis*. It occurs in eutrophic water and ponds with a high organic load, with a temperature superior to 20 °C. This infection is responsible for high mortality rates. *Branchiomyces spp.* invasion is usually localized in the blood vessels of fish gills (carps, trouts, eels). In eels, *Branchiomyces* fungus may disseminate to the gut, the pericardiumand the spleen. Disease is transmitted by fungal spores, but transmission course is still unknown. Gill blood vessels contamination is responsible for vessels obstruction, ischemia, haemorrhages and subsequent gill necrosis. This progression is rapid.

Concerning mycotic infections in aquarium fishes, it must be assumed that fungal agents are ubiquitous in aquatic habitats. Fungus spores are always present within the aquarium, and can become invasive when fish damage its gills or skin. Even vegetal used as enrichment elements of the aquarium environment are fungal vehicles (*Pythium* spp.). Fungal infections are typically secondary infections. Whitish mycelia growing on fish body surface is the main physical signal observed. Most of the anti-fungal drugs available, including biocides, are able to provide adequate protection for aquarium fish. Anti-fungal medicines are usually the best option to fight aquarium infections, but generally good hygiene practices are very relevant as preventive strategy. Commerce of aquarium fish must always be supported by appropriate sanitary documentation.

VIRAL DISEASE OF FISH

Among fish infectious disease, those with more devastating consequences are caused by virus. Once a viral illness is in progress in a fish farm, disease course cannot be changed using medication or any other health care treatment. Therefore, it is important to implement measures to allow preventing secondary infections, to maintain water and environmental quality, to regulate temperature and adequate diet practices. Most relevant viral fish diseases are: epizootic haematopoietic necrosis; infectious haematopoietic necrosis; spring viraemia of carp; viral haemorrhagic septicaemia; infectious salmon anaemia; red sea bream iridoviral disease; koi herpesvirus disease; and limphocystis disease . Most of these diseases are international reportable diseases.

EPIZOOTIC HAEMATOPOIETIC NECROSIS (EHN)

This illness includes systemic clinical or subclinical infections.

Aetiology: Epizootic haematopoietic necrosis virus (EHNV) belongs to *Ranavirus* genus, *Iridoviridae* family. EHNV is very resistant to hard environmental conditions, including drying and freezing, being able to survive in the environment for years.

Susceptible Fish Species: Since EHNV detection in Australia in 1986, the virus has been found in rainbow trout (*Oncorhynchus mykiss*), macquarie perch (*Macquaria australasica*), redfin perch (*Perca fluviatilis*), mosquito fish (*Gambusia affinis*), mountain galaxias (*Galaxias olidus*) and silver perch (*Bidyanus bidyanus*). Analogous necrotizing diseases have been described in sheatfish (*Silurus glanis*), catfish (*Ictalurus melas*), turbot (*Scophthalmus maximus*), black bullhead (*Ameirus melas*) and pike (*Esox lucius*) [43, 44].

Goldfish (*Carassius auratus*) and common carp (*Cyprinus carpio*) are naturally resilient to these infections [45].

Major Clinical Signs and Lesions: Most frequent signals of the clinical form of the disease include distended stomach, darkening of skin, fins and gills haemorrhages, swollen kidney and spleen [45]. In aquaculture farms, tanks or ponds, fish stop eating and juveniles (<25 mm) become unsettled due to equilibrium loss. Deaths arise mostly in summer, during a brief period; afterwards, infection may not be observed for a long time, which seems to indicate a possible subclinical evolution. In rainbow trout mortality rates are lower, as disease seems to be less serious in this fish species, primarily disturbing small fingerlings with less than 125 mm.

Diagnosis: Clinical diagnosis is not accurate because there aren't specific clinical signs. Suspicions begin when fish are found dead, with skin, fin and gill lesions, and detection of deficient management measures, including overpopulation and deficient water quality.

In histophatological examination, unspecific acute necrosis lesions can be observed in the liver, kidney and spleen, but also in heart, pancreas, intestines, gills and pseudobranch [46].

EHNV grows in many fish cell lines, being the golden standard for virus isolation. Virus identification, after growth in cell culture, may be attained by immune or molecular methods, including ELISA and PCR. It is important to refer that protocols that apply polyclonal antibodies may show false-positive results with ranaviruses, except with Santee Cooper ranaviruses [47].

Sanitary impacts: Like almost all viral illness, there is no efficient treatment. Also vaccination is not available.

Epidemics are related with deficient water characteristics, including not regulated temperature (11 °C to 17 °C for rainbow trout; > 12 °C for redfin perch). To reduce infection impact in farmed rainbow trout it is crucial to maintain low stocking rates and adequate water quality. Other EHNV vectors comprise boats, including nets and further equipment, fish bait, and birds.

Strategic key for EHN control is good husbandry practice. Risk management needs to focus on farm protection, fish health maintenance, avoiding the infected fish introduction, biosafety measures against aquatic birds, and blocking international commerce of fish from infected zones. This is an OIE reportable fish disease.

INFECTIOUS HAEMATOPOIETIC NECROSIS (IHN)

IHN can affect fish reared in fresh or sea water. When it occurs on farms, it may be responsible for elevated mortality rates and severe economic losses.

Aetiology: IHN is caused by a Novirhabdovirus of the *Rhabdoviridae* family. IHNV is very susceptible, being highly labile to high temperatures, acid treatments and ether; however, it may persist for a month in the environment, especially in cooler temperatures and in the presence of high levels of organic material. Susceptible fish species: IHNV was found in members of the *Salmonidae* family: rainbow trout (*Oncorhynchus mykiss*), chinook salmon (*O. tshawytscha*), sockeye (*O. nerka*), chum (*O. keta*), amago (*O. rhodurus*), masou (*O. masou*), coho (*O. kisutch*), atlantic salmon (*Salmo salar*), cutthroat trout (*O. clarki*), brown trout (*S. trutta*), chars (*Salvelinus namaycush, S. alpinus, S. fontinalis* and *S. leucomaenis*) and ayu (*Plecoglossus altivelis*). IHN has also been found in non-salmonids fish, like European eel (*Anguilla anguilla*), pike (*Esox lucius*), shiner perch (*Cymatogaster aggregata*), herring (*Clupea pallasi*), cod (*Gadus morhua*), sturgeon (*Acipenser transmontanus*) and tube-snout (*Aulorhychus flavidus*). Virus has also been reported in wild fish [48]. Fryes of several salmonid species are the most vulnerable. Adult fish are usually more resilient to these infections, but differences in individual susceptibility have been described. Rainbow trout and Chinook and sockeye salmons fryes may express mortalities near 100%.

Major Clinical Signs and Lesions: Virus replication in capillaries, hematopoietic tissues and in the kidney is responsible for its destruction and development of clinical signs. Depending on fish species, rearing ecological conditions and virus strain, outbreaks of IHN may be expressed by hyperacute or chronic clinical forms. IHNV infection promotes generalized oedema and haemorrhage, being responsible for high mortality rates due to osmotic balance impairment. Most insidious clinical signs and lesions are lethargic or hyperactivity, darker skin, exophthalmus, stomach swelling, haemorrhage in several locations, including skin, internal organs, fins, skull and lateral line, visible anaemia and scoliosis.

Losses in acute outbreaks may be higher than 90% [49]. In chronic cases, fishes in several phases of infection can be present in the pond.

Diagnosis: Clinical signals conjugued with epizootic data may lead to disease diagnosis, although only laboratorial procedures allow confirmation. For laboratorial tests, the optimal tissue materials for sampling are the spleen, kidney, heart or encephalon, and also ovarian fluid and milt. Whole small fries (<4 cm long), can also be tested. Histopathological examinations may reveal unspecific necrotic areas in the kidney, spleen and liver, and intranuclear and intracytoplasmic inclusions can be observed in the pancreas. The pathognomonic IHN lesion is the submucosal eosinophilic granular cells necrosis.

IHNV detection is conventionally performed using cell cultures, and confirmation methods include immunological (neutralization, ELISA) or molecular techniques (PCR, hybridization, sequencing).

Sanitary Impacts: Disease is most severe when water temperature falls below 10 °C. IHNV was already reported in North America, Asia and Europe. Fish reservoirs include diseased fish and healthy carriers, both from cultures or from the wild. IHNV is disseminated by direct contact or by ingesting contaminated food. It can also be transmitted by contaminated semen and eggs, and invertebrate vectors have implicated in some cases [49]. One strategic measure for IHN control is to avoid virus exposure through the implementation of strict hygiene practices policies and the establishment of biosafety measures. Another relevant measure is vaccination, and autogenous and recombinant vaccines are available in North America [50, 51].

IHN is an OIE reportable fish disease.

SPRING VIRAEMIA OF CARP (SVC)

SVC is a very relevant infectious disease affecting fish of the *Cyprinidae* family, mainly common carp. Disease is widespread in Europe, where it causes significant morbidity and mortality. SVC is an official notifiable disease according to the OIE health code.

Aetiology: SVC is promoted by several subtypes of Rhabdovirus carpio (SVCV), belonging to genus *Vesiculovirus, Rhabdoviridae* family [52]. The virus is easily inactivated by formalin, glycerol, ozone and both acid and alkaline and lipid solvents. Physical treatments like heating > 60 °C for 15 minutes, gamma radiation, UV light, extremes of pH are also effective. Virus infectivity remains in tap water and mud for at least six weeks. Susceptible fish species: Fish of *Cyprinidae* family are the main targets of SVCV. Besides common carp (*Cyprinus carpio*), SVCV has been observed in crucian carp (*Carassius carassius*), bighead carp (*Aristichthys nobilis*), silver carp (*Hypophthalmichthys molitrix*), tench (*Tinca tinca*) and sheatfish (*Silurus glanis*).

Major Clinical Signs and Lesions: Signs are not specific: equilibrium loss, erratic swimming, exophthalmos, ascitis, oedema and organs haemorrhage. Virus causes haemorrhage and inflammation of the swim bladder leading to abdominal distension, lethargy, lateral swimming and sinking. Affected fish may also appear pale, exhibiting a darker skin and fins with point haemorrhages. They may congregate in slow moving areas. In clinically affected fish, long thick mucoid traces appearing from the vent will be noticed. Fish are often anaemic due to viral attack to haematopoietic tissues. Affected fish show kidney, spleen and liver tissues destruction, leading to haemorrhage, loss of water-salt balance and impairment of immune response. Higher mortality arises during spring when the temperature of the water is between 10 and 17 °C. This seasonality responsible for higher frequencies in 9-12 and 21-24 months old fish.

Diagnosis: SVC can be diagnosed in carp combining clinical signs, virus isolation in cell culture and molecular methods. Necropsy and histopathological examination of affected fish reveals splenomegaly with reticuloendothelial hyperplasia and enlarged melanomacrophage centres. Liver becomes hyperaemic with multifocal necrosis and degeneration. Kidney tubules are often clogged with casts and cells are vacuolated and degenerate due to virus multiplication. Pericarditis is also observed. Haemorrhage and organ swelling with oedema is characteristic. There may be white-grey patches of tissue and organ necrosis. Blood vessels may show oedematous perivasculitis. Swim bladder lamina changes into a discontinuous monolayer with dilated vasculature and lymphocyte infiltration. Electron microscopy reveals virus particles in the fish cells nuclei. Definitive diagnosis can be achieved by virus isolation, direct immunofluorescence (IF) or ELISA. Virus neutralisation can be used as an alternative to virus isolation. Nested PCR and RT- PCR methods are available, being the former more sensitive. Sampled fish for laboratory examinations should be transported under refrigeration or on ice but never frozen. It is important to note that fish bodies decomposing will seldom produce results. All organs, including kidney, spleen and encephalon, should be immediately sent to laboratory analysis. SVCV isolation attempt must be performed within 24 hours after organ collection. Neutralizing antibodies against SVCV cross-react with other homologous virus when different serological techniques are used (sero-neutralization, immunofluorescence, immunoperoxidase or ELISA). Cross-reaction is higher with pike fry rhabdovirus (PFR), suggesting a phylogenetic relation between these two viruses.

Sanitary impacts: SVC has been detected in *Cyprinidae* in Europe, North and South America, Russia and Asia. Virus transmission route is horizontal, by faecal casts, urine and gill mucus. It is also transmitted by some bloodsucking mechanical vectors like louse, leech, crustacean and annelids. It usually enters *via* gills and replicates here before dissipating to the swim bladder. Vertical transmission, through eggs, is possible but less frequent. Waterborne transmission is believed to be the main infection route. Feral carp is more vulnerable than farmed animals. International trade of goldfish and other ornamental fish (guppies, zebrafish and pumpkinseed) without appropriate boarder control may be a route for virus introduction in endemic zones. International fish commerce from infected areas must be blocked. Regular physical and chemical equipment disinfection, careful fish handling to reduce stress and dead fish safe disposal can reduce outbreaks severity. Disease occurs typically in water < 18 °C, so is predominant during spring. Some experimental efforts have been developed to prevent virus infection through DNA vaccines and immunostimulatory therapeutics.

VIRAL HAEMORRHAGIC SEPTICEMIA (VHS)

VHS is described generally as one of the most devastating fish diseases worldwide. That is the reason why VHS is an OIE reportable fish disease and the scientific community as well as the other stakeholders (fisheries managers, aquaculturists, and traders) must be aware of some special compulsory regulatory requirements associated with this pathogen.

Aetiology: The agent is the Viral Hemorrhagic Septicemia Virus (VHSV) , genus *Novirhabdovirus*, another member of *Rhabdoviridae* family. Rhabdoviruses are bullet-shaped viruses that contain a single-stranded RNA genome, belonging to the same family as the Rabies Virus. However, it is extremely important to realize that VHSV is an exclusive fish pathogen and do not poses any health risk to humans. Four genotypes have been described. Virus may survive in the aquatic environment, depending on the water quality and the presence of organic materials. Freezing temperatures will not succeed to inactivate all virus particles, but will reduce more than 90% virus infectivity or titers. Susceptible fish species: The list of fish species that have been reported as vulnerable to this virus infection is very extensive. Some authors argue that all fish species are susceptible to this viral infection. VHS has been described in multiple fish species belonging to the following families: Anguillidae (*Anguilla anguilla*); Centrarchidae (*Micropterus salmoides, M. dolomiteu, Pomoxis nitromaculatus, Lepomis macrochirus, L. gibbosus, Ambloplites rupestris*); Salmonidae (*Salmo* spp., *Oncorhynchus* spp., *Salvelinus* spp., *Thymallus* spp., *Coregonus* spp.); Lotidae (*Lota lota, Enchelyopus cimbrius*); Cyprinidae (*Pimephales notatus, Notropis atherinoides, N. hudsonius*); Catostomidae (*Moxostoma macrolepidotum, M. anisurum*); Merlucciidae (*Merluccius productus*); Moronidae (*Dicentrarchus labrax, Morone chrysops*); Pleuronectidae (*Limanda limanda, Platichthys flesus Pleuronectes platessa, Hippoglossus hippoglossus, Reinhardtius hippoglossoides*); Scianidae (*Aplodinotus grunniens*); Gadidae (*Gadus* spp., *Trisopterus* spp., *Melanogrammus aeglefinus, Merlangius merlangus, Micromesistius poutassou*); Gobiidae (*Pomatoschistus minutos, Neogobius melanostomus*); Ictaluridae (*Ictalurus punctatus, Amieurus nebulosus*); Clupeidae (*Clupea* spp., *Sprattus sprattus, Sardinops sagax, Dorosoma cepedianum*); Esocidae (*Esox masquinongy, E. niger*); Percopsidae (*Percopsis omiscomacys*); Percidae (*Perca flavescens, Sander vitreus*); Sebastidae (*Sebastes inerrmis, S. schlegelii*); Serranidae (*Epinephelus akaara*); Scophthalmidae (*Scophthalmus maximus*), Sparidae (*Acanthopagrus schlegelii, Pagrus major*) and Soleidae (*Solea senegalensis*). VHSV natural reservoirs are the diseased fish and the carriers. Individual variation in susceptibility can be observed, due to including genetic variability and fish age, the younger fish have higher susceptibility. In general, VHS mortality cases occur in older fish who have never been in contact with the disease.

Major Clinical Signs and Lesions: VHS may have two clinical forms. The acute form expresses higher death rates. Fish can present anemic gills, dark skin, ascitis, exophthalmos, irregular behaviour, haemorrhages in the eyes, skin, intestines and muscles, and congestion and necrosis in the kidney, hematopoietic organs and in the hepatic parenchyma. The chronic form of the disease has lower and prolonged death rates. Fish develop lethargy, anemic gills, dark skin, exophthalmus, and swollen abdomen and internal organs, including the spleen, liver and kidneys.

The virus can destroy endothelial cells of the blood vessels, and for this reason the vessels are then unable to retain blood and haemorrhage occurs. While haemorrhage is a common lesion associated with VHS, it can also occur as a result of fish diseases caused by other viral, bacterial, parasitic diseases and also by water quality, toxic or mechanical injuries. Diagnosis: VHS signals are observed by the Icthyopathologist for diagnosis orientation. The fact that not all fish species and that not all fish in a group of the same species are present with the same VHSV signs is a further complication. VHSV definitive diagnosis must be based on laboratory testing of appropriate samples collected from infected fish, adequately preserved. Target sampling organs are the kidney, heart, spleen, and also the brain in the chronic form.

The current officially accepted method for VHSV diagnosis is a two-step procedure. The first step is screening the virus using fish cell cultures. In case of virus activity in cell cultures (cytopathic effect), the virus culture is used for subsequent procedures. To identify the virus, the second step of the process involves using testing methods specific for VHSV. In this step, infected cell culture material is prepared and assayed by RT- PCR using specific primers for VHSV. VHS is an OIE reportable illness.

Sanitary impacts: The disease is not treatable. To combat VHS occurrence, an effort has to be made to assess how the virus can be transmitted as well as how it might be transmitted. It is crucial to avoid trading infected fish. Disease may occur in fishes from all ages. Healed fish develop defensive immunity and fish surviving to epizootics events will become long-term carriers.

Waters with deficient characteristics, high feeding rate and fish density, and other viral, bacterial, mycotic or parasitic diseases are able to influence fish susceptibility to VHS, as well as bad management practices, such as inadequate high water temperature, low quality of feeding, overpopulation and frequent handling s. Several immunostimulants, such as yeast beta-glucans and probiotics, can be used to prevent dissemination, as well as eggs disinfection. A commercial vaccine is not yet available.

INFECTIOUS PANCREATIC NECROSIS (IPN)

IPN is an extremely infectious disease of intensive farming young salmonids, with high economic losses.

Aetiology: IPN is caused by a Birnavirus (IPNV), *Birnaviridae* family. Isolates display wide antigenic diversity, being divided in two serogroups, being observed that most strains belong to serogroup A. There are variable degrees of virulence among different isolates.

Susceptible Fish Species: IPN can affect the rainbow trout (*Oncorhynchus mykiss*), brook trout (*Salvelinus fontinalis*), brown trout (*Salmo trutta*), Atlantic salmon (*Salmo salar*) and Pacific salmon (*Oncorhynchus* spp.). This virus has also been related to subclinical infections in nonsalmonids, including pike (*Esox lucius*), yellowtail (*Seriola quinqueradiata*), dab (*Limanda limanda*), halibut (*Hippoglossus hippoglossus*), turbot (*Scophthalmus maximus*), Atlantic cod (*Gadus morhua*) and loach (*Misgurnus anguillicaudatus*). Other affected fish families include Anguillidae, Atherinidae, Bothidae, Carangidae, Cotostomidae, Cichlidae, Clupeidae, Cobitidae, Coregonidae, Cyprinidae, Esocidae, Moronidae, Paralichthydae, Percidae, Poecilidae, Sciaenidae and Soleidae .

Older fish seem be more resistant.

Major Clinical Signs and Lesions: IPN clinical expression is not quite specific, being characterized by sudden outbreaks with high death rates that may reach 90%, depending on virus strain specific virulence and infective dose exposure, host specie, stage vulnerability and environmental conditions. Signs include erratic swimming, skin darker pigmentation, ascites, exophthalmus, acute catarrhal enteritis, cylindrical mucoid faece, anemia, and haemorrhage in the intestines, mainly in pyloric caeca.

Diagnosis: Clinical and epizootological data may alert for IPN. Fish necropsy may help and allow collecting material for laboratory procedures. Diagnosis is performed using histopathology, revealing necrotic areas in the pancreas, gut and kidneys, and also hyalinic degeneration of the skeletal muscle; immunological methods, and virus isolation using cell cultures.

Sampled fish materials are the entire alevin, if body length is inferior than 4 cm, entire intestines and kidney if the fish body length is between 4 cm and 6 cm; the liver, kidney and spleen for larger size fish, including also the ovarian fluid for asymptomatic or apparently healthy fish.

Sanitary Impacts: IPN has a worldwide distribution, being present in North America, South America, Europe and Asia, with the exception of Oceania. IPN transmission can occur horizontally through water or vertically through eggs.

Control procedures are based application of sanitary measures, including averting the use of fertilized eggs from infected farms or overpopulation, and promoting the use of good quality water. There are no available effective vaccines, and eggs surface disinfection is not completely effective.

INP is also an obligatory notifiable disease, integrating the official list of OIE illnesses enforcing international notification and market blockage.

INFECTIOUS SALMON ANAEMIA (ISA)

ISA affects the Atlantic salmon (*Salmo salar*). First observed in Norway, it already has been described in Canada, Chile, United Kingdom and the USA.

Aetiology: The disease is transmitted by a low virulence orthomyxo-like virus (ISAV), a single species from *Isavirus* genus of the *Orthomyxoviridae* family. Several strains have been described, being the European strain and the North American strain the most common. Virus transmission may occur directly *via* diseased fish and their excretions, or indirectly *via* equipments and handlers or *via* a vector, the crustacean sea louse (*Lepeophtheirus salmonis*). The virus survives in salted water.

Susceptible Fish Species: Atlantic salmon (Salmo salar) is the single fish species described to be vulnerable to this disease, developing visible clinical signs. Infection is mainly observed in sea, but disease outbreaks in fresh waters.

Major Clinical Signs and Lesions: Disease is characterized by anaemia, ascites, hemorrhagic eyes, petechial peritoneum, congestion in several organs, such as the liver and spleen, and liver and kidney necrosis.

Diagnosis: ISAV can be detected through histopathology, immunological and molecular methods, or using cell cultures.

Sanitary Impacts: Few environmental factors have been assessed as determinant for ISA outbreaks. In subclinical animals, some stress elements, including anti-parasitic therapeutics or bacterial diseases may be responsible for ISA infections and outbreaks, 2-3 weeks afterwards. ISAV reservoirs are not completely elucidated, but disease spreading has been related to subclinical carrier smolts, fish slaughtering establishments and fish industries.

Evidence for ISAV in wild salmonids was described. ISA spread could be effectively controlled if general regulatory measures were implemented, especially regarding the international movements of marketed fish: compulsive health control procedures (active surveillance). transport protocols and sanitary controls on fish farms. ISA is an official reportable fish disease for some regional fish health organizations.

KOI HERPESVIRUS DISEASE (KHVD)

KHVD is a viral disease responsible for high morbidity and mortality rates in common carp (*Cyprinus carpio*) [53]. It is a notifiable fish disease with mandatory consequences in global trade regulations.

Aetiology: Koi herpesvirus KHV is designated as Cyprinid herpesvirus 3 (CyHV3) from the *Alloherpesviridae* family. KHV survives in hot waters at temperatures of 23-25 °C for 4 hours, but it is inactivated by UV radiation, temperatures > 50 °C and several biocides, including iodophor, benzalkonium chloride, ethyl alcohol and sodium hypochlorite.

Susceptible Fish Species: KHVD has been detected in cyprinids, including koi carp, common carp, goldfish (*Carassius auratus*) and grass carp (*Ctenopharyngodon idella*) [52]. It affects animals from several age groups [53]. After infection, fish remain carriers for life [55].

Major Clinical Signs and Lesions: Although disease incubation period is 3 days, clinical signs are observed only after one to three weeks. Diseased fish signs include erratic swimming and respiratory difficulties. Internal signs include the development of adhesions and internal organs swelling [53]. Infected fishes can develop secondary infections with ectoparasites or bacteria, showing specific signs.

Diagnosis: KHVD does not express pathognomonic lesions. Diagnosis is dependent on virus detection or isolation. External signs may include skin, gills and fins discoloration, altered epidermidis texture, altered mucus production in skin and gills, enophthalmia, skin and fins haemorrhages and fin destruction. Histopathology may reveal anemic gills and severe necrosis and inflammation, abdominal adhesions, kidney enlargerment, sometimes exhibiting petechial haemorrhages, and necrosis in the kidney, spleen, pancreas, liver, brain, intestines and mouth.

Definitive KHV diagnosis may be performed by molecular (PCR) or immunological (ELISA, VN) techniques [56]. Virus isolation and identification by PCR to identify carriers can be achieved from fresh tissues from euthanized fish, including gill, kidney and spleen from diseased animals or intestine (gut) and encephalon from subclinical fishes, or from blood, feces, mucus and gill biopsies [55]. ELISA can be performed using blood samples.

Sanitary Impacts: Disease has been detected in many worldwide regions. KHVD has been described in Europe, Asia, South Africa and North America. KHV transmission may be horizontal *via* infected fish or water contaminated with feces, urine and mucus, or vertical *via* eggs. Control and prevention strategies include avoiding exposure to diseased fishes, quarantine, eggs disinfection, and application of hygiene and biosafety measures [56]. An efficient vaccine is not available.

KHV is included in the World Organization for Animal Health (OIE) fish disease list that blocks international trade since January 2007.

RED SEA BREAM IRIDOVIRAL DISEASE (RSIVD)

RSIVD is a relevant disease for marine aquaculture fish, being responsible for high mortality rates, from 20 to 60%.

Aetiology: RSIVD is promoted by the red sea bream *iridovirus* (RSIV), belonging to *Megalocytivirus* genus, a recently proposed genus within *Iridoviridae* family [57].

SGIV is inactivated at 56 °C for 30 minutes; is sensitive to ether and chloroform; is inactivated by formalin (0.1%); is stable in tissue freezed at -80 °C. Susceptible fish species: Since the first RSIVD outbreak recorded in *Pagrus major*, the virus has been detected several species of cultured marine fish in Japan, including species belonging to the orders Perciformes , Pleuronectiformes and Tetra-odontiformes. Susceptible species include the yellow fin sea bream (*Acanthopagrus latus*), crimson sea bream (*Evynnis japonica*), Japanese amberjack (*Seriola quinqueradiata*), greater amberjack (*Seriola dumerili*), yellowtail amberjack (*Seriola lalandi*), a hybrid of yellowtail amberjack and Japanese amberjack (*S. lalandi* × *S. quinqueradiata*), striped jack (*Pseudocaranx dentex*), northern bluefin tuna (*Thunnus thynnus*), Japanese Spanish mackerel (*Scomberomorus niphonius*), Japanese jack mackerel (*Trachurus japonicus*), chub mackerel (*Scomber japonicus*), Japanese parrotfish (*Oplegnathus fasciatus*), spotted knifejaw (*Oplegnathus punctatus*), cobia (*Rachycentron canadum*), snubnose pompano (*Trachinotus blochii*), chicken grunt (*Parapristipoma trilineatum*), crescent sweetlips (*Plectorhinchus cinctus*), Chinese emperor (*Lethrinus haematopterus*), spangled emperor (*Lethrinus nebulosus*), largescale blackfish (*Girella punctata*), rockfish (*Sebastes schlegeli*), croceine croaker (*Pseudosciaena crocea*), Hong Kong grouper (*Epinephelus akaara*), convict grouper (*Epinephelus septemfasciatus*), Malabar grouper (*Epinephelus malabaricus*), longtooth grouper (*Epinephelus bruneus*), orange-spotted grouper (*Epinephelus coioides*), yellow grouper (*Epinephelus awoara*), greasy grouper (*Epinephelus tauvina*), brown-marbled grouper (*Epinephelus fuscoguttatus*), giant grouper (*Epinephelus lanceolatus*), a sea bass (*Lates calcarifer*), Japanese sea perch (*Lateolabrax japonicas, Lateolabrax* sp., *L. barramundi*), hybrid of striped sea bass and white bass (Morone saxatilis × *M. chrysops*), largemouth bass

(*Micropterus salmoides*), bastard halibut (*Paralichthys olivaceus*), spotted halibut (*Verasper variegatus*) and torafugu (*Takifugu rubripes*). Regarding ISKNV infection, Chinese perch (*Siniperca chuatsi*), red drum (*Sciaenops ocellatus*), flathead mullet (*Mugil cephalus*) and *Epinephelus* sp. Generally, juveniles have higher susceptibility.

Major Clinical Signs and Lesions: Clinical signs are not pathognomonic, although lethargy concomitant with severe anaemia, petechiae in the gills and spleen enlargement may support diagnosis. Changes in fish behavior also help diagnosis, namely swim inactively and abnormal and conspicuous respiratory exercises due to anemia. Histopathology may reveal the presence of basophilic distended cells in several organs, including the spleen, heart, kidney, liver and gills. These may react to anti-RSIV MAb when immunohistochemistry testing is used. Target organs and infected tissues to be selected for sampling are the spleen, kidney, heart, intestine and gills. Tissue samples can be pooled but no more than five from juvenile fish (< 3 cm). Samples must be preserved, by storing at 4 °C to be used within 24 hours or at -80 °C for longer periods. Most appropriate organs for immunofluorescence (IF) smears are the spleen, heart, kidney, liver and intestine. Fish bodies showing signals of advanced decomposition will not be suitable for any method. RSIV and ISKNV isolation is performed using specific cell lines. Spleen or kidney tissues from diseased fish are also suitable samples. Uninfected and RSIV-infected cell monolayers have to be used as negative and positive controls, respectively. Corroborative diagnostic criteria are based on the confirmation of RSIV or ISKNV presence following development of viral cytopathic effect (CPE) and virus identification using antibody-based antigen detection and/or molecular methods (PCR). Samples used for indirect fluorescent antibody testing include acetone-fixed infected cell monolayers that have developed CPE.

Sanitary Impacts: Depending on host fish species, fish age, water temperature and other culture conditions, mortality rates may range between 0% and 100%. Morbidity is unknown. RSIVD caused by RSIV and ISKNV has been reported in Japan, China, Taipei, Hong Kong, Korea, Malaysia, Philippines, Singapore and Thailand.

Outbreaks have been reported mostly in summer when water temperature arises above 25 °C. The main route of RSIVD transmission is horizontal *via* water. RSIV vertical transmission has not yet been investigated. A number of general husbandry practices are used to reduce RSIVD-associated losses. These include: introducing pathogen-free fish; implementing hygiene practices on farms; implementing of practices to increase water quality; avoiding overpopulation and overfeeding to decrease stress. An effective RSIVD formalin-inactivated vaccine is commercially available in Japan.

RSIVD is included in the OIE list of fish diseases where international notification is mandatory.

Many other virus infections have been reported in fish. Some have specific local impacts or are not responsible for high economic losses. Those are the cases of "Lymphocystis Disease", "Herpesvirus disease of Salmonids", "Epithelioma papillosum" (Fish Pox) and "Viral encephalopathy and retinopathy".

LYMPHOCYSTIS DISEASE (LCD)

LCD is a chronic viral disease that affects oceanic and freshwater fishes. It is promoted by Lymphocystivirus (LCDV), an iridovirus member of the *Iridoviridae* family. LCD is expressed by the growth of nodules, especially on the fins, skin and gills.

LCD has been detected globally, and it affects several species. In general, it does not cause high mortality rates, but affects fish growth and reduces fish commercial value. LCD has been described in oceanic and freshwater fish species, used as ornamental fish or food. It affects teleosts, including *Centrarchidae, Centropomidae, Chaetodontidae, Cichlidae, Cyprinodontidae, Gobiidae, Lophidae, Osphronemidae, Pleuronectidae, Pomacentridae, Pseudo-chromidae, Sciaenidae, Scophtalmidae, Serranidae,* and many others. LCD does not affect catfish (*Siluriformes*), goldfish, koi, barbs, danios (*Cyprinidae*) or salmonids (*Salmonidae*).

Besides the most obvious and characteristic LCD nodules, clinical signs include exophthalmia, erractic swimming and breathing difficulties. Histologically, nodules are clustered groups of greatly enlarged cells (fibroblasts), enclosed in a thick membrane. Subclinically infected fish do not show visible lesions, but may have affected internal organs, including the eye and spleen.

LCD spreads horizontally *via* direct contact with infected fishes, and vertical transmission seems not to occur. LCD is usually an self limited disease, with lesions disappearing in a few weeks. However, affected fish cannot be sold or consumed during this time.

HERPESVIRUS DISEASE OF SALMONIDS

This disease has been primarily described in the rainbow trout.

Transmission is believed to be direct. Clinical signs include lethargy, protruding gills, mucoid feces, exophthalmos, ascitis, anemia, eyes and fins haemorrhages, multifocal necrosis of the myocardium, liver, kidney and intestines. Syncytial cells around pancreas acinar cells are considered to be a pathognomonic signal.

Prevention and control can be achieved by avoiding contact with diseased fishes.

PARASITIC DISEASE OF FISH

Many fish are definitive and intermediate hosts of several parasites. Most of these parasites threat fish life; some have relevant economic impacts; some are responsible for direct hazards to icthyophagic humans and carnivores. These are the main reasons why fish parasites are so relevant [58]. Fish from natural ecosystems or aquaculture systems may be affected by many different species of macro or micro parasites. All these parasites are Eukaryotic organisms belonging to different kingdoms (Animalia, Chromoalveolata, Cnidaria) and phyla (Annelida, Acanthocephala, Nematoda, Platyhelminthes, Arthropoda) [58].

Fish parasitic Annelida belongs to a large phylum of segmented worms, which includes many species including ragworms, earthworms and leeches. The majority of these species are leeches, which have been associated with freshwater, marine brackish and aquarium fish. Most annelids found in marine environment are mostly blood-sucking parasites, while most freshwater species are predators. They have suckers at both body ends, helps them to move like inchworms.

Parasites from *Piscicolidae* family are found in fish surfaces, including in the skin, mouth, branchial, cloaca and caudal fin [58]. Most frequently reported leech on fish is Piscicola geometra, already described in freshwater fish from Europe and Asia.

Other leeches species can be locally abundant: *Hirudo medicinalis (*in *Esox lucius* and *Cyprinus carpio*), *Actinobdella* sp. (in *Barbus* spp.), *Hemiclepsis marginata* (in *Cyprinus carpio*), *Hirudinea* sp. (in *Platichthys flesus*), *Pontobdella muricata* (in *Raja* spp. and *Torpedo marmorata*) and *Cystobranchus respirans* (in *Rutilus* sp.).

Leeches have slim and flexible bodies and vigorously swim to reach the preys. They damage fish skin and adjacent soft tissues, allowing blood to flow into the parasite digestive tract. These are not host specific and the produced damage eliminate depends upon the number of annelids present. Smaller fish can be very wounded and even die due to extreme parasite infestation.

Regarding phylum *Acanthocephala*, this is a group of parasites comprising of worms with 2 to 20 mm in length, with an anterior proboscis with several hooks. Their body is divided into a presoma, which includes the proboscis and associated structures, and a cylinder-shaped trunk. Acanthocephalans life cycles are intricate, as they include several hosts (birds, crustacean, molluscs and aquatic mammals) for both developmental and resting stages. Acanthocephalans have been found in marine and freshwater fish especially in Salmonids and blue fish. Most frequent species are *Acanthocephalus clavula, A. lucii, A. anguillae, A. minor, Acanthogyrus* sp., *Echinorhynchus borealis, E. clavula, E. salmonis, E. truttae, Neoechinorhynchus rutili* and *Pomphorhynchus laevis.*

Usually, infected fish exhibit no external abnormalities, except slimness. Intestines of diseased fishes can be hemorrhagic. Histologically, parasites proboscis can appear firmly attached to the mucosa of the intestines. Depending on infestation intensity, affected fish may exhibit emaciation, lethargy and anaemia, resulting in death. Morbid effects on fish host are low, being the damage extent related to the proboscis penetration depth which may vary with fish hosts, ranging from mild to severe, with the development of granuloma and fibrosis, peritonitis and systemic alterations.

Control methods have not been developed because the parasite is considered to be low pathogenic. It is important to control the population of intermediate hosts by removing food and faeces remnants. These parasites are susceptible to fenbendazoles.

In this group of Metazoan parasites, those having more relevance belong to phylum Nematoda. Nematodes include both roundworms and horsehair worms. They affect oceanic, salted and freshwater fishes, both in their adult and larval stages of their life cycle [2]. They usually affect the gut but, sometimes, larvae migrations can reach several organs with severe damage. Some species are responsible for high levels of mortality. A peculiar group of nematodes which may have a potential deleterious effects to other animals (definitive ou intermediary hosts) include the genera *Anguillicoloides, Philometra, Phoacascaris, Skrjabillanus, Anisakis, Pseudoterranova, Contracaecum, Eustoma* and *Huff-manela*. Ingestion of infected raw fish meat having *Anisakis, Pseudoterranova* or *Contracaecum* larvae, represents a threat to humans, for causing digestive symptoms and alergies.

Adults and larvae of nematodes are regularly found in fish worldwide. Common infestation signs include body deformation, lethargy, hemorrhages, enteritis, equilibrium loss, erratic swimming, gills ulceration, distressing of fins and development of skin nodules. They also cause anorexia and anaemia, as they feed on host nutrients, tissues, sera and blood.

Nematode diagnosis can be performed by histopathology and molecular techniques (PCR). Prevention of human transmission can be achieved by freezing the infected fish lower than -20 °C for 24 hours, cooking above 60 °C or salting; however, it is important to refer that dead parasites can induce immunitary response in humans due to the release of immunogenic products.

Huffmanela is another genus of Nematoda which parasites fishes. They can affect the skin, mucosa, muscles, swimbladder and intestines of several fish species, including elasmobranchs and bony fishes. These parasites life cycle is not completely elucidated. It has been shown that females deposit their eggs in the hosts' tissues, where they continue to mature, forming masses that appear as noticeable black areas in the muscles or fish organs, being a problem for fish commercialization, as fish are refused for consumption and are not allowed to be placed in the market (Fig. **1**).

Fig. (1). Aspect of *Trisopterus minutus* muscle affected by *Huffmanela* sp. parasitism (Courtesy of Dr. Miguel Mendes).

An effective therapeutic for controlling fish nematodes is not available, although flubendazol, mebendazole, levamisole, trichlorphon and triclabendazole have been applied, especially in aquaria collections and aquaculture.

Other Metazoan group that is an important fish parasite is Cestoda, a taxonomic group from Platyhelminthes phylum, in which the mature parasite is found in the vertebrates intestines. The life cycle of these parasites have intermediate phases, in which they affect several regions of the vertebrate and invertebrate hosts' bodies. Cestodes, "tapeworms" are ribbon-shaped, with their bodies separated into small fragments designated proglotids. Diagnosis depends upon parasite detection within the fish intestines or of cysts in its body. Diseased fish show anorexia, anemia, pale skin, and secondary infections are frequent. Small amounts of plerocercoids can be detected in several major organs, including the brain, heart, spleen, kidney, gonads or muscles, with distressing consequences.

There are several species that parasite fishes, including adult cestodes living in the intestine (*Protocephalus* spp., *Ligula* spp.) and cyts (plerocercoids) detected in internal organs and muscles of fish (*Diphyllobothrium* spp.).

Diphyllobothrium latum can be found in fish (intermediary hosts) and mammals (definitive hosts). The adult tapeworm may live in the intestine of canids, felines, bears, pinnipeds and mustelids and can also infect humans accidentally. *D. latum* has been found in Scandinavia, Russia, Baltic regions and North America,. Other members of the *Diphyllobothrium* genus include *D. klebanovskii,* that infects the Pacific salmon, *D. dendriticum* (the salmon tapeworm), *D. pacificum, D. cordatum, D. ursi, D. lanceolatum, D. dalliae, D. nihonkaiense,* and *D. yonagoensis*. All these species are able to infect humans. In fact, human diphyllobothriosis represents an emerging infectious disease in regions where the habits of eating raw or undercooked fish are recent. Symptoms of human diphyllobothriosis are usually mild, including weight loss, fatigue, abdominal ache, diarrhoea, nausea and constipation. Almost 80% of affected humans do not show symptoms, remaining un detected for many years. In some chronic infections, diphyllobothriosis originates a serious vitamin B12 deficiency, since Diphyllobothrium absorbs almost all hosts' B12 intake, leading to severe anaemia, spinal cord degeneration and neurological symptoms. The most effective measure to prevent human transmission is the implementation of accurate fish health inspection procedures, prior to fish introduction in the market. Fish is declared unfit for consumption if plerocercoid cysts are observed in the fish flesh.

Trematoda class parasites are also found in fish, some of them have zoonotic potential and economical impacts. Trematodes are commonly known as flukes, and belong to Platyhelminthes phylum.

Trematodes, or "flattened oval worms", are usually 1 mm to few centimetres in length. They present one or two suckers, one adjacent to the mouth, and the other (when present) on the parasite base. They may be divided in two main subclasses: *Aspidogastrea* and *Degenea*. Recently, a third class, *Monogenea*, has been described, in spite of its phylogenetic similarity with Cestoda class. *Monogenea* include parasites which whole life cycle occurs on a single host, and contains two major genera, *Dactylogyrus* and *Gyrodactylus*. Dactylogyrus sp. is a fish parasite with 0.2 to 2.0 mm length, seven pairs of marginal hooks, one pair of hooks at the opishaptor, and two to four pigmented spots on the anterior region, designated as "eyes" or "eye spots". All dactylogyrus are oviparous with no uterus. *Dactylogyrus* is frequently present on *Cyprinidae* gills. Infections incidence varies according to the seasons, being more frequent in autumn and winter, as there is a relation between water temperature and infection severity; also, fish spawning period seems to predispose to infection. Infected *Cyprinidae* become lethargic, swim near surface, and has a decrease in appetite; they show inflamed gills, excessive mucous secretion and accelerated respiratory movements. In serious infections, *Dactylogyrus* may cause gills haemorrhaging, leading to secondary infections and death. Heavily infected fish are also anorectic, showing respiratory distress and erratic behavior, including leaping outside the water.

World trade of infected carps and ornamental goldenfish may be responsible for parasite dissemination. *Gyrodactylus* spp. comprises of a group of 8 species of small parasites, with a maximum length of 0.4 to 0.5 mm, relevant for aquarium fish health. Some species have been incriminated in very significant economic losses in Salmon aquacultures, especially *Gyrodactylus salaris*, commonly known as Salmon fluke. This parasite resembles a minuscle leech, and has been related with the decrease of Atlantic salmon populations in Norway, having been inadvertently introduced in this area in the 1970s. By 2001, salmon native populations in Norwegian rivers had been virtually wiped out, as a consequence of intense biocides treatments. Other species that can be parasitized include rainbow trout (*Oncorhynchus mykiss*), grayling (*Thymallus thymallus*), Arctic char (*Salvelinus alpinus*), North American brook trout (*S. fontinalis*), North American lake trout (*S. namaycush*) and brown trout (*Salmo trutta*) [61].

It binds to fishes using the haptor, a specialized attachment organ with sixteen hooks located at the posterior end. It feeds by attaching its anterior extremity to the host using cephalic glands, releasing the pharynx and a digestive liquid containing proteolytic enzymes, able to dissolve the salmon skin, which is then absorbed by the parasite. In large infestations, parasites action generates big wounds, damaging the fish epidermis and promoting secondary infections. If not treated, mortality rates of susceptible farmed Atlantic salmon can reach 100%. Mortality rates in other susceptible species are usually low or not recorded.

G. salaris is an extremely dangerous parasite and efficient sanitary control measures must be implemented to avoid contamination of salmonids aquacultures. Since 2009, OIE included this disease as a targeted parasitosis for specific surveillance and has defined methodologies to declare gyrodactylosis free zones. With that goal, regulatory measures for diagnostic/detection methods have been established, which also required international certification before trade. Gyrodactilosis control strategy also includes application of biocide treatments to fish at risk.

Another very important group of trematodes relevant as fish parasites belongs to subclasse *Digenea*. These "flatworms" present a syncytial tegument and two suckers. Adults can be found in the digestive tract of fishes and in the internal organs of vertebrates, including humans. They have an intricate life cycle, including consecutive stages of larvae, interchanging asexual and sexual steps and hosts changes, before adult development in its final target host. Fish can be primary or intermediate hosts.

The most common and relevant Digenea trematodes belong to the genera *Diplostomum, Opistorchis, Clonorchis, Paragonimus and Posthodiplostomum*. Diagnosis of Digenea infections is dependent upon genus and species identification. *Diplostomum* parasites do not pose a serious threat to fish life. Adults can be found in the intestine of fish-eating birds, such as herons, seagulls and cormorants. Their body has 0.3 to 0.5 cm in length, and is separated into two regions, a compressed anterior extremity and a tubular and thinner posterior extremity. Eggs are released and disseminated through bird faeces to the water. After three weeks, eggs hatch originating free-swimming ciliated miracidia that invade the first intermediary host, the aquatic snails (*Lymnaea* sp.), by penetrating the hepatopancreas. Miracidia become a sporocyst, which is then divided in one or more sporocysts that are able to release numerous cercariae into the water. Then, cercariae invade a second intermediary host, generally small fishes (*Cyprinidae, Percidae*) through the fins, skin, gills, cornea or brain. Primary host fish eat these small fishes, becoming infected, being then consumed by fish-eating birds.

Posthodiplostomum spp. trematode life cycle is identical, although cercariae infectivity persists for only 24 hours after the discharging of the aquatic snail. Cercaria penetrate the fish scales through the scale pocket, irritating the skin and originating congestion and haemorrhages. Larvae disseminate to fishes visceral organs one to three hours after penetration, and metacercariae can be found in several organ, including the liver, kidney, heart and spleen. Metacercariae skin invasion consequences include the deposition of melanin, responsible for the designation of "black spot disease". Fin lesions are known as "black grub", while small white or yellow stains in internal organs are known as "white grubs" or "yellow grubs".

Opistorchidae, particularly *O. felineus*, represent a quite different health problem. *O. felineus* and *O. viverrini* are trematodes which infect the icthyophagic mammals liver, being known as cat liver flukes. Regarding its life cycle it includes three hosts. The first intermediary hosts are *Bithynia inflata*, *B. troschelii* and *B. leachii*, snails that can be found in freshwater; the second ones are freshwater fishes; and the final ones are fish-eating animals, including cats and humans. Opisthorchiasis may vary from asymptomatic to severe disease, affecting several organs, including the liver, pancreas and gall bladder. Disease symptoms include fever, sickness, gastrointestinal disturbances, skin rash and anaemia. Disease outcome depends upon prompt diagnosis and treatment with praziquantel, and a late response may be responsible for the development of cirrhosis and liver cancer. Control may be achieved *via* physical examination of fish and systematic thermal treatment of fish used for consumption.

Another important *Digenea* is the very small parasite *Clonorchis sinensis*, which adults lanceolate body measures approximately 11 to 20μm by 3 to 4.5μm. They have an anterior operculum, a posterior lump, a translucent and brownish color and they are all hermaphrodites. They infect humans who eat raw or undercooked freshwater *Cyprinidae* fish having metacercaria encysted on muscles. Acid-resistant cyst enables the metacercariae to survive in the human gastric acids, allowing to reach unharmed the small intestine. Afterwards, it reaches the human liver *via* circulatory blood. Infection severity depends upon the number of metacercaria ingested, and may range from mild asymptomatic infections to severe acute disease, which symptoms include fever, diarrhea, abdominal ache, liver swelling and jaundice. Dissemination to the gall bladder may cause cholecystitis, as flukes obstruct the bile duct lumen, where they feed on bile. This parasite has also been related with the development of liver and bile duct cancer.

This agent, the Chinese liver fluke, is the third most predominant worm parasite worldwide, being found in China, Taiwan, Japan, Korea, China, Vietnam and Russia.

Diagnosis is achieved by the detection of eggs in human feces by microscopic examination, immunological (ELISA) or molecular (PCR, real-time PCR) methods.

Drugs used for treatment comprise triclabendazole, albendazole, praziquantel, bithionol, levamisole and mebendazole .

Paragonimus westermanii is another *Digenea* with zoonotic relevance. It is a lung fluke that attacks human lungs after eating raw or undercooked infected crab or crayfish. A less frequent condition of paragonimiasis occurs when the parasite travels to the central nervous system, being responsible for a serious clinical problem.

P. westermani can be found in several Asian countries, including Japan, China, Philippines, Vietnam, Korea, Taiwan and Thailand ; *P. africanus* can be found in Africa, while P. mexicanus has been reported in Central and South America. They can infect humans who eat raw or undercooked shellfish (ceviche, kilawin, bagoong). To avoid these infections, freshwater crabs or crayfish should never be eating raw, being allowed to cook for at least 10 minutes at 63 °C. Paragonimiasis is an acute infection with low fever, with cough, abdominal pain and discomfort that may evolve to a chronic disease with symptoms similar to bronchitis or tuberculosis. Assymptomatic or mild infections are also described. Effective treatment for paragonimiasis is available. Artropoda are another Metazoan phylum relevant as fish parasites. The list of fish parasites species is very extensive. Most frequent Crustaceae subphylum known as fish parasites belong to genera *Lerneocera, Sphyrion, Lernae, Argulus, Caligus, Lepeophtheirus* and

Sarcotaces. More than 3 hundred species belonging to *Cymothoidae* family have also been associated with fish. The copepod *Lernaeocera branchialis*, known as "cod worm", is an oceanic fish parasite, mostly found in North Atlantic. It evolves from a small pelagic crustacean larvae to a copepodid larvae, ranging from 2-3 millimetres in size, and then to a large copepod, with more than 4 cm long in the adult stage. *L. branchialis* is an ectoparasite, whose life cycle has many stages, including motile and sessile periods. It has two hosts: it infects a *Pleuronectidae* or a *Cyclopteridae* as a secondary host, and another where it parasites "cods" or other gadoids as primary hosts. It is a pathogen with a negative impact on commercial fishing and mariculture of *Gadidae*, including cod aquaculture. In this fish species, it is responsible for reducing the efficacy of fishfeed use, delaying gonads development. It may also be observed up to 30% in weight loss, increased mortality due to exposed injuries with blood loss, obstruction of vessels or aorta and secondary infections.

Another copepod frequently found in *Scorpaenidae* and *Sebastidae* in North Atlantic is *Sphyrion lumpi*. Water lice, known as "anchor worms", are other copepods that can also have a special role as fish parasites. Most common fish lice belong to genera *Lepeophtheirus, Caligus* and *Argulus*. These genera are external parasites that damage their hosts by sucking their blood, causing localized skin and soft tissue direct damage. Blood volume reduction affects fish activities due to general weakness and also decreases the growth rate. These parasites can also lead to fish death, which may also be due to secondary bacterial and viral skin or musculature infections (indirect effects). There are many species of freshwater lice, like Argulus coregoni, *Argulus foliaceus* and *Argulus japonicus* . Freshwater lice do little harm to their fish hosts and are not hazardous for humans. However, the wounds they make can allow microbes' infections and, consequently, promote fish death. They can be identified as two black stains located adjacent to the eyes, after detailed examination. *Argulus coregoni* and *A. foliaceus* epidemics have been described in rainbow and brown trout fisheries in Scotland since 1994.

The parasites must not be mistaken with the marine lice *Lepeophtheirus salmonis* and *Caligus* spp. that infect trouts and salmons. Sea lice are Copepoda from *Caligidae* family, being the most widely distributed marine parasite in salmon industry. Sea lice cause physical and enzymatic damage at their binding and suckling sites, resulting in abrasion injuries, which gravity depends on the sea lice and fish species, lice concentration, fish age and well-being. Consequences of lice infestation include variations in mucus production, skin injury, blood and fluids loss, electrolyte changes, increase of cortisol production, immune-depression, and growth and performance reduction.

Vaccination could be an environmentally friendly alternative for sea lice control; however, research on vaccines development is still at an early stage. Recent studies have suggested that subolesin/akirin/my32 are good antigen candidates for arthropod infestations control, including sea lice.

Several species from Amphipoda and Isopoda orders are also relevant as fish skin Artropodes, are known as "sea fleas" and "fish fleas". "Sea fleas" are found in seawater, even at 5 thousand meters depths, and can attach themselves to fish and eat through their skin. They appear as creamy and translucent, and some European species are also tiny, and difficult to be spotted. Nevertheless, it is important to refer that their size may vary from a few millimetres to 25 cm (*Anilocra gigantea*). These parasites bite fish and crustacean, usually at the base of fins. Bites are often very itchy, being responsible for the development of red lesions and blisters. Some particular fisheries procedures promote these infections. In "long line", for example, captured death fishes remain in water for many hours, being prone to "sea fleas" attack, which may ingest a very significant portion of the fish, even all the flesh.

Protozoa are another major group of fish parasites. These microscopic organisms with eukaryotic structures belong to Chromalveolata kingdom (ex- Chromista). Fish protozoa may be found on the fish surface, within the gills, and/or internal organs or muscles. When infecting the gills, they promote respiratory distress that can evolve to death in the presence of extra environmental stress factors. Protozoan parasites present only on the skin, fins or scales are not generally lethal, except when a simultaneous secondary bacterial infection occurs; nevertheless, they are quite stressful.

The most frequent fish protozoan parasites belong to two major phylums: Ciliphora and Apicomplexa. Ciliphora are ciliated parasites up to 2 mm long, with cilia attached to the external wall, responsible for undulant movement. Cilia occur in all group members sometimes only in one life cycle phase, and can be used for swimming, swarming, connecting, feeding and feeling. Ciliates can be found almost in fresh, salted and oceanic waters.

Most important fish parasites Ciliates are *Ichthyophthirius multifilis, Trichodinia* spp., *Ambiphyra* sp., *Ichthyobodo* sp. and *Chilodonella* sp. *I. multifilis* is one of the most frequent parasites of freshwater fishes, promoting a parasitic disease known as "Ich" or "white spot". It is a major problem for aquarists. The mature parasite has a diameter around 1 mm and can be found on the fish gills and skin.

Ciliate mature stage is known as trophont, characterized by a horseshoe nucleus and a cilia covered exterior. It has a direct life cycle, but it can also be found outside the host. The "trophont" is positioned inside the fish host epidermis, where it encysts and splits into hundreds of immature forms, the tomites, theronts or trophozoites, until it leaves the fish. The tomites penetrate the fish skin and gills to end the life cycle. Registered signals include respiratory distress, abnormal hiding, resting on the bottom, flashing, upside-down swimming near surface and rubbing and scratching against surfaces or objects. Gills and cutaneous infections promote epidermis hyperplasia.

Subclinical infections have also been reported. It is not easy to detect a trophozoite attached to the gills. Once *Ichthyophthirius multifilis* is introduced into a large fish aquaculture plant, it is difficult to control dissemination due to its fast life cycle and peculiar life stages. If not controlled, fish mortality rate may reach 100%. With treatment, disease can be controlled but the cost is high, due to fish loss, labour and applied chemicals. Formalin, malachite green, or combinations of the two, have been described in the past as the best treatments. However, nowadays these biocides are not allowed to be applied for treating fish intended for consumption due to safety issues, but in the aquariums their use is still possible. "Ich" prevention is based on general hygiene practices for fish production or for aquariums, such as: never introduce fish without health assurance or free of all signs of disease; never use fish from tanks with dead or diseased fish; newly acquired fish must be placed in quarantine space for at least two weeks; never use ornamental plants that were previously kept in a fish tank, as plants also need at least 4 days of quarantine; fish exhibiting any signal of the disease must be immediately removed to a quarantine tank and treated; temperature, pH or ammonia levels fluctuations must be avoid, as well as other stressful events; fish must be fed with controlled feed; tanks must never be overstocked; water characteristics must be regularly controlled and water must be frequently renewed.

Trichodina are also ciliates that can be found on fish surface. They are characterized by the presence of a ring of interlocking cytoskeletal denticles, which provides support and allows for adhesion to surfaces, including fish tissue. This group is formed by three genera: *Trichodina, Trichodonella* and *Tripartiella*. They all present a dish shape with a diameter of approximately 0.1 mm, being surrounded by cilia.

Trichodina species live in salted, estuarine or freshwater ecosystems. Some species are preferentially located on the gills, some on the skin, and others parasitize on both locations. Trichodina species reproduce by parthenogenesis, through binary fission. Most species are host specific and presumably spread through direct fish to fish contact, incidental contact between susceptible host fish, and through water. Infected fish may show respiratory and osmoregulatory difficulties and fin and skin ulcers.

Clinically infected fish may 'flash' in the attempt to scratch off the organisms. Superficial white punctiform lesions appear on the body surface and the fins may become frayed. Scales may be loose and concomitant opportunistic bacterial invasions may lead to secondary infections causing skin ulcerations and erosions. Diagnosis is not easy, depending upon the parasite detection or histopathological analysis.

Microscopic examination of mucus obtained from the skin surface may reveal rapidly spinning ciliated organisms, with typical morphology; gill, fin and skin biopsies can also be examined. Chemical treatments are relevant for aquarium fish, using formalin, salt, potassium permanganate or copper sulphate baths. As the life cycle is direct, a single treatment is usually adequate. General good hygiene practices are the only strategy to prevent *Trichodina* infections.

Ichthyophonus hoferi, previously classified as a Fungi of the *Zygomycetes* class, is a protozoan parasite that has been incriminated in internal organs and muscles infections in rainbow trout (*Oncorhynchus mykiss*), chinook salmon (*O. tshawytscha*), Pacific herring (*Clupea pallasii*), yellowtail (*Seriola quinqeradiata*) and other Pacific fish. *Ichthyophonus* spp. have binucleate hyphal bodies that grow into thick walled spherical multinucleate bodies, with a diameter of 20 to 125 µm. The multinucleate spherical bodies extend the hyphal bodies and produce spores. Fish become infected by ingesting the multinucleated spherical bodies, which germinate in stomach and penetrate its mucosa. Disease outbreaks were signaled in cultured rainbow trout in Japan (fed on raw herring) and in wild Chinook salmon in North America. It affects wild, aquaculture and aquarium fishes, promoting a chronic disease characterized by spine deformation (scoliosis, lordosis), exophthalmos and granulomatous injuries on several organs, such as the skin, heart, liver, spleen, kidney or muscle.

For diagnosis, suspect lesions macerates may be cultured on Sabouraud's dextrose agar. Control procedures comprise eliminating diseased fish and not include raw fish products in the diet. This parasite is not infectious to humans. There are sessile ciliated protozoan organisms of *Ambiphrya* genus, with a tubular body that ranges from 60 to 100 mm. They bind to fish skin or gills, being responsible for the development of chronic infections characterized by respiratory distress. Diagnosis depends upon identification of scrapings on skin or gill scrapings or by histopathology. Another relevant fish parasitic protozoon is *Ichthyobodo* sp., previously known as *Costia* sp. This parasite may be found on fish gills and skin. All freshwater fish species are susceptible to *Ichthyobodo* spp. The most frequent species is *Ichthyobodo necator*, protozoa of the *Bodonidae* family, with a pyriform body with a size of 8 to 13 mm. This parasite is frequently found feeding mostly on fish dead cells. It is not dangerous for healthy fishes, due to their effective immune system. It has a very short life cycle, with two forms. The feeding form has an oval form and binds to the fish gills and skin. The free, non-feeding form has an oval body and can swim due to the presence of 2 (rarely 4) unequal flagella. *Ichthyobodo necator* is a classic opportunist parasite, waiting for weakened hosts to attack, triggering mass infestations and leading to high mortality rates. During their lives, *Cyprinidae* and *Salmonidae* can develop resistance to this parasite, but juveniles are very susceptible. In fact, I. necator is extremely dangerous for juveniles and leads inevitably to the death of the offspring. *I. necator* attaches to the debilitated host *via* the attachment plate and feeds through the cytostome and the cytopharyngeal canal protruding into the host cell. Infected fish will flash or scrape against surfaces or objects and in more extreme cases they stop eating (anorexia) and gasp at the waters' surface. A typical infection sign is an excessive mucus production ("blue slime disease"). Heavily infested fish exhibits skin haemorrhagic lesions, skin and gills epithelial hyperplasia, fins ulceration and erosion. Diagnosis depends upon protozoa detection within affected fish tissues, using cytology or histopathology. The typical flagella are observed in a fresh microscope slide. Identification of the *Ichthyobodo* species requires morphological observation using Diff-Quik stain and molecular genomic analysis. New alternative therapeutics control methods are under development, as traditional formalin treatment is now forbidden in fish intended for food [62]. Spironucleus salmonicida lives in fish intestines and may also produce abscesses in fish muscles. This flagellate protozoa is pyriform to oval with a sharp posterior end, usually being 10-12 mm long to 6-8 mm wide. Occasionally, rounded individuals are found. Taxonomically, Spironucleus spp are now positioned in the Excavata kingdom, Metamonada phylum and *Diplomonadida* order. Spironucleus spp. are characterized by having three pairs of long flagella, located in the anterior extremity of the body. Flagella originate from the blepharoplast at the axostyles anterior top. They can multiply by binary division, or go through schizogony in the ceca or intestine cells.

Most common clinical signs of infected fishes are malnutrition and emaciation. Diagnosis depends on parasite observation *via* cytology or histopathology, performed using of cecal or duodenal scrapings.

Chilodonella sp. is a motile protozoan able to invade fish skin and gills. It has a ciliated heart-shaped body. Its dimensions range from 30 to 70 mm in length and 20 to 40 mm wide and its surface is covered with cilia, moving in a characteristic slow spiral. In the posterior end they present a macronucleus, surrounding by or enclosing several micronucleus. Unlike other ciliates, *Childonella* sp. contains numerous mini somatic chromosomes in its macronucleus, and three chromosomes in its micronucleus. Some species, including *Childonella uncinata*, may promote contamination by mosquito larvae and specific bacteria, like *Klebsiella* spp.. Although *Childonella* sp. is primarily a free-living freshwater organism, inhabitant of fish mucous and secretions, it has been incriminated on *Salmonidae* death promoted by respiratory and osmoregulatory distress.

Fish with chilodonelliasis show increased mucous production, respiratory distress swollen gills, altered behavior and anorexia. Diagnosis depends upon analyses of fresh biopsies of infected tissues, tested either by cytology or histopathology. They are 0.5 to 0.7 mm long, with typical shape and parallel bands of cilia. *Chilodonella* sp. has been incriminated in fish death since 1902. Some epizootic episodes may have been triggered by overcrowding, environmental gill diseases, and/or microscopic mineral particles injuring the fry's gills. Some authors stated that the application of 100 ppm hydrogen peroxide for 30 minutes during three days can lead to successful treatment in aquarium fish. Another relevant group of not ciliated Protozoa, able to parasite fish, belongs to the Apicomplexa phylum and includes parasites usually known as "coccidia". Coccidians are usually vertebrates' parasites, being found in the intestines, natatory bladder, kidney, spleen, roe and testis. Coccidian lifecycle is complex, involving merogony, gametogony and sporogony.

Many *Eimeria* coccidians species has been described as fish parasite: *Eimeria aurati* in goldfish (*Carassius auratus*); *Eimeria baueri* in crucian carp (*Carassius carassius*); *Eimeria leucisci* in barbel (*Barbus barbus bocagei*); *Eimeria lepidosirenis* in lungfish (*Lepidosiren paradoxa*); *Eimeria rutili* in Europena chub (*Leuciscus cephalus*) and in Iberian nase (*Chondrostoma polylepis*); *Eimeria vanasi* in blue tilapia (*Oreochromis aurea*); *Eimeria anguillae* in European and Pacific eels; *Eimeria daviesae* in gobi (*Gobius kessleri*); *Eimeria percae* in yellow perch (*Perca flavescens*); *Eimeria variabilis* in seebull (*Myoxocephalus bubalis*); *Eimeria gadi* in haddock (*Melanogrammus aeglefinus*) and *Eimeria sardinae* in harring (*Clupea harengus*). *Eimeria* spp. are ubiquitous of the vertebrates gut epithelium, in which they complete their development. Impacts of coocidiosis on fish health are not completely assessed. There are other groups of parasites which can cause disease in fishes, previously classified as Protozoan, and that are nowadays included in different kingdoms, like Animalia (*Myxozoa*) and Fungi (*Microsporia*). Myxosporea belong to the Phylum Cnidaria, subphylum Mixozoa, and are 10 μm to 20 μm parasites of tissues and organ cavities, being able to produce spores.

Life cycle may require a vertebrate (fish) and an invertebrate host (annelid or polychaete worm); in each host the parasite develops its own sexual and asexual stages. Many myxosporidians species infect the skin, gills, muscle and viscera, where they form nodules or cysts, depending on the host and infected organs. Severe infestations can cause death. Each parasite is species and organ specific.

One of the most relevant myxozoan parasites in fish is *Myxobolus cerebralis*, which is one of the most pathogenic parasites, responsible for major economic losses. They are responsible for the "whirling disease" in *Salmonidae* [63], a chronic illness frequently referred in the USA, although recent outbreaks have been reported in Europe, Russia, South Africa and New Zealand. It feeds on the axial skeleton cartilage of wild and aquaculture salmon and trout. Disease signs include skeletal distortion, neurological injury, distressed swimming, feeding difficulties and increased vulnerability to predators. In the early stages of infection fish can elimate *Myxobolus* from their skin *via* their immune system, but this response depends on the fish species [64]. Mortality rate is up to 90% for fingerlings, and survivors are deformed, acting as reservoirs.

M. cerebralis can affect many salmonid species, including Atlantic salmonids (*Salmo salar*), Pacific salmonids (*Oncorhynchus* spp.), the char (*Salvelinus* spp.), the grayling (*Thymallus thymallus*); and the huchen (*Hucho hucho*).

To confirm diagnosis, *Myxobolus cerebralis* parasites must be observed histologically within the axial skeleton cartilage.

Myxozoan parasites are normally present in wild and captive fish. They do not affect their hosts when fish and environment are in equilibrium. When any kind of stressing situation occurs, including handling, deficient water quality or overpopulation, parasites develop and disease arises. This is the case of *Henneguya ictaluri*, the aetiological agent of "proliferative gill disease" (PGD), a known parasitosis of the channel catfish, *Ictalurus punctatus* invades the gills, resulting in respiration and osmoregulatory distress. Its life cycle includes an intermediary host, named Dero digitate, which is a tiny oligochaete worm.

Diagnosis depends on parasite detection in the gills.

Another myxosporean parasite, *Henneguya salminicola*, usually infects salmonids from North America, Canada and North Europe. Fish responds by enclosing parasites into cysts inside the muscles. *Henneguya* has a complex lifecycle with two hosts, a *Salmonidae* that discharges spores after reproducing, which invade a secondary host, an invertebrate tubificid worm. The cycle is completed when these worms release an infective form of the parasite as juvenile fishes migrate to the ocean, where they remain until the following reproduction season.

Henneguya infection seems to be asymptomatic. However, fish production may become compromised due to the market refusal of salmon, harboring the visible white cysts and with flesh emaciation. It must be stressed that *Henneguya* is harmless for public health, as it doesn't affect humans. Other fishes in which *Henneguya* has been reported include pike, perch, catfish, bream and char.

Very similar to *Henneguya* cysts are those produced by *Kudoa* spp., excepted for the fact that the myxosporean parasite is exclusively found in marine fish. *Kudoa* has a global dissemination, infecting several fish species, including *Sparidae, Pleurinectidae, Salmonidae, Clupeidae, Gadidae* and *Scombridae* (Fig. **2**).

Salmonid ceratomyxosis is another myxosporidian parasitosis, promoted by *Ceratomyxa shasta*, prevalent in the Pacific Northwest. It infects wild and aquaculture salmons, where it can be found in several organs, including the intestines, gall bladder, spleen, liver, gonads, kidney, heart, gills and muscles. Disease signs include emaciation, fatigue, anorexia, skin darkening, ascitis, exopthalmia, haemorrhagic vent, swollen abdomen and kidney abscesses. These signs differ between salmonid species, and also depend on host life stage. Subsequent intestinal damages and concomitant secondary infections may lead to death.

A presumptive diagnosis can be performed by observation of trophozoites in the intestines, or by histopathology.

Some North American States classify ceratomyxosis as a reportable disease.

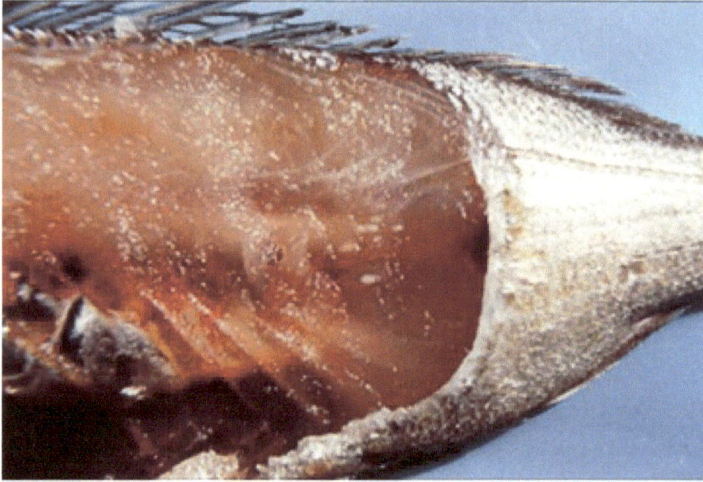

Fig. (2). Pseudocyst of *Kudoa thyrsites* in *Pagellus acarne* muscle (original). *Kudoa thyrsites* spores are typical stellate in shape, with four valves and four polar capsules. K. thrysites life cycle is not completely elucidated. It has been hypothesized that it has an indirect life cycle involving marine invertebrates [59]. Upon infection by the actinospore, the sporoplasm migrates to fish muscles where it forms a white pseudocyst, in which the developing spores are present. Kudoa and other myxozoan species are responsible for important economic losses, by causing post-mortem "myoliquefaction", characterized by extensive flesh, rendering the fish unmarketable. It is not infective to humans. Prevention of *K. thyrsites* infections is impossible because it occurs in open water netpens. Currently, there are no available treatments.

CONCLUDING REMARKS

Many physical, chemical and biological agents are incriminated in fish diseases. At the present, biological aetiologies have major impacts on fish health and fisheries economy. Microbial and parasitic diseases of fish are not only responsible for significant economic losses in aquaculture, but also for international blockage of global market, to avoid disease dissemination. Some even have zoonotic potential. In this text are listed species with significant sanitary impacts and major consequences to the global market, having substantial negative consequencies on the economic rentability of fish farms and commerce [1]. More detailed descriptions dedicated to some diseases are justified by considering in international regulatory laws. It was also taken into consideration the fact that some fish diseases are not adequately studied. In addition to the microbiological and parasitic determinantes mentioned, many others can affect fish life. The presence of predators and symbiotic animals in a specific aquatic environment, like other fish, aquatic mammals or birds, are also relevant to microbial and parasitic diseases evolution. Whenever treatments are available, it is crucial to evaluate their application, based on regulamentary or compulsory legislation; available treatments for aquarium fish may be not allowed for fish produced in aquaculture systems intended for human consumption. Aquaculture fish production submitted to antimicrobial treatments need to respect withdrawal periods and residues official safety levels before being introduced to the market. When fish classified as unfit and its sub products cannot be recycled for fish consumption, the food chain would remain contaminated. Prevention is always more efficient than treatments in reducing the negative impacts of fish microbial and parasitic diseases.

CONFLICT OF INTEREST

The author confirms that author has no conflict of interest to declare for this publication.

ACKNOWLEDGEMENTS

Declared none.

REFERENCES

[1]   Roberts RJ. Fish Pathology. Chichester, UK: John Wiley & Sons, Ltd 2012; p. 514.
       [http://dx.doi.org/10.1002/9781118222942]

[2]   Paperna I, Zwerner DE. Parasites and diseases of striped bass, *Morone saxatilis* (Walbaum) from the
       lower Chesapeake bay. J Fish Biol 1976; 9: 267-87.

[3]   Organisation Mondiale de la Santé Animale (OIE).. Code sanitaire pour les animaux aquatiques
       Available from: http://www.oie.int/fr/normes-internationales/code-aquatique/acces-en-ligne/.

[4]   Hubbert RM. Bacterial diseases in warmwater aquaculture. In: Shilo M, Sarig S, Eds. Fish culture in
       warm water systems: problems and trends. Boca Raton, Florida, USA: CRC Press 1989; pp. 179-94.

[5]   Barton BA, Iwama GK. Physiological changes in fish from stress in aquaculture with emphasis on the
       response and effects of corticosteroids. Annu Rev Fish Dis 1991; 1: 3-26.
       [http://dx.doi.org/10.1016/0959-8030(91)90019-G]

[6]   Faisal M, Abdelhamid HS, Torky H, Soliman MK, Abu Elwafaa N. Distribution of *Aeromonas
       hydrophila* in organs and blood of naturally and experimentally infected *Oreochromis niloticus*. J
       Egypt Vet Med Assoc 1984; 44: 11-20.

[7]   Ruangpan L, Kitao T, Yoshida T. Protective efficacy of *Aeromonas hydrophila* vaccines in nile tilapia.
       Vet Immunol Immunopathol 1986; 12(1-4): 345-50.
       [http://dx.doi.org/10.1016/0165-2427(86)90139-X] [PMID: 3765355]

[8]   Darunee S-O, Muroga K, Nakai T. A case of *Edwardsiella tarda* infection in cultured coloured carp *Cyprinus carpio*. Fish Pathol 1984; 19: 197-9.
[http://dx.doi.org/10.3147/jsfp.19.197]

[9]   Kitao T, Aoki T, Tawara K, Kumada K, Shiomitsu K, Fukudome M. On an edwardsiellosis in tilapia. Abstract, Annual meeting Jap Soc Sci Fish. April 1980; 80.

[10]   Miyashita T. *Pseudomonas fluorescens* and *Edwardsiella tarda* isolated from diseased tilapia. Fish Pathol 1984; 19: 45-50.
[http://dx.doi.org/10.3147/jsfp.19.45]

[11]   Lio-Po G, Wakabayashi H. Immuno-response in tilapia *Sarotherodon niloticus* vaccinated with *Edwardsiella tarda* by hyperosmotic infiltration method. Vet Immunol Immunopathol 1986; 12(1-4): 351-7.
[http://dx.doi.org/10.1016/0165-2427(86)90140-6] [PMID: 3765356]

[12]   Sakata T, Hattori M. Characteristics of *Vibrio vulnificus* isolated from diseased tilapia. Fish Pathol 1988; 23: 33-40.
[http://dx.doi.org/10.3147/jsfp.23.33]

[13]   Muroga K, Yo Y, Nishibuchi M. Pathogenic *Vibrio* isolated from cultured eels -I. Characteristics and taxonomic status. Fish Pathol 1976; 11: 141-5.
[http://dx.doi.org/10.3147/jsfp.11.141]

[14]   Bruno DW. Prevalence and diagnosis of bacterial kidney disease (BKD) in Scotland between 1990 and 2002. Dis Aquat Organ 2004; 59(2): 125-30.
[http://dx.doi.org/10.3354/dao059125] [PMID: 15212278]

[15]   Austin B. Progress in understanding the fish pathogen *Aeromonas salmonicida*. Trends Biotechnol 1997; 15: 131-4.
[http://dx.doi.org/10.1016/S0167-7799(97)01026-3]

[16]   Tobback E. Early pathogenesis of *Yersinia ruckeri* infections in rainbow trout (*Oncorhynchus mykiss, Walbaum*). PhD thesis, Faculty of Veterinary Medicine, Ghent University, Ghent, Belgium 2009; 155.

[17]   Woo PT, Bruno DW. Fish diseases and disorders Viral, Bacterial and Fungal infections. New York, USA: CabInternational 2011; p. 930.
[http://dx.doi.org/10.1079/9781845935542.0000]

[18]   Daskalov H, Austin DA, Austin B. An improved growth medium for *Flavobacterium psychrophilum*. Lett Appl Microbiol 1999; 28(4): 297-9.
[http://dx.doi.org/10.1046/j.1365-2672.1999.00522.x] [PMID: 10212443]

[19]   Bernardet JF, Nakagawa Y. An introduction to the family *Flavobacteriaceae*. In: Dworkin M, Falkow S, Rosenberg E, Schleifer KH, Stackebrandt E, Eds. The Prokaryotes: a Handbook on the Biology of Bacteria. New York, USA : Springer 2006; pp. 455-80.

[20]   Eppinger M, McNair K, Zogaj X, Dinsdale EA, Edwards RA, Klose KE. KloseDraft KE. Genome sequence of the fish pathogen *Piscirickettsia salmonis*. Genome Announc 2013; 1(6): e00926-13.
[http://dx.doi.org/10.1128/genomeA.00926-13] [PMID: 24201203]

[21]   Amin NE, Abdallah I, Faisal M. Easa MEl-S, Alaway T, Alyan SA. Columnaris infection among cultured Nile tilapia *Oreochromis niloticus*. Antonie van Leeuveen 1988; 54: 509-20.

[http://dx.doi.org/10.1007/BF00588387]

[22]   Mauel MJ, Miller DL. Piscirickettsiosis and piscirickettsiosis-like infections in fish: a review. Vet Microbiol 2002; 87(4): 279-89.
[http://dx.doi.org/10.1016/S0378-1135(02)00085-8] [PMID: 12069766]

[23]   Gaggero A, Castro H, Sandino AM. First isolation of *Piscirickettsia salmonis* from Coho salmon, *Oncorhnchus kisutch* (Walbaum), and rainbow trout, *Oncorhynchus mykiss* (Walbaum), during the freshwater stage life cycle. J Fish Dis 1995; 18: 277-9.
[http://dx.doi.org/10.1111/j.1365-2761.1995.tb00303.x]

[24]   Neubrand E, Enriquez R, Henriquez C. Studies on the pathology and transmission of *Piscirickettsia salmonis* in coho salmon (*Oncorhynchus kisutch*). European Association of Fish Pathologists. In: Seventh International Conference 'Diseases of Fish and Shellfish'.; Palma de Mallorca, Spain. 1995.55

[25]   Inglis V, Roberts RJ, Bromage NR. Salmonid Rickettsial Septicaeamia. In: Inglis V, Roberts RJ, Bromage NR, Eds. Bacterial diseases of fish. Oxford, UK: Blackwell Scientific Publishing 1993; pp. 245-54.

[26]   Noga EJ, Wright JF, Pasarell L. Some unusual features of mycobacteriosis in the cichlid fish *Oreochromis mossambicus*. J Comp Pathol 1990; 102(3): 335-44.
[http://dx.doi.org/10.1016/S0021-9975(08)80022-9] [PMID: 2365849]

[27]   Colorni A. A systemic mycobacteriosis in the european sea bass *Dicentrarchus labrax* cultured in Eilat (Red Sea). Isr J Aquacult 1992; 44: 75-81.

[28]   Knibb W, Colorni A, Ankaoua M, Lindell D, Diamant A, Gordin H. Detection and identification of a pathogenic marine Mycobacterium from the European seabass *Dicentrarchus labrax* using polymerase chain reaction and direct sequencing of 16S rDNA sequences. Mol Mar Biol Biotechnol 1993; 2(4): 225-32.
[PMID: 8293073]

[29]   Lescenko P, Matlova L, Dvorska L, *et al.* Mycobacterial infection in aquarium fish. Vet Med Czech 2003; 48: 71-8.

[30]   Kusuda R, Kawai K. Bacterial diseases of cultured marine fish in Japan. Fish Pathol 1998; 33: 221-7.
[http://dx.doi.org/10.3147/jsfp.33.221]

[31]   Shamsudin MH, Tajima K, Kimura T, Shariff M, Anderson IG. Characterization of the causative organisms of ornamental fish mycobacteriosis in Malaysia. Fish Pathol 1990; 25: 1-6.
[http://dx.doi.org/10.3147/jsfp.25.1]

[32]   Fischer O, MAtlovA L, Bartl J, DvorskA L, MelichArek I, PavlA-k I. Findings of mycobacteria in insectivores and small rodents. Folia Microbiol (Praha) 2000; 45(2): 147-52.
[http://dx.doi.org/10.1007/BF02817414] [PMID: 11271823]

[33]   Wayne LG, Kubica GP. Genus *Mycobacterium* Lehmann and Neumann 1896, 363AL. In: Sneath PHA, Mair NS, Sharpe ME, Holt JG, Eds. Bergey's Manual of Systematic Bacteriology, 2. Baltimore, USA: The Williams & Wilkins Co. 1986; pp. 1436-57.

[34]   Guerrero C, Bernasconi C, Burki D, Bodmer T, Telenti A. A novel insertion element from *Mycobacterium avium*, IS1245, is a specific target for analysis of strain relatedness. J Clin Microbiol 1995; 33(2): 304-7.

[PMID: 7714183]

[35]   Tung M-C, Shin-Chu C, Shin-Shyong T. General septicaemia of streptococcal infection in cage cultured tilapia, *Tilapia mossambica*, in southern Taiwan. Fish Dis Res 1985; 12: 95-105.

[36]   Eldar A, Bejerano J, Bercovier H. *Streptococcus shiloi* and *Streptococcus difficile*: two new streptococcal species causing a meningoencephalitis in fish. Curr Microbiol 1994; 28: 139-43. [http://dx.doi.org/10.1007/BF01571054]

[37]   Iida T, Wakabayashi H, Yoshida T. Vaccination for control of streptococcal disease in cultured yellowtail. Fish Pathol 1982; 16: 201-6. [http://dx.doi.org/10.3147/jsfp.16.201]

[38]   Kusuda R, Kawai K, Masui T. Etiological studies on bacterial pseudotuberculosis in cultured yellowtail with *Pasteurella piscicida* as a causative agent. On the serological properties. Fish Pathol 1978; 13: 79-83. [http://dx.doi.org/10.3147/jsfp.13.79]

[39]   Yasunaga N, Hatai K, Tsukahara J. *Pasteurella piscicida* from an epizootic of cultured red sea bream. Fish Pathol 1983; 18: 107-10. [http://dx.doi.org/10.3147/jsfp.18.107]

[40]   Nizan S, Hammerschlag E. First report of pasteurellosis in freshwater hybrid tilapia (*Oreochromis aureus* x *O.niloticus*) in Israel. Bull Eur Assoc Fish Pathol 1993; 13: 179-80.

[41]   Lilley JH, Callinan RB, Chinabut S, Kanchanakhan S, MacRae IH, Phillips MJ. Epizootic Ulcerative Syndrome (EUS) Technical Handbook. Bangkok: The Aquatic Animal Health Research Institute 1998; p. 88.

[42]   Jensen BB, ErsbA,ll AK, Ariel E. Susceptibility of pike *Esox lucius* to a panel of Ranavirus isolates. Dis Aquat Organ 2009; 83(3): 169-79. [http://dx.doi.org/10.3354/dao02021] [PMID: 19402450]

[43]   Gobbo F, Cappellozza E, Pastore MR, Bovo G. Susceptibility of black bullhead *Ameiurus melas* to a panel of ranavirus isolates. Dis Aquat Organ 2010; 90(3): 167-74. [http://dx.doi.org/10.3354/dao02218] [PMID: 20815324]

[44]   Langdon JS. Experimental transmission and pathogenicity of epizootic haematopoietic necrosis virus (EHNV) in redfin perch, *Perca fluviatilis* L., and 11 other teleosts. J Fish Dis 1989; 12: 295-310. [http://dx.doi.org/10.1111/j.1365-2761.1989.tb00318.x]

[45]   Reddacliff LA, Whittington RJ. Pathology of epizootic haematopoietic necrosis virus (EHNV) infection in rainbow trout (*Oncorhynchus mykiss* Walbaum) and redfin perch (Perca fluviatilis L). J Comp Pathol 1996; 115(2): 103-15. [http://dx.doi.org/10.1016/S0021-9975(96)80033-8] [PMID: 8910739]

[46]   Cinkova K, Reschova S, Kulich P, Vesely T. Evaluation of a polyclonal antibody for the detection and identification of ranaviruses from freshwater fish and amphibians. Dis Aquat Organ 2010; 89(3): 191-8. [http://dx.doi.org/10.3354/dao02198] [PMID: 20481086]

[47]   European Food Safety Authority. Scientific opinion of the panel on AHAW on a request from the European Commission on aquatic animal species susceptible to diseases listed in the Council Directive

2006/88/EC. EFSA J 2008; 808: 1-144.

[48]   Bootland LM, Leong JC. Infectious hematopoietic necrosis virus. Fish Diseases and Disorders: Viral, Bacterial and Fungal Infections. Oxon, UK: CAB International 1999; 3.

[49]   Kurath G. Biotechnology and DNA vaccines for aquatic animals. Rev - Off Int Epizoot 2008; 27(1): 175-96.
         [PMID: 18666487]

[50]   Lorenzen N, Lorenzen E, Einer-Jensen K, Heppell J, Davis HL. Genetic vaccination of rainbow trout against viral haemorrhagic septicaemia virus: small amounts of plasmid DNA protect against a heterologous serotype. Virus Res 1999; 63(1-2): 19-25.
         [http://dx.doi.org/10.1016/S0168-1702(99)00054-4] [PMID: 10509712]

[51]   Falk K, Aspehaug V, Vlasak R, Endresen C. Identification and characterization of viral structural proteins of infectious salmon anemia virus. J Virol 2004; 78(6): 3063-71.
         [http://dx.doi.org/10.1128/JVI.78.6.3063-3071.2004] [PMID: 14990725]

[52]

[53]   Haenen OL, Way K, Bergmann SM, Ariel E. The emergence of koi herpesvirus and its significance to European aquaculture. Bull Eur Assoc Fish Pathol 2004; 24: 293-307.

[54]   Michel B, Leroy B, Stalin Raj V, *et al.* The genome of cyprinid herpesvirus 3 encodes 40 proteins incorporated in mature virions. J Gen Virol 2010; 91(Pt 2): 452-62.
         [http://dx.doi.org/10.1099/vir.0.015198-0] [PMID: 19846671]

[55]   Eide K, Miller-Morgan T, Heidel J, Bildfell R, Jin L. Results of total DNA measurement in koi tissue by Koi Herpes Virus real-time PCR. J Virol Methods 2011; 172(1-2): 81-4.
         [http://dx.doi.org/10.1016/j.jviromet.2010.12.012] [PMID: 21185329]

[56]   St-Hilaire S, Beevers N, Joiner C, Hedrick RP, Way K. Antibody response of two populations of common carp, *Cyprinus carpio* L., exposed to koi herpesvirus. J Fish Dis 2009; 32(4): 311-20.
         [http://dx.doi.org/10.1111/j.1365-2761.2008.00993.x] [PMID: 19236553]

[57]   Noga EJ. Lymphocystis. Fish disease: diagnosis and treatment. Iowa: Wiley-Blackwell, Ames 2010; pp. 171-3.

[58]   Mashego SN. A seasonal investigation of the helminth parasites of Barbus species in water bodies in Lebowa and Venda, South Africa PhD Thesis, University of the North, Sovenga, South Africa 1982; 191.

[59]   Varela MC. Parasitas e Parasitoses em Piscicultura. In: Ordem dos Médicos VeterinMários, Ed. Lisboa, Portugal 2005; p. 285.

[60]   Hochberg NS, Hamer DH. Anisakidosis: Perils of the deep. Clin Infect Dis 2010; 51(7): 806-12.
         [http://dx.doi.org/10.1086/656238] [PMID: 20804423]

[61]   Bakke TA, Harris PD, Hansen H, Cable J, Hansen LP. Susceptibility of Baltic and East Atlantic salmon *Salmo salar* stocks to *Gyrodactylus salaris* (Monogenea). Dis Aquat Organ 2004; 58(2-3): 171-7.
         [http://dx.doi.org/10.3354/dao058171] [PMID: 15109139]

[62]   Scientific Opinion on risk assessment of parasites in fishery products, EFSA Panel on Biological

Hazards (BIOHAZ). EFSA Journal 2010; 8: 1543.

[63]   Suzuki K, Misaka N, Sakai DK. Efficacy of green tea extract on removal of the ectoparasitic flagellate *Ichthyobodo necator* from chum salmon, *Oncorhynchus keta*, and masu salmon, *O. masou.* Aquaculture 2006; 259: 17-27.
[http://dx.doi.org/10.1016/j.aquaculture.2006.05.004]

[64]   Hedrick RP, El?"Matbouli M. Recent advances with taxonomy, life cycle, and development of *Myxobolus cerebralis* in the fish and oligochaete hosts. Am Fish Soc Symp 2002; 29: 45-53.

[65]   Lom J, DykovA I. *Microsporidian xenomas* in fish seen in wider perspective. Folia Parasitol (Praha) 2005; 52(1-2): 69-81.
[http://dx.doi.org/10.14411/fp.2005.010] [PMID: 16004366]

CHAPTER 5

Introduction to Anaesthesia and Surgery in Fish

Nuno Pereira[1,2,3,4,*]

[1] *Oceanário de Lisboa, Esplanada D. Carlos I, 1990-005, Lisboa, Portugal*

[2] *Faculty of Veterinary Medicine. Universidade Lusófona de Humanidades e Tecnologias, Lisboa, Portugal*

[3] *ISPA, Instituto Universitário. Ciências Socias, Psicológicas e da Vida, Lisboa, Portugal*

[4] *IGC, Instituto Gulbenkian de Ciência, Oeiras, Portugal*

Abstract: Handling and manipulation of fish, requires, almost without exception, sedation or anaesthesia due to integument fragility and health as well as welfare considerations.

Integument adaptations to fish immersed life such as thin epidermis covered with a protective layer of mucous, intradermic scales and slender hypodermis must be protected during physical manipulations. Also, as sentient animals, fish must be spared the eventual pain and stress caused by husbandry, research and veterinary procedures.

Fish anaesthesia and surgery have become common procedures in ornamental fish industry, public aquaria and research, and in the case of anaesthesia and aquaculture.

Surgery is one of the reasons to anesthetize fish as these animals can undergo a variety of surgical procedures, from simple cutaneous interventions to more sophisticated intracoelomic and even cardiac or hepatic surgeries.

Keywords: Analgesia, Anaesthesia, Anaesthesia monitoring, Fish, Fish surgery techniques, Sedation.

ANAESTHESIA

Despite persistent controversy, there is a robust data supporting the concept of

---

* **Address correspondence to Nuno Pereira:** Oceanário de Lisboa, Esplanada D. Carlos I, 1990-005, Lisboa, Portugal; E-mail: npereira@oceanario.pt.

Manuela Oliveira, Fernando Bernardo, Joana I. Robalo (Eds.)

fish being capable of nociception or even pain perception. Observation of nociceptors in teleost fish, behavior and physiological responses to pain that are controlled by analgesics, are strong arguments to consider the alleviation of pain and distress in fish undergoing surgery or other potentially painful procedures [1, 2].

Nevertheless, partial or total absence of peripheral afferent neurons with unmyelinated axons (C-fibers: responsible for the "delayed dull pain associated with noxious stimuli") in elasmobranchs can cast some doubts on their nociception or pain perception capacities. On the other hand, the presence of opiate receptors in elasmobranchs can point to the existence of some pain perceptions [3 - 5].

For those with a more skeptical approach to this discussion, prudency recommends that a precautionary principle should be followed. Therefore, when in doubt, it is preferable to consider that fish pain perception is a possibility and to implement measures to prevent and alleviate pain [4, 6, 7].

Furthermore, avoidance of unnecessary stress mitigates homeostasis disturbances and prevents serious health and economic consequences caused by secondary and tertiary stress responses [8].

Sedation and anaesthesia (Table **1**) are necessary to alleviate stress and eventual pain in husbandry, clinical and surgical procedures in aquaculture, ornamental fish industry, public aquaria and research. Procedures like weighing, sorting, vaccination, transport, clinical examination, surgery and euthanasia should be performed under different depths or stages of anaesthesia [1, 4, 9, 10].

Table 1. Definitions of sedation, narcosis, hypnosis, general and surgical anaesthesia [Adapted from 4, 11 and 12].

Sedation	Preliminary state of anaesthesia characterized by depression of the central nervous system with drowsiness, generally unaware of its surroundings, dulled sensory perception and possibly with some analgesia but responsive to painful stimulation. No gross loss of sensory perception and equilibrium.
Narcosis	Drug-induced state of deep sleep from which a patient cannot be easily aroused. Analgesia may be present or not.

(Table 1) contd.....

General anaesthesia	Reversible and controlled depression of the central nervous system with loss of sensory perception, hypnosis (sleep-like state), analgesia, suppression of reflex activity and relaxation of voluntary muscle. In this state, the animal is not arousable by noxious stimulation.
Surgical anaesthesia	General anaesthesia in a state/plane that provides unconsciousness, muscular relaxation, and analgesia enabling painless surgery.

In fish, as in mammals, birds and reptiles, various stages of anaesthesia (Table **2**) can be identified indicating the depth of anaesthesia, that is influenced by the anaesthetic drug chosen, dose and length of exposure. Other factors may also contribute to the depth of anaesthesia, induction and recovery time, namely fish species and size, besides also water parameters including temperature, pH, salinity and hardness [4, 6, 9, 13]. Depending on the species and induction time, these stages can be more or less obvious, being more evident in slower inductions [13].

Table 2. Stages of anaesthesia [Adapted from 1, 4, 6, 13 - 16].

Stage	Plane	Description	Signs	Comments
0		Normal	Voluntary swimming Normal reaction to visual and tactile stimuli, equilibrium, muscle tone and respiratory rate	
1	1	Light Sedation	Voluntary swimming Small loss of reaction to visual and tactile stimuli Normal equilibrium, muscle tone and respiratory rate	In transports can reduce stress and physical trauma
	2	Deep sedation	Absence of voluntary swimming Normal equilibrium Muscle tone and respiratory rate with small decrease No reaction to visual and tactile stimuli Only aware of gross stimulation Still responds to postural changes	Good plane for close visual observation and for minimal manipulation

(Table 2) contd.....

Stage	Plane	Description	Signs	Comments
2	1	Light narcosis / excitement phase	Partial loss of equilibrium Muscle tone decrease Respiratory rate increased and/or irregular Increased reaction to visual and tactile stimuli Still weak responses to postural changes	Higher risk of physical injury or escape / jump from container or aquarium
	2	Deep narcosis	Total loss of equilibrium Muscle tone decrease Respiratory rate decreasing to almost normal Reaction to strong visual and tactile stimuli Lack of responses to postural changes Heart rate normal	Good plane for weighing, measuring, external sampling and blood sampling (in some species of elasmobranchs no chemical anesthesia is required - done under tonic immobility). Avoid painful procedures as analgesia may be present or not.
3	1	Light anaesthesia	Total loss of equilibrium Total loss of muscle tone Respiratory rate decreases Reaction to tactile stimuli – only to deep pressure Lack of responses to postural changes Decrease heart rate	Minor surgical procedures like fin biopsies and gill biopsies
	2	Surgical anaesthesia	Total loss of equilibrium Total loss of muscle tone Respiratory rate very low Absence of reaction to visual or tactile stimuli Lack of responses to postural changes Slow heart rate	Major surgical procedures
4		Medullary collapse	Total loss of equilibrium Flaccid muscle tone Apnea – absence of respiratory rate which can be followed in several minutes by cardiac arrest if anesthesia depth is not decreased	Stage used in euthanasia by anesthetic overdose

There are several physical and chemical methods of fish anaesthesia. The most commonly utilized is the inhalation anaesthesia (Table **3**).

Table 3. Main methods of sedation or anaesthesia in fish [Adapted from 4, 13, 15, 17 - 20, 60, 71].

Method		Comments
Physical	Hypothermia	Severe water temperature reduction immobilizes fish. Direct contact between fish and ice should be avoided. While some reduction in sensitivity to stimulation is achieved, there is at least an incomplete analgesia. Useful in species like Atlantic salmon to produce a sedation that allows prolonged transport. Not appropriate in painful procedures. In zebra fish, there is some discussion regarding the use of hypothermia as a humane euthanasia method. The absence of analgesia and the formation of ice crystals in tissues after rapid cooling are the main debatable issues. Ice crystal formation seems to be absent but the analgesia proves continues in doubt.
	Electrical anaesthesia / electronarcosis	Achieves sedation and eventual anaesthesia with analgesia for a short period. Useful in certain situations for capture of wild fish, handling and minor procedures. Can be dangerous for handlers. Induces tetany and there is a certain incertitude regarding modes of action. Can be applied in fresh and salt water fish. Strong muscle convulsions induced by these methods result frequently in, not always fatal, a internal lesions like spinal injuries and associated hemorrhages, especially in salmonids. Lesions occurrence varies according to the applied electro narcosis technique. Other side effects reported are gill or vent hemorrhages and asphyxia.
	Tonic immobility(TI) / hypnosis in elasmobranches	Induced by placing the animal swiftly in a ventral-dorsal position (upside-down) for a period of time. It's defined as a "coma-like stasis" or as a reversible immobile state where animals are aware of their environment. TI has an unknown mode of action; several terrestrial and aquatic animals can experience it and can last from seconds to hours in unrestrained animals. Some elasmobranchs species can experience tonic immobility with relaxed muscle tone and with the presence of deep rhythmical ventilations by buccal pumping. It's an innate behavior in those species and can be a strategy to cope with the presence of predators. TI can be useful to restraint wild or captive elasmobranchs for non-painful procedures (*e.g.* external examination, biometry, tube feeding, blood collection, skin scraping, and gill visualization). It is a stressful experience, triggering primary and secondary stress responses. An initial short-term ventilatory inefficiency is compensated by an increase of ventilation rates that is part of the primary stress response. In terrestrial animals, it is associated with an apparent insensitivity to painful stimulation. TI can be preceded by a brief period of excitement especially in batoids. Possible physiological effects associated with tonic immobility: Blood pressure depression and bradycardia that can be mitigated by branchial irrigation. Blood acidosis associated with an increase of carbon dioxide blood concentration. Hyperglycemia. Blood electrolytes – Increase of Na, Ca and Mg. Decrease of K. TI is not directly associated with variations in blood lactate or bicarbonate concentrations. If preceded by a prolonged period of exercise, these alterations can be exacerbated due to an increase of lactate and a decrease in bicarbonate blood concentrations. TI induction can be dangerous so it should be avoided in fragile individual or for a prolonged time. The ventilatory depression is exacerbated in animals that depend solely in ram ventilation. For these species and also for the ones that can use buccal pump ventilation, artificial branchial irrigation by pumping water into the mouth decrease a the ventilatory depression and the onset of blood acidosis due to rise of carbon dioxide blood concentration. (Brooks 2011)

(Table 3) contd.....

Method			Comments
Chemical	Inhalation	Anaesthetic drug in aqueous solution	Immersion – induction and maintenance.
			Induction by direct application to gills.
			Maintenance by artificial ventilation (for longer procedures or when fish have to be kept out of water for more than 5 minutes).
		Using gases	Gaseous anaesthesia is rarely used and is possible in fish with aerial respiration like Catfish, one-gilled eels, Synbranchidae and fish with modified lungs. More common is the use of diluted carbon dioxide for fish sedation or euthanasia. Acidosis that is observed with this method can be attenuated with addition of sodium bicarbonate as a buffer. When used for euthanasia, it should be followed by bleeding to achieve death. Anaesthesia and analgesia are feeble so there are concerns regarding animal welfare.
	Parenteral		Intravenous, intraperitoneal and intramuscular

INHALATION OR IMMERSION ANAESTHESIA

Contrary to terrestrial animals, where the volatile anaesthetic is delivered in the lungs by air, in fish the anaesthetic is diluted in water that flows between the branchial lamellae delivering the drug to the blood stream by rapid diffusion at the gill epithelium [4, 9, 13].

Gills are exposed, either by immersing the fish in water with anaesthetic or by direct gill application of the dissolved anaesthetic. Due to concentration gradient, the anaesthetic is absorbed through the gill's epithelium. To reduce/diminish anaesthesia, gills are exposed to water without the anaesthetic and the anaesthetic is gradually eliminated from the gills and eventually by the kidney and skin [4, 6, 13].

Some obligate and facultative air breathing fish can have prolonged and inconsistent inhalation anaesthesia inductions [4].

INDUCTION

Before anaesthesia, whenever possible, fish should fast at least one feeding cycle to decrease the risk of regurgitation and the possible gill epithelial damage. This fasting reduces the oxygen demand, the elimination of ammonia and water quality deterioration during anaesthesia particularly in recirculatory inhalation systems [6, 13, 20].

Stressed fish can have faster anaesthesia absorption through the gills with larger

amount of anaesthetic reaching the blood flow resulting in deeper and prolonged anaesthesia [21 - 23]. Capture procedures (*e.g.* crowding fish for pre anaesthesia netting in aquaculture) should therefore be carried out as smoothly as possible to minimize stress [4].

Induction should be done in a container with the same water and quality parameters of the fish aquarium water. The anaesthetic can be dissolved in the induction chamber prior to the introduction of fish, when working with known species. When facing any incertitude in the desired dose, the anaesthetic can be gradually dissolved with the fish already placed in the container. After reaching the desired anaesthesia plane, the procedure should be quickly performed with the fish immersed, yet in the induction chamber or out of the water (*e.g.* clinical examination, radiographies, fin clipping, cutaneous mucous scraping, gill biopsy, blood and urine collection or tagging) (see Fig. **1**).

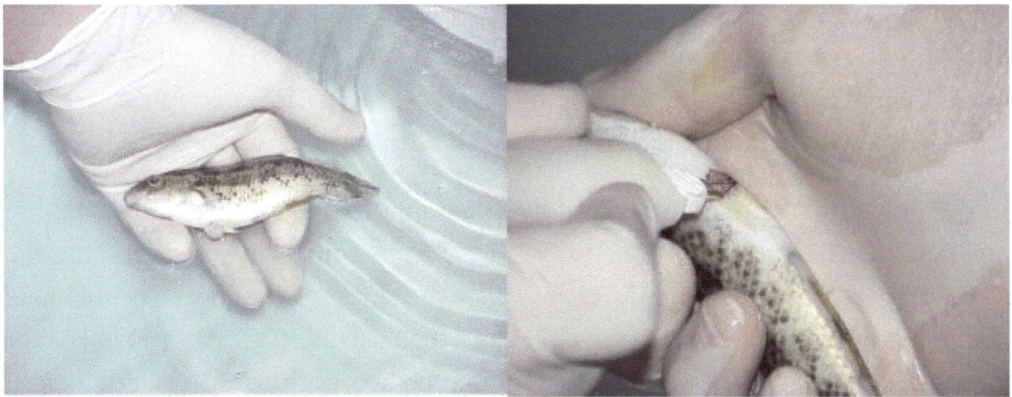

Fig. (1). Brief procedures like this cutaneous mass ablation by thermocautery can be performed with an anesthesia without maintenance (Courtesy of Oceanário de Lisboa).

To capture fish in large aquaria or in the wild, anaesthesia can be delivered directly to the gills. This can be achieved with plastic sprays or atomizer bottles. Normally the dose of the anaesthetic is superior than that used in a standard immersion induction. In some situations, the higher concentrated anaesthetic can also be applied directly in the mouth. This method can at least sedate the animal to facilitate further handling. It can also be useful to avoid contact between eggs and the dissolved anaesthetic in striping procedures [4].

If fish are maintained in the induction water, anaesthesia depth continues to increase due to a cumulative action involving drug dose and exposure time (*e.g.* Ms222 brain and muscle concentration continue to increase even if concentration of anaesthetic in the water is unchanged after induction). So for longer procedures, water anaesthetic concentration must be regulated to control the anaesthesia depth [4]. Also, because fish shouldn't be manipulated for more than 5 minutes when out of the water (depending on the species), in procedures requiring extended anaesthesia or longer manipulation out of the water, a system for inhalation anaesthesia with artificial ventilation should be used [6].

Starting from the induction and throughout the anaesthetic procedure, a supplementation of oxygen by aeration with an air stone is recommended because the manipulation and some of the anaesthetics can cause hypoxia.

Using oxygen instead of air also decreases considerably the risk of supersaturation, or gas bubble related disease. So, near a saturation level, dissolved oxygen will control anaesthesia related hypoxia and, especially elasmobranchs, will mitigate the drop of blood pH (acidosis) due to the increased carbon dioxide level in the blood [6, 9, 15].

A slow induction can increase stress by extending the phase of excitability that potentiate the risk of fish injury. It is possible to minimize stress, increasing the anaesthetic dose simultaneously rising also the anaesthetic risk. To attenuate induction stress, reduce anaesthetic doses and increase anaesthesia safety. Two or more anaesthetics can be used in combination. While the first or initial drug induces sedation, the second drug will induce anaesthesia (*e.g.* metomidate, ms222 in Atlantic cod and halibut; diazepam and quinaldine sulphate in gilthead sea bream (*Sparus aurata*) and European sea bass (*Dicentrarchus labrax*) [6, 13, 22].

This methodology is as useful in fish, as it is in mammals. An interesting report of the use of etomidate administrated intramuscularly as a sedative followed by hypotonic immobility and chemical inhalation anaesthesia with eugenol in a sandbar shark (*Carcharhinus plumbeus*) provides an example of combined anaesthesia [24].

Table 4. Systems for anaesthetic delivery with artificial ventilation [Adapted from 4, 6, 9, 13, 16, 21, 25] (see Fig. 3).

The fish is usually positioned over a wet foam surface adapted according to its anatomy. A humid material like towel paper can be placed between the animal and the foam. This platform will contribute to maintain the skin moisten and to drain flowing water. The animal will receive aerated water in the mouth that will pass through the gills exiting at the opercula or branchial slits. Ideally, two small tubes or a bifurcated one should be placed inside the buccal cavity, directed to both gills. One of these tubes can periodically be used to maintain the skin wet. The flow direction should respect the normal direction (normograde) of water at the gills to preserve gill lamellae integrity and to allow the counter-current mechanisms of gas exchange at the secondary lamella to occur, maximizing anaesthetic exchange also. However, a retrograde flow can be used without major problems in special cases, such as in oral surgeries, or when anatomical features hamper placing a tube in the buccal cavity. Retrograde flow can also be used when particular body positions hamper the water entering the buccal cavity and reaching the gills (*e.g.* rays in dorsal ventral positions). In these cases, the water can be administrated at the opercula or spiracle. The author also uses this approach to shortly reinforce the anaesthesia plane. Water containers, with or without anaesthesia, can be placed either in a higher plane than the fish, relying on gravity to produce the flow of water or below the specimen, resorting to water pumps. In both systems, water recirculation can be applied, either when one sole recipient containing water with anaesthetic or two recipients, one with and another without anaesthetic, are used. In both cases, water is transported by tubes and carried into the fish. Water flow rate can be controlled by a valve or any alternative device. When two recipients are used, a three-way-valve allows controlling the anaesthetic delivery.
Non-recirculation systems
These systems are very easy to create and are mostly suitable for small fish. Adapted intravenous fluid bags or other small containers can be used and water delivered by a modified given set with an optional three-way-valve to control water plus anaesthetic flow. In such systems (water movement depending solely on gravity), flow can reach 250ml/min, which is adequate for fish up to 250g Syringes can also be used for water with anaesthetic administration but it is not as practical. Water that leaves the fish opercular cavity is drained into a container and not reused.
Recirculating systems
The main difference regarding these systems that are more adjusted to larger fish, is that the water, with or without anaesthetic, that leaves the fish opercular cavities is reused. Normally, the use of electric water pumps is incorporated to force water movement with a recommended flow rate of 1-3 l/min. Insufficient flow can contribute to hypoxia, rise of plasma carbon dioxide levels and may lead to deficient anaesthesia delivery or prolonged anaesthetic elimination. Excess flow can damage the thin gill lamellar epithelium and cause gastric or intestinal dilatation. Anaesthesia can be continued for hours but in prolonged procedures, problems concerning the maintenance of water temperature and water quality (rising of ammonia and fecal products) may require the use of cooling or heating equipment and water changes. Another important concern when using recirculating systems for prolonged procedures is the maintenance of the skin moisture.

MAINTENANCE

If necessary, anaesthesia can be prolonged, preferably in a recirculatory inhalation system (Table **4**) (see Figs. **2 - 5**). These systems, of variable sophistication, use either water pumps or gravity to deliver water to the gills. If the fish is out of the water, which is often the case, its body surface should be kept wet. The person in charge of the anaesthesia and monitoring should be also responsible for maintaining the skin moist. Depending on the anaesthetic, type of procedure, fish species and personal experience and preference, maintenance anaesthetic concentration can be reduced, generally to half of the induction dose or the same, if in an inhalation anaesthesia system that can control the delivered anaesthesia [6]. The author uses normally the same concentration for anaesthesia maintenance (MS222) and if the anaesthetic delivery isn't adequate to reach the desired anaesthesia deep or length, the concentration of the anaesthetic is adjusted, either by adding new anaesthetic or by diluting the previous concentration.

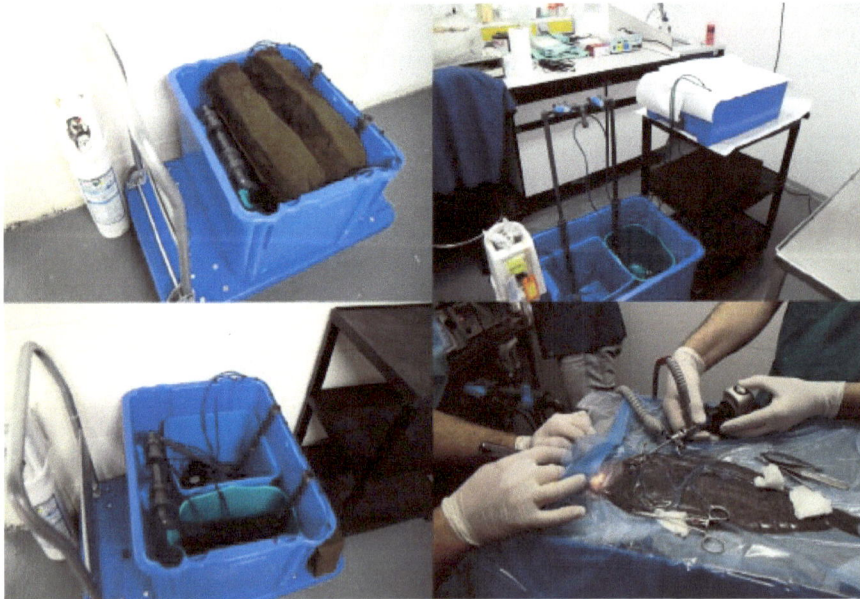

Fig. (2). Prolonged procedures like this laparoscopy to a common sole (Solea solea) need longer maintenance anesthesia periods so there is the need of using a system for anesthetic delivery with artificial ventilation. In these systems, valves can control what water is delivered to the gills, with or without anesthetic, to achieve and maintain the desired anesthesia phase (System designed and assembled by Hugo Batista and laparoscopy performed by Dr. Rui Bernardino) (Images courtesy of Oceanário de Lisboa).

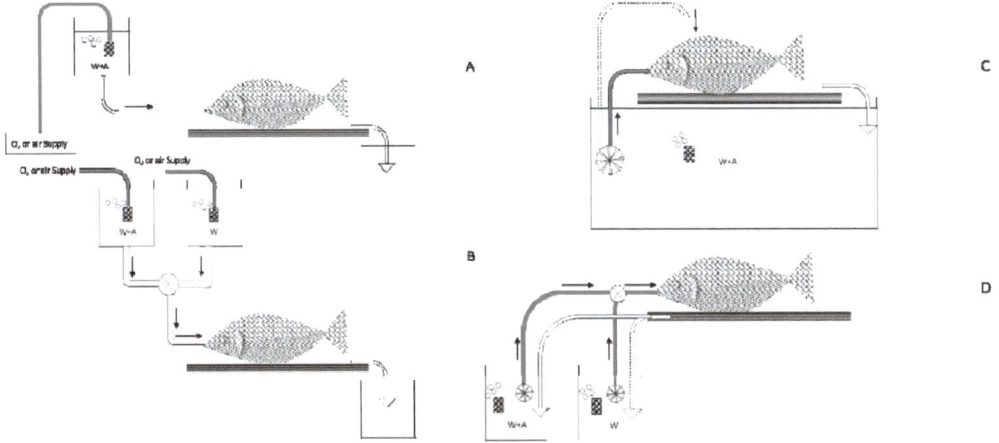

Fig. (3). Systems for anaesthetic delivery with artificial ventilation. Non-recirculation systems (A and B) and recirculating systems (D and C) [Adapted from 4, 6, 9, 13, 16, 21, 25].

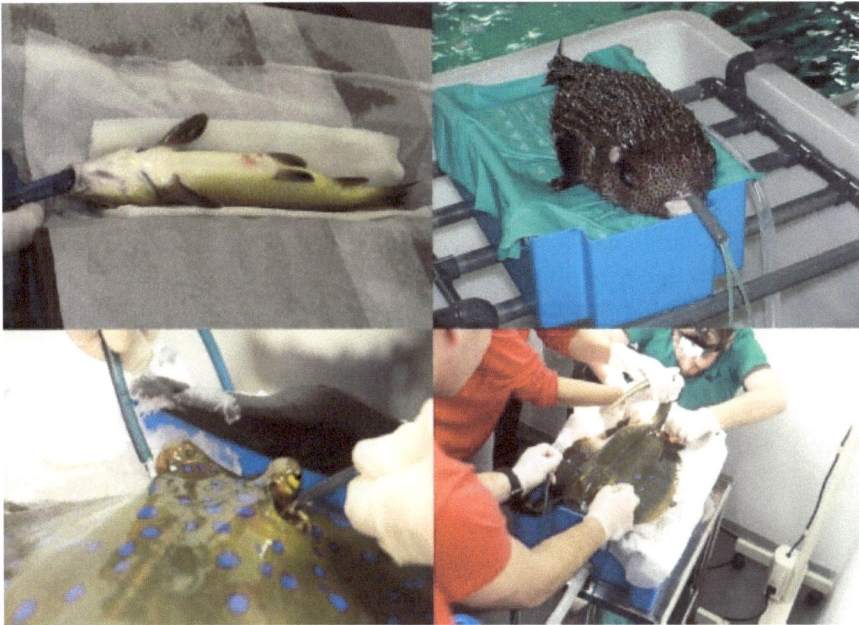

Fig. (4). With most of the teleosts and sharks, two small tubes can be used to deliver aerated water with or without anesthesia to the mouth and gills ((upper left corner). However, in some cases like this spotted porcupinefish (Diodon hystrix), these tubes must be protected from the teeth (upper right corner). In the lower images a bluespotted ribbontail ray (Taeniura lymma) is being anesthetized to perform an abortion due to fetal deaths and the tubes are placed at the spiracles to deliver water with or without anesthesia to the gills (Courtesy of Oceanário de Lisboa).

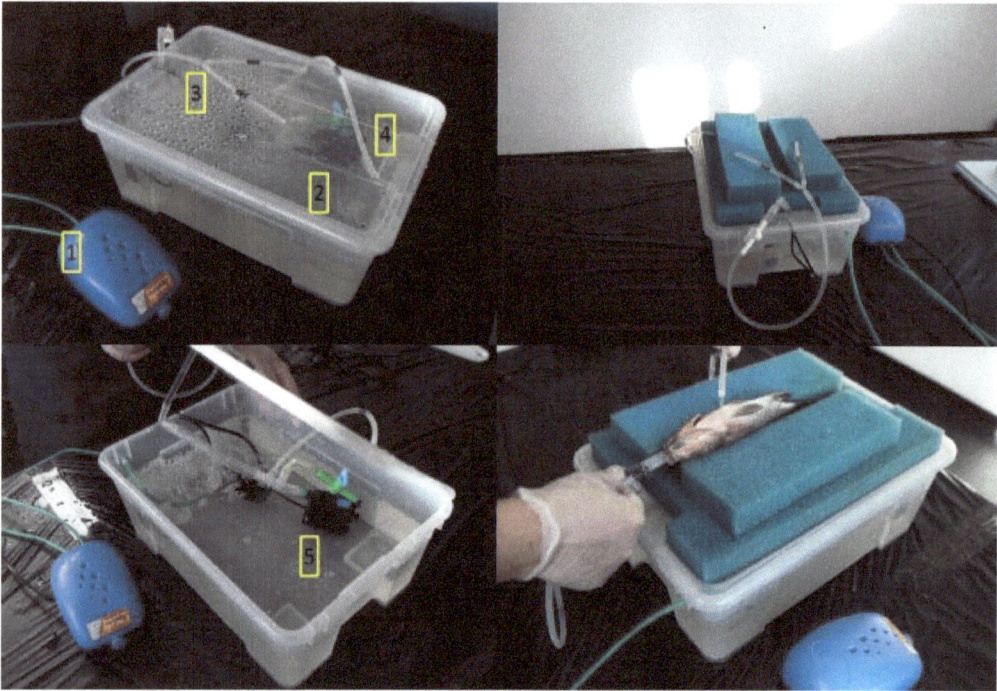

Fig. (5). This type of recirculatory system for anesthetic delivery (System C in Fig. **3**) reaches anesthesia depth enough to perform gonadectomies in Mozambique tilapia (Oreochromis mossambicus). Despite the fact that the anesthetic concentration is generally the same and is not very practical to change this concentration, is this species, surgical anesthetic plane can be maintained for 20 to 30 minutes (1 – air supply;2 – plastic container with a perforated cover;3 – tubes to deliver water with anesthetic to th oral cavity and gill;4 – tube to maintain the fish skin moisted;5 – pump) (Courtesy of Boga/Ciimar fish research facility).

RECOVERY

To finish the anaesthesia, water without anaesthetic should be delivered to the gills, in an anaesthesia delivery system or in a recovery aquarium. In this last case, the fish should be placed, if possible, in an area with higher water flow, facing the current.

Depending on the anaesthetic, species, individual fish and water temperature, awakening can vary in swiftness. Generally, it is faster in higher temperatures, as should be the induction time. In the recovery period, which may take minutes to hours, whenever possible, fish should be maintained in a dim light and kept protected from external stimulation like abnormal vibrations, intense light and intrusive tank manipulation or observation, which could be perceived as stressors.

During this period, there is an increased risk of physical injury that should be minimized by having an adequate recovery aquarium with no potentially damaging physical structures [4, 6, 9].

The risk of post MS222 and eugenol anaesthesia hypoxemia should be reduced by maintaining the anaesthesia recovery water with normal or slightly elevated dissolved oxygen values [6].

After anaesthesia recovery, the fish should be transferred to the tank of origin only when completely awakened to prevent injuries or attacks by eventual cohabitants and should stay under observation for up to 24 hours or more to prevent or resolve any problem that may occur. It should be borne in mind that a fish can take more than 7 days to regain homeostasis after anaesthesia [14].

MAIN ANAESTHETICS USED IN FISH

Inhalation Anaesthetics

MS222, benzocaine, clove oil eugenol, isoeugenol and 2-phenoxyethanol are amongst the most used inhalation anaesthetics in fish [1, 4, 9, 10, 26]. Due to budgetary concerns, legal restraints or eventual side effects to fish and humans, a growing research effort is underway to study options for these common anaesthetics [27]. New products, like essential oil of *Lippia alba*, Menthol, *Condalia buxifolia* methanolic extract and other substances isolated from plants, are being researched in known or novel ornamental and aquaculture species in countries such as Brazil [28 - 32].

One should choose an anaesthetic and gain experience with its use in one or more species. Recommended doses for an increasing number of drugs are available in the literature with indication for fish species, mainly related to aquaculture and research [4, 10, 26]. Regarding Ms222 and ketamine plus medetomidine it is possible to find data for a larger list of ornamental species (teleosts and elasmobranchs doses), due to public aquaria veterinary medicine expertise [9, 15, 33, 34].

If working with an anaesthetic or a fish species for the first time, for which no data is available, it is advisable to previously test the anaesthestic in one or a small number of individuals [23].

MS222 and Benzocaine

Tricaine methane-sulphonate (Synonyms: MS222, Tricaine mesilate, Metacaine) (Table **4**), a sodium channel blocker that inhibits the conduction of nervous impulses, with a unknown action in the central nervous system, is the most common anaesthetic used in research and in fish medicine [9, 23, 35]. It is available as a 100% pure white powder, fresh and salt water-soluble (< 11% w/v), acidic in aqueous solution and is used both in teleosts and elasmobranchs [4, 15, 23]. MS222 is also used in aquaculture procedures and is legally approved in various countries [23], although in recent fish anaesthetic research in countries like Brazil, it is being replaced by less expensive drugs, like clove oil/eugenol and benzocaine [4, 10, 27, 31, 36].

Benzocaine is chemically related with Ms222 and can have some benefits regarding Ms222 in few specific situations (Tables **5** and **6**).

Table 5. MS222 [Adapted from 4, 6, 9, 13 - 15, 17, 20, 23, 26, 35, 37 - 42, author personal communication, (https://www.pharmaq.no/sfiles/3/48/4/file/uk-ie-pl-v1_2013-04-10.pdf). *degree day: a unit that is a product of time and temperature (*e.g.* 70 degree days means a withdrawal time of 7 days at 10°C or if at 12°C it will be 5.8 days).

Good safety margin, decreasing at warmer water with low hardness due to an increase in potency, in higher water temperature induction and recovery times can be smaller. Therapeutic index also decreases in smaller fish.

Larger animals can have slower responses to anaesthesia due to smaller relation between gill surface and body dimensions. Unhealthy fish and higher stock density (<80g/l) can have higher anaesthetic risks. Repeated anaesthesia can reduce induction times.

Liposoluble drug with a slower recovery time in older, fatter, gravid fish or prolonged exposure to the drug because there is an accumulation in fat tissues.

Doses for induction and maintenance may differ and vary with species and personal experiences. Maintenance dose can be changed during anaesthesia. This applies to other anaesthetic drugs. Generally, maintenance dose is half the induction dose. With Ms222, the maintenance dose can be the same of the induction or even increased if in an appropriated system for inhalation anaesthesia (that can control the anaesthetic delivered to the fish).

It is possible to store in stock solutions (*e.g.* 10g/L in sealed dark bottle). Stability time of these solutions varies according references, from 12 hours to 3 months. Oily residues and color change are related to possible potency reduction. It is best to use fresh prepared solutions.

(Table 5) contd.....

MS222 solution in water has a low pH (up to 2.8 in fresh water). Less dramatic pH variations in sea water causing gill, epidermal and corneal aggressions and osmoregulation disturbances - can be buffered with sodium bicarbonate (1:1 or 1:2) up to 7 to 7.5. Other buffers can be used like imidazole, sodium hydroxide or Tris-buffer. Use of unbuffered MS222 can jeopardize external protozoa histological observation.

Example of stock solutions in zebrafish: 400 mg tricaine powder plus 97.9 ml DD water plus ~2.1 ml 1 M Tris (pH 9). Adjust pH to ~7. Store in freezer (there is no information regarding frozen MS222 stability). Use as an anaesthetic dilute 4.2 ml Ms222 stock solution in 100 ml tank water.

Buffered solutions have quicker induction time, more consistent maintenance period and quicker recovery.

Absorbed and eliminated primarily by the gills (in some species also by the skin) by diffusion and osmotic pressure gradient. It's partially eliminated by the kidney and bile after hepatic and kidney metabolism.

Induction, depending on the dose amongst other factors, can be quick, less than one minute. Recovery time is often quick, few minutes, but occasionally can be more prolonged, up to 6 hours.

Doses, according to species, for sedation range between 7 to 50mg/l (ppm) and for anaesthesia between 30 to 350mg/l (ppm). For Syngnathidae, lower doses are recommended ranging from 50 to 75ppm. For unreported species, range between 50 to 90ppm maintained. In elasmobranchs doses range between 50 to 125ppm. Zebrafish normal dose range is between 160 to170 ppm.

To sedate or anesthetize by direct application to the gills, a higher dose up to 1gr/ml can be used, but always well tamponated and can be used in conjugation with a dye as methylene blue to see if the solution passes through the gills.

It is not effective in some species (*e.g.* Gulf Mexico sturgeon) and dangerous in the following *Apistogramma ramiezi, Balantocheilus melanopterus, Etroplus surrantensis, Melanotaemia macculochi, Monodactylus argenteus, Phenacogrammus interruptus and Scatophagus argus* (Pharmaq leaflet).

Possible side effects:

Hypoxemia associated to hypoventilation Hypercapnia and acidosis

Tachycardia or bradycardia Hyperglycemia with latter hypoglycemia

Elevation of potassium, magnesium, hemoglobin, hematocrit, catecholamines cortisol and lactate Electrolyte disturbances due to urinary output increase can last 7 days after anaesthesia.

When used in field conditions, the hazards of releasing fish before passing the withdraw time should be considered. Can decrease sperm motility so avoid presence of ms222 during external fertilization

Despite tissue levels, almost absent 24 hours after anaesthesia, there is withdrawal time for human consumption: 21 days (USA and Norway); 10 days (New Zealand); 5 days if water > 10°C(Canada); 70 degree days*(UK) Handling hazards:

Risk of skin and respiratory tract irritation

Risk of reversible retinal toxicity to handlers / avoid systemic absorption

Handlers should avoid contact with skin, eyes and respiratory tract and use safety glasses.

Legal

Permitted in the UK, Italy (vaccination and research purposes), Spain and Norway. Approved in the United States only on Ictaluridae, Salmonidae, Esocidae and Percidae

Clove Oil, Eugenol and Isoeugenol

Clove oil is obtained by distillation of two species of plants namely *Syzygium aromaticum* and *Eugenia caryophyllata*, and has several active principles: phenolic eugenol (85-95%); isoeugenol and methyleugenol. Clove oil can be purchased with no veterinary prescription and is inexpensive (Table **7**). Eugenol is also available in a concentration of nearly 100%, avoiding the presence of methyleugenol, a suspected carcinogenic substance [4, 17, 26].

Table 6. Benzocaine [Adapted from 1, 4, 9, 26, 43].

Insoluble in water, so it should be dissolved in acetone, ethanol or propylene glycol
Stock solution – 100g/L of acetone or ethanol in sealed dark bottle, if possible, should be refrigerated for 1 year. It is not necessary to add a buffer.
Good safety margin. Less at higher temperatures but not affected by water hardness or pH. No consequences in reproduction and minor ones in fed appetite and growth.
Fat soluble drug: slower recovery in older, fatter, gravid fish or prolonged exposure to the drug. Recovery time: 3-15 minutes.
Doses, according to species, can range between 25 to 50mg/l (ppm) and in higher doses in higher water temperatures.
Useful for species sensitive to MS222 (*e.g. Morone saxatilis*)
Possible side effects similar to MS222, with the exception of water low pH consequences. Can have immunosuppressant effects and hypoxemia associated with hypoventilation
Eliminated from water by activated carbon. Without any treatment, breakdown in water occurs in 4 hours. Handlers should avoid contact with skin, eyes and respiratory tract and use safety glasses.
Despite evidence of quick anaesthetic tissue elimination, the withdrawal time for human consumption in USA is 21 days.

Table 7. Clove oil, eugenol and isoeugenol [Adapted from 1, 4, 6, 17, 24, 26, 35, 37, 43 - 47].

Clove oil / Eugenol
Non-soluble in water mainly in cold temperatures. Need to be previously dissolved in 95% ethanol in a 10% stock solution that is effective for 3 months
Compared with MS222 induction time, it is usually faster but recovery tends to be longer. Good anaesthesia consistence compared with other anaesthetics, regardless of the temperature.
Reported a narrow safety margin in red pacu Piaractus brachypomus compared with MS222. Southern stingrays can be sensitive to eugenol.
Analgesic properties are unproved in fish. Inexpensive.
Possible side effects: Hypoxemia
Hypercapnia and acidosis
Less cortisol elevation compared with MS222 Increased catecholamine, hematocrit and glycemia
Hypoventilation and respiratory failure especially at higher doses can be caused by an oily coating of the gill epithelia interfering with the gaseous exchanges or by a hypothesized neurotoxic effect
Risk of mild gill necrosis after repeated exposure to low dose of eugenol Anecdotal reports of death in Acanthuridae fish.
Repeated anaesthesia can have a negative effect on the growth of rainbow trout
Handlers should avoid contact with skin, eyes and respiratory tract and use safety glasses.
Doses, according to species, can range between 25 to 150mg/l (ppm) and in higher doses in higher water temperatures (*e.g.* 100-120 for induction and 40ppm for maintenance). Sedation can be achieved with doses ranging from 2-5ppm. In sand tiger sharks - Carcharhias taurus a dose of 30ppm of eugenol (99% pure) for induction and 15ppm for maintenance is reported. This animal was premedicated with etomidate 1mg/kg/IM.

(Table 7) contd.....

Isoeugenol
Active ingredient in Aqui-S Indicated for harvesting and fish transportation. More effective than clove oil. Good safety margin. Possible side effects: Depression of cardiovascular system Increased catecholamine and hematocrit Stress alleviation can be inefficient Southern stingrays can be sensitive to Aqui-S. Inexpensive. Doses, according to species, can range from 10 to 60mg/l (ppm)

2-Phenoxyethanol

2-Phenoxyethanol does not have significant advantages regarding other drugs but is inexpensive so it is still used as a sedative for handling procedures like transport and as anaesthetic for surgery [4, 26] (Table **8**).

Table 8. 2-Phenoxyethanol [Adapted from 1, 4, 26].

Colorless or straw-colored oily liquid and it is water soluble if well mixed. Anaesthetic solution is effective for 3 days and has some bactericide and fungicidal properties Inexpensive Recovery can be abrupt Possible side effects: Elevation of cortisol, catecholamine, glycaemia and hematocrit Hypoventilation, hypoxia, hypercapnia and acidosis Blood pressure depression Deficient analgesia More potent at lower temperatures Humans / handlers: harmful if swallowed, inhaled or absorbed through the skin. May cause reproductive defects. Severe eye and skin irritant. Prolonged exposure can cause a neuropsychological syndrome Doses can range in teleosts from 60 to 600ppm but in species with no information regarding this anaesthetic administration should be initiated with 100 to 300ppm.

Oxygen Narcosis in Elasmobranchs

Some elasmobranch species can be immobilized for minor procedures by being exposed to a flow of 100% oxygenated water across the gills. This effect is probably related with the rise of plasma carbon dioxide due to a respiratory depression. If prolonged, this method can cause dangerous acidosis [15].

Injectable Anaesthetics

Parenteral or injectable anaesthesia in fish (either by intramuscular or more rarely intracoelomic or intravenous administrations) is useful when induction by inhalation anaesthesia is not feasible. This can be the case of elusive or large fish in big exhibition aquaria and of some Scrombidae species like tuna that cannot endure confinement. Several available injectable anaesthetics (Table **9**) can be administrated by hand, pole syringe or submersible dart gun.

Ideally, the intramuscular injection site should be the dorsolateral musculature, between the dorsal fin and the lateral line, avoiding the coelomic cavity. A retrograde drainage of the injected liquid is a common occurrence, especially in elasmobranchs. To minimize drainage, several precautions are recommended such as, deep injections with the needle inclined to administer the injected liquid away from the skin injection puncture, leaving the needle with the syringe or dart in place for some time after the administration, or when possible, pressing the orifice with a finger or hand for a few seconds. As in all injections, and whenever possible, the needle should enter the skin in a 45° cranially orientated degree to minimize skin damage and to spare scales as much as possible.

Induction can last longer in sedentary animals due to less muscular movements and the concurrent decreased blood circulation.

Intraperitoneal injections have a higher risk in damaging intracoelomic organs if performed without previous fasting and in a non-immobilized fish. To minimize this risk, the needle angle of approach should be smaller than in intramuscular injections. Intraperitoneal injections can have unpredictable induction times due to different serosa permeability to drugs.

Intravenous injections, which should only be performed in immobilized fish, provide quick induction and usually a shorter period of anaesthesia compared to inhalation anaesthesia. The more accessible blood vessels are the caudal aorta and vein. In teleosts and in elasmobranchs, during prolonged procedures, these caudal tail vessels can be catheterized by a needle. In larger sharks, catheterization can be performed by a spinal needle (with a removable stylet) that is inserted in the ventral midline caudal to the cloaca near the first or second anal fins, depending on the species anatomy. In more prolonged or in painful procedures, parenteral administration of anaesthetics can be followed by immersion anaesthesia to extend or deepened the anaesthesia. In addition, injectable anaesthetics usually require artificial ventilation, especially in animals depending solely on ram ventilation (*e.g.* some shark species and tunas) [6, 9, 13, 15, 20, 24, 26].

Table 9. Injectable anaesthetics [Adapted from 4, 6, 9, 15, 24, 33, 48 - 54].

Ketamine hydrochloride / Medetomidine
Ketamine alone can give an incomplete anaesthesia, apnea episodes and extended recovery with excitement and in sharks, muscle spasms are reported. Ketamine safety margin is reduced at higher temperatures. Association with α2-agonists will mitigate theses effects. Medetomidine has replaced xylazine in these associations with ketamine. There are several reports with this association: i – consistent anaesthesia in bonito(ketamine 4 mg/kg + medetomidine 0.4 mg/kg) and Pacific mackerel (ketamine 53–228 mg/kg + medetomidine 0.6–4.2 mg/kg) but with much higher dose of ketamine in mackerel (up to 50x); ii – induction of light anaesthesia plane but with cardio-respiratory depression in Gulf of Mexico sturgeon (*Acipenser oxyrinchus de soti*) (ketamine 3-7 mg/kg + medetomidine 0.03-0.07 mg/kg); iii – in sturgeon hybrid *Acipenser naccarii* female × *Acipenser baerii* male (ketamine 4 mg/kg + medetomidine 0.04 mg/kg), intravenous administration resulted in adequate anaesthesia for husbandry and artificial reproduction procedures. Other doses are reported in a few teleosts with doses ranging from ketamine 6 to 42 mg/kg + medetomidine 0.122 to 0.246 mg/kg. This combination is highly species dependent and in most cases it only gives a mild sedation, which is enough to facilitate capture and eventual anaesthesia deepening by inhalation anaesthesia. In elasmobranchs, this combination is reported in several species and in non-reported ones, a starting dose is recommended: medetomidine 0.09-0.10 mg/kg + ketamine 4-5 mg/kg. Results are species related and can vary also according to previous muscle activity and stress. Induction time range can go from 5 minutes to 20 minutes and some species do not reach complete anaesthesia despite re-dosing. Medetomide is reversed by atipamizole (5 to 6 times the injected medetomidine). The anaesthesia must not be revised too soon because fish will be only under ketamine action, which can be stressful and increase the risk of physical trauma during recovery.

Propofol
This anaesthetic provides a rapid induction of anaesthesia in Gulf of Mexico sturgeon but with cardio-respiratory depression that can be mitigated by artificial ventilation. It can be associated with mortality in African lug fish (*Portopterurs annectens*). It is a safe anaesthetic in spotted bamboo sharks (*Chiloscyllium plagiosum*) without cardio-respiratory depression and with a rapid induction and a recovery time reaching 1 hour. The normal administration mode is intravenous but there are recent reports of inhalation anaesthesia with this drug in *Carassius auratus* and in silver catfish *Rhamdia quelen*.

Tiletamine / Zolazepam
In teleosts, this association can be useful for capture of fish in large aquaria or at the sea. It can be also used in species that have inconsistent anaesthesia with inhalation drugs like MS222. There is a report regarding intramuscular sedation/anaesthesia in more than 20 species of captive sea water fish. The response can be highly species, weight and temperature dependent. Bearing this in mind, dose for sedation ranges from 5 to 10 mg/kg and for anaesthesia can reach 20mg/kg. No mortalities were registered but recovering time reached up to 72 hours, uneventfully. Use of smaller doses in some cold-water species and in fish larger than 3 kg will attenuate the problem with prolonged recoveries. Doses can be decreased until 3mg/kg in these cases (*e.g.* some species of the Muraenidae family). This association can be concentrated in a small volume, which can facilitate drug administration. In nile tilapia (*Oreochromis niloticus*), only the intravenous administration, not intramuscular, induced consistent anaesthesia and in sharks, this combination caused irritability, rapid swimming and biting.

(Table 9) contd.....

Etomidate
This non-barbiturate hypnotic and non-analgesic drug, as metomidate, is described extensively as an immersion anaesthetic. An interesting application is reported in elasmobranchs as a sedative 1mg/kg/IM to facilitate shark capture (sand tiger shark, *Carcharhias taurus*). It can be followed by tonic immobility induction allowing non-painful procedures or by inhalation anaesthesia for surgery. Etomidate recovery period is not long and administration of corticosteroids is recommended due to depression of cortisol synthesis associated to this drug.

Other anaesthetics can be consulted in the vast available literature [4, 9, 20, 33, 34, 55].

Anaesthesia Monitoring

Depth of anaesthesia is typically monitored by observation of ventilatory rate (opercular or branchial slits movement), body movement or by the response to external physical stimulation. Other parameters, such as equilibrium, caudal fin strokes, and jaw tone can be useful for monitoring induction or superficial sages of anaesthesia (Table **2**). To check these parameters during a surgery, a transparent surgical drape is helpful. More refined monitoring methods are possible and depending on the species, type of procedure and availability of equipment, heart frequency can be assessed by a Doppler probe (audible assessment), ultrasonography and, less frequently by ECG [6, 13, 20, 39].

A Doppler probe can be placed on the ventral surface within the area of the heart projection, normally between the pectoral fins or just caudal to the opercula or gill slits. Other options for probe placement are: inside the opercula, in the caudal direction, under the gill arches or the dorsal face of the tongue.

The pulse oximeter probe (of two basic types, transmission and reflectance) can be placed in the dorsal surface of the tongue, deep into the anus or cloacal opening (reflectance probe) or on the fins (transmission probes). It can be helpful to assess heart rate but is not as reliable to control blood gases saturation. Heart rate reading's accuracy OF Pulse oximeter can be confirmed by ultrasonography [24, 39, 56].

Electrolytes, blood biochemical (*e.g.* glucose, lactate and bicarbonate), blood gases and pH control can be important, especially in elasmobranch prolonged surgeries and can be done by a portable blood gas analyzer [9, 24].

Anaesthesia Accidents and Complications

Even when factors that influence anaesthesia are carefully controlled, sporadic anaesthesia accidents and complications occur and must be faced.

One possible complication is a prolonged anaesthesia recovery, especially with a low ventilatory rate. In such situation, the elimination of the inhalation anaesthetic can be fastened by increasing the heart rate and blood circulation in the gill capillaries. Due to the existence of a ventilator/cardiac reflex, this will be achieved by enhancing the flow of water (without anaesthetic) that enters in the mouth and passes though the gills [4].

This increased water flow can be obtained by gently (exaggerated flux can damage the gills) propelling water into the mouth with a syringe, a rubber enema pump or a tube connected to a water pump and, if the fish dimensions allow it, it can be pulled forward against the water [6, 9]. In the author's experience, in teleosts, gentle atraumatic (for the fish and the handler) physical stimulation of the oral cavity can be an adjuvant measure to stimulate autonomous ventilation.

Bear in mind that older or fatter fish recovers slowly from anaesthesia induced by liposoluble drugs due to the body lipids accumulation [4].

During anaesthesia, apnoea or depressed ventilation should be controlled and reversed by maintaining the oxygen aerated water flow to the gills and by stopping temporarily the supply of water with dissolved anaesthetic. In fish, heart failure occurs normally only several minutes after the respiratory failure, especially if the animal has oxygen supplemented water flowing through the gills providing a safety interval to allow normalization of the ventilatory rate [9].

If assisted ventilation and anaesthetic-free water flow are not enough to reverse apnoea, doxapram (respiratory stimulant, 5mg/kg IV, IP, IM or 7,5mg/kg applied topically at the gills) can be useful although elasmobranchs may become excited and aggressive [26, 57 - 59].

In more debilitated or shocked fish, corticosteroids (dexamethasone 1-2 mg/kg IM, IV or IP), epinephrine and fluid therapy can be of help [26, 34]. The author finds methyl prednisolone sodium succinate, 2-4 mg/kg IM, efficient in these situations. The dose range for this last drug can be risen up to 10mg/kg and may also be administrated intravenously [60].

In cardiac asystole, heart digital massage applied unilaterally or bilaterally in the caudal ventral areas of the opercular cavity, can be performed in teleosts of appropriated size and adequate opercula anatomy. Adrenaline 1:1000 0.2-0,5ml IP, IM, IV, and IC may also be indicated in this situation [61].

In elasmobranches, blood pH regulation is somewhat frail. Blood acidosis, with increased lactate and carbon dioxide levels, can occur and is frequently related to strenuous exercise or ventilatory depression. In anaesthetic procedures involving tonic immobility, particularly in prolonged procedures, ventilatory depression with or without bradycardia, induces a rise in carbon dioxide and a decrease in blood pH. To mitigate these alterations, assisted ventilation with oxygen supplemented water flowing through the gills, imperative in ram obligate ventilator sharks, should be available and may also control hypoxemia.

Hypoxemia can occur both in teleosts and elasmobranchs even if the animal exhibits apparently normal ventilation. Bradycardia and blood pressure decline that occur with most anaesthetics can cause a decrease in oxygen blood concentration. This can be worsened by the hemodynamic alterations caused by erythrocytes swelling, which reduces capillary blood flow in the gills. Ideally, respiratory efficiency can be monitored by blood gas analysis (*e.g.* iStat) [4, 33].

The risk of lactic acidosis in elasmobranch (higher in pelagic species and in large animals) following anaesthesia of animals recently captured may be reduced by shortening the period of pursuing. Minimizing physical exercise prior to anaesthesia, maintaining adequate absorption of oxygen and excretion of carbon dioxide by the gills and providing a constant flow of water, are three important measures to mitigate the onset of blood acidosis [24, 33, 58, 60].

In marine elasmobranchs, fluid therapy must mimic as much as possible its plasma osmolarity characterized by high concentrations of urea, NaCl and trimethylamine oxide [60].

During prolonged anaesthesia, fluid therapy with elasmobranch-ringer solution reduces the risk of tissue hypoperfusion and lactate production and reinforces blood buffer capacity. If fluid therapy with elasmobranch-ringer solution is not enough to correct the blood acidosis, other available options may be used but blood pH evaluation is recommended (Table **10**).

Administration of sodium acetate or sodium bicarbonate can provoke a life threatening alkalosis; therefore, these treatments should be monitored by blood pH, lactate and bicarbonate analytical control [6, 24, 33, 58, 60] (see Fig. **6**).

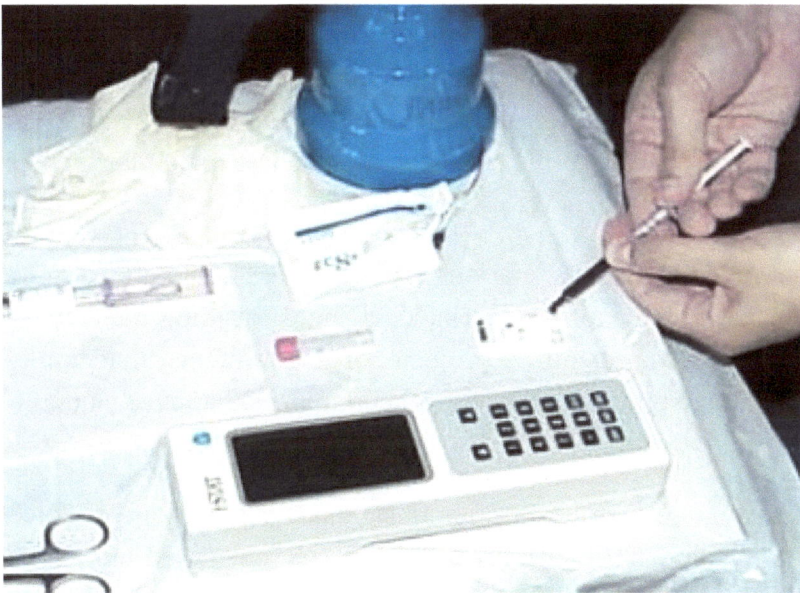

Fig. (6). A portable blood gas analyzer (ideally providing readings of blood pH, electrolytes, blood biochemical parameters like glucose, lactate, bicarbonate and blood gases like oxygen and CO2) to assess acid-base status can be a valuable tool to control blood acidosis treatment efficiency (*e.g.* i-STAT) (Courtesy of Oceanário de Lisboa).

Table 10. Pharmaceutical prevention and treatment of blood acidosis in elasmobranchs [Adapted from 24, 58 and 60].

Elasmobranch-ringer solution
Sodium chloride (10 g/l), NaHCO3 (0.1 g/l), and urea (26 g/l) in LRS to reach an osmolality of 960 mOsm/kg. Sterilized by a 22mm filter prepared no more than 24 hours prior to administration and stored at 48°C. It may be administered intravenously or intraperitoneally. Dose is calculated according to weight and a dose of circa 0,075ml/kg/min is reported in a sandbar shark (*Carcharhinus plumbeus*). If constant evaluation of the acid-base status is available, rate of the dose can vary according to the need. There are alternative formulations described in the literature.

(Table 10) contd.....

Sodium acetate
40 mEq/L of sterile water or 0,45% salinity 20–30 ml/kg/h IV or IP. It is safer than sodium bicarbonate, with less risk of provoking alkalosis.
Sodium bicarbonate
Reported doses are: 1 mEq/l/IV or 0.088 mg/kg /IV-IP/diluted in isotonic saline or 5%dextrose sterile solutions / very gradual administration by fractionated administration to diminish the risk of iatrogenic alkalosis.

ANALGESIA

Nociception is now accepted in fish although pain perception can still cause debate [1, 7]. Research in teleost fish associates painful events with behavioral disturbances and a humane treatment of fish should be concerned in sparing these animals' unnecessary pain or distress [1]. Nevertheless, pain management in fish, related or not with surgery, is yet poorly investigated [35].

In addition to ethical considerations, researchers should avoid or minimize physiological and behavior pain related changes, as they can alter scientific results. Additionally, scientific projects or experiments that imply animal manipulation must be approved by ethical committees that have animal welfare in full consideration.

Compared with other groups of animals, analgesic options (Table **11**) and strategies to handle pain in fish are still a relative uncharted territory and more research is necessary to minimize both the suffering of fish as well as the deleterious consequences of painful procedures.

Table 11. Fish analgesics [Adapted from 1 - 3, 58, 60, 62 - 64].

Butorphanol 0,4mg/kg/IM
A single dose prior to anaesthesia recovery in koi carp showed slightly better recovery from surgery and the absence of side effects. In chain dogfish (*Scyliorhinus rotifer*), this drug did not show potential as an analgesic.
Ketroprofen 2mg/kg
In koy carp, it seems to reduce post-surgical muscle damage and can provide analgesia in goldfish (*Carassius auratus*).
Lidocaine 1mg/kg/IM
In rainbow trout, it was effective in reducing behavior and physiological pain related alterations.

(Table 11) contd.....

Carprofen 1-5mg/kg/IM
It has controlled effects of noxious stimulation in carp but causes a decrease in activity and in appetite as a side effect.
Morphine 5-50mg/kg
It can be an effective analgesic in rainbow trout, flounder and goldfish. This drug has slower excretions rate in fish (half-life 37 hours with total elimination time of 56 hours) and can be a promising surgery fish analgesic because one injection should be enough to provide a durable anaesthesia. Another advantage is that, at least in trout, there are no alterations in behavior, feeding and physiology.
Tramadol
An opioid drug, showed analgesic properties in common carp (*Cyprinus carpio*) and further studies could establish this drug as a valid analgesic option for fish.

One of the main problem is the often need, in most of the available options, of more than one administration to reach prolonged analgesia. The decision of retreatment may depend on several factors such as, the presence of behavioral signs possibly related with pain (if possible humane endpoints should be established) [63], potential side effects of the analgesic, and, when animals are not easily manipulated, it should be considered if handling stress doesn't surpass the potential drug advantages.

Surgery

With the development and increasing technical sophistication of public aquaria medicine and fish research, surgery overcomes an initial resistance to undergo even simple surgical procedures. The major difference between ornamental fish medicine and fish research, is that in the last scenario, surgeries are performed by the researchers, mostly non-veterinarians. The gap of communication that still exists between fish veterinarians and researchers should be reduced and veterinarian knowledge related with feasible surgical aseptical techniques, anaesthesia, analgesia and surgical procedures should be transmitted to researchers to optimize scientific outcomes and improve animal welfare. This collaboration between fish veterinarians and researches was already addressed by some workers [65, 66] regarding telemetry tag implantations but should be extended to other fields of laboratory fish research [35, 67]. Nevertheless, veterinarian can also gain from these collaborations, since surgery in research fish (biomedical, fisheries and aquaculture related) had an early development and contributed to the recent sophistication in ornamental fish surgery [35]. Fish research groups can benefit from veterinarian consultancy in anaesthesia and surgery but also in issues like the design and implementation of health monitoring programs.

In ornamental fish medicine and laboratory fish research, surgical procedures have generally different objectives but the necessary surgical technical knowledge is common (see Figs. **7** and **8**). The aim of this introductory text is not to review in detail the published surgical procedures in fish, but rather to address general surgical principles. Some surgical procedures that can be performed in fish are listed in Table **12**.

Table 12. Surgical procedures in fish [Adapted from 6, 7, 24, 35, 37, 54, 67 - 82].

Ablation of external or internal masses. Resolution of anal and ovarian prolapse. Catheterization for fluid collection, drug delivery or blood pressure measurements (dorsal and ventral aorta, curvier duct, gall bladder). Endoscopy (gastrointestinal and intra coelomic). Exploratory celiotomy. Eye surgery (cataracts, enucleation, corneal treatment and suture, eye prosthesis). Foreign bodies extraction (endoscopy or intra celomic surgery). Gonadectomy. Implantation of microchip transponders. Intracardiac perfusions. Laparoscopy and gastric biopsies, Lipid keratopathy (green moray eels), Microsurgery (*e.g.* hepatic related blood vessels). Neurology research procedures (brain injections and cannulations). Odontology. Orthopedic surgery (mandibular fractures in moray eels). Pneumocystectomy/buoyancy problems. Pseudobranchectomy. Surgery of skin ulcers or lacerations. Telemetry implantation. Tissue or organ ablation (hypophysectomy, isletectomy, hepatic lobules, heart apex, kidney, gill and fin biopsies).

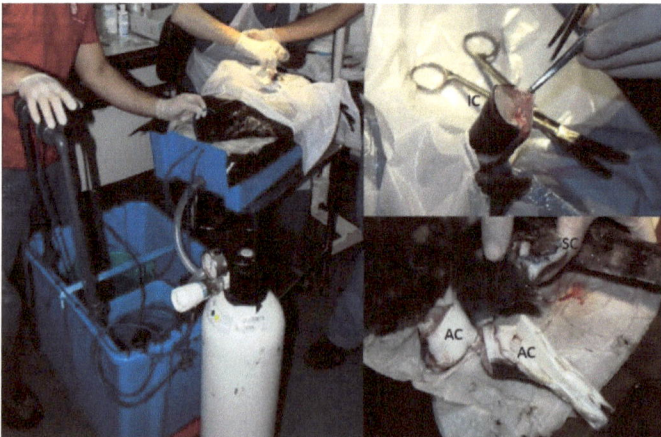

Fig. (7). Public aquaria surgeries have generally a curative goal, as in the case of this undulate ray (*Raja undulata*) with a multiple clasper fracture that went through a unilateral clasper amputation (IC – incisioned clasper; SC – sutured amputated clasper caudal extremity; AC – amputated clasper) (Courtesy of Oceanário de Lisboa).

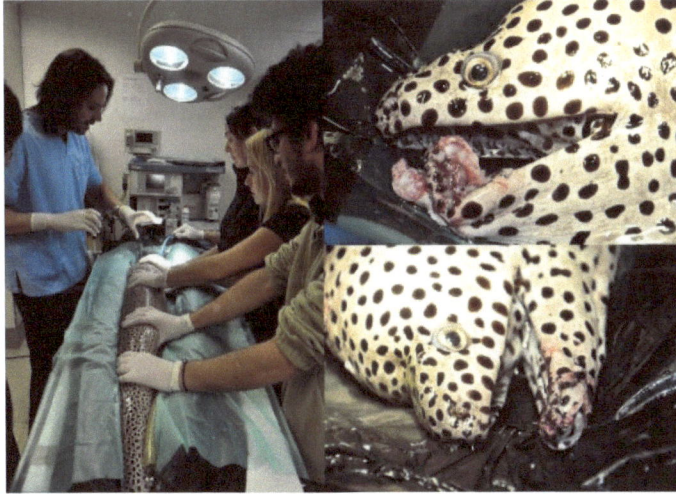

Fig. (8). Orthopedic surgery in an exposed mandibular fracture in a laced moray (*Gymnothorax favagineus*). This animal was anesthetized with tiletamine/zolazepam intramuscularly. (Courtesy of Oceanogràfic of the City of Arts and Sciences of Valencia).

Aseptic Surgery Technique in Fish / Adaptations

In fish surgery, it is almost impossible to obtain an effective aseptic surgical field. Amongst the several differences regarding terrestrial animal surgery, the omnipresence of water is the major obstacle to maintain a surgical environment free of microorganisms [39].

Nevertheless, in some particular interventions like intracoelomic procedures or eye surgery, an aseptic surgical field can eventually be reached. This rigorous asepsis is even more doubtful if surgery involves bowel incisions or contaminated wounds [35]. Also, field surgeries such as telemetry implantation can hardly be performed under total asepsis. Fish surgery should therefore be considered a clean-contaminated procedure.

Antibiotherapy can be considered, in light of this situation or concept, but should never be an expedient to replace the implementation of basic aseptic surgical measures.

Choice of antimicrobial agents should, whenever possible, respect previous bacterial identifications and antibiogram results (antimicrobial susceptibilities) for the particular situation. This choice of the antibiotic for fish that are being released in the wild, should consider withdrawal periods and environmental impact [66].

Antibiotic administration should be perioperative (ante, intra or post-surgery) and if there is just the possibility of one administration (*e.g.* difficult capture or release in the wild) a good option is an intraperitoneal injection of enrofloxacin (10mg/kg) that should maintain therapeutic levels for 5 days [37]. If the fish is accessible for several days, injections of broad-spectrum antibiotics like ceftazidime 30mg/kg/IM/every 3 days or enrofloxacin 5-10mg/kg/IM/each 48 hours can be an option [6, 37]. The author also prescribes regularly a post-surgery prolonged flumequine immersion (50ppm/8-10 days).

Before surgery, cleaned surgical instruments should be sterilized and not just simply disinfected. Preferably, sterilization should be done by autoclaving. Cold chemical methods such as immersion in 2% glutaraldehyde (*e.g.* CidexPlus, for at least 10 minutes) should be used only as a last resource, in situations such as multiple telemetry implantation procedures in field surgeries. Ideally, each surgery should have a separate set of instruments. With cold chemical sterilizers, instruments should be thoroughly washed with sterile water to eliminate residues before using the instruments in the fish. Simply washing or disinfecting surgical instruments before surgery should be discouraged and furthermore, disinfection with alcohol, benzalkonium chloride or chlorhexidine does not provide sterilization. Intracoelomic implants (*e.g.* telemetry or drug delivery implants) should be sterilized by gas or chemical cold sterilization, and never by autoclaving. Overall, improper aseptic technique can lead to microbial contamination and post-surgical infections, inaccurate research results and compromise animal welfare [35, 66].

Pre-surgical Procedures

In ornamental fish medicine, a preoperative medical assessment should be performed. Diagnosis and prognosis can be clarified with additional diagnostic exams like radiography, ultrasonography and blood analysis [37, 40]. However, these pre-surgical analytical profiles (blood biochemistry and hemogram), a common procedure in general veterinary surgery, are not often performed in fish

[24, 82]. Health status is a major surgical prognostic factor (*e.g.* septic, cachectic and anaemic fish will have a poorer prognosis) [40]. In the clinical examination, microscope observation of fin, gill biopsies and cutaneous mucous, can rule out subclinical infections which may be later exacerbated due to surgical stress. In research scenarios, even a brief medical evaluation is often absent partly because it is assumed that fish are healthy [35] and also because a veterinary clinician is not always available.

Fish should be placed in a wet foam surgical bed with its shape adapted to the anatomy of the patient [39]. In fusiform fish, foam with a V shape is recommended and foam with a small incision is appropriate for smaller fish like zebrafish. In pleuronectiform fish and in rays, a flat surface is preferable, which in the case of fish dorsal recumbency, should have a space to accommodate and protect the eyes.

As described in the anaesthesia section, fish should fast before surgery to prevent regurgitation, with potential gill lamella clogging (sestonosis) or epithelial damage, and to avoid water quality deterioration, ammonia build-up, particularly in prolonged surgeries [20]. This fasting can also facilitate the visualization of intra-coelomic structures, that otherwise will be hidden by the food containing enlarged gastro intestinal tract (*e.g.* gonadal assess in the tilapia is practically impossible with a full stomach and intestine).

Stress should be minimized before surgery, favoring a quiet induction and recovery [4, 6].

When handling fish, skin and the cutaneous mucous layer should be preserved because they are important components of the nonspecific immune system [37, 83]. Failure in preserving these elements increases the risk of secondary infections and osmotic disturbances. Only in the case of large scales it is necessary to carefully remove, with tissue forceps, a row along where the surgical incision is to be made. In fish with small scales (*e.g.* salmonids) removal of the scales prior to the incision is not necessary [37]. It should be noticed that scales are embedded in the dermis, so they should be spared as much as possible. Because fish only have a vestigial hypodermis, its disruption exposes the underlying muscle.

Whenever possible, in-water transfers or use of plastic bags instead of nets to handle fish and the use of latex gloves also minimize handling stress [6, 55].

Disinfection of the surgical field also preserves the cutaneous mucous layer. Alcohol use and surgical scrubbing should be avoided [57]. The area should be gently cleaned with diluted iodopovidone (1:20) or chorhexidine (1:40) [37].

Due to the absence of eyelids, eyes should be protected from physical damage and desiccation with a sterile eye gel lubricant whenever the fish is emerged. In prolonged procedures, protection from light with moist gauze calms the fish and avoids retinal damage [20, 61].

Sterile surgical drapes are essential to minimize surgical field contamination [37] and they are often disregarded in research surgeries, mainly in field procedures. A compromise can be achieved, using clean disinfected, and not sterile drapes when researchers need to perform numerous surgical procedures in a short length of time.

Classic surgical towel drapes are unsuitable for fish surgery because they contribute to dry the skin, damaging the protective cutaneous layer.

Surgical drape clamps, even small, should be avoided. Instead, a sterile gel lubricant applied between the fish and the plastic drape can help secure it in place [39]. These plastic transparent drapes also contribute to maintaining the skin moist, allowing good observation of the patient, anaesthesia monitoring and helping prevent water from entering the surgical site [37].

Surgical instruments are similar to those used in mammals and avian surgeries and in smaller fish, microsurgery or ophthalmological instruments are quite useful. Also, some adaptations can be useful, such as the use of abdominal atraumatic retractors to open the opercula (see Fig. **9**).

Head loups with mounted illumination are useful for visualizing small fish or small anatomical structures. These loups will also facilitate the surgical work deep inside the coelomic cavity [37].

Surgical microscopes are rarely used but can help when doing cornea microsurgeries or lens related procedures. Adapted binocular lens can substitute quite efficiently a proper surgery microscope.

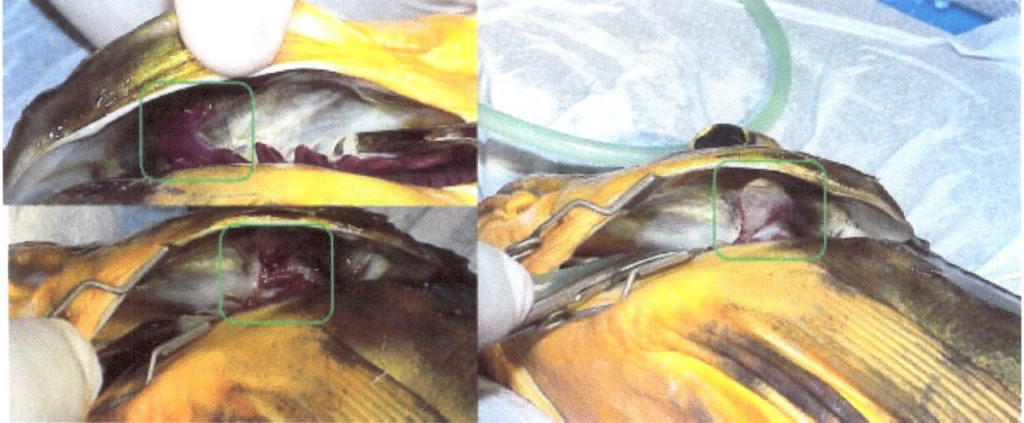

Fig. (9). To perform a pseudobranchectomy in a large-scaled gurnard (*Lepidotrigla cavillone*) the access to the interior of the opercular cavity is facilitated by the use of a small rodent cheek retractor. The pseudobranchea is visible inside the green square at the left images. At the right it was already absent (Courtesy of Oceanário de Lisboa).

Surgical Procedures

During fish surgery, there are particular aspects that should be considered because they differ substantially from terrestrial animal surgery.

All handling of fish tissues should be minimized and performed carefully to reduce trauma and inflammation. Skin is rarely mobile so there is often a need to rely on second intension healing [35].

In celiotomies, the standard surgical approach is along the ventral midline. The incision should be performed cautiously to avoid damaging internal organs because the coelomic muscle wall can be confused with the intestine due to their similar colours [35]. In some cases, a pelvic girdle osteotomy is necessary and may be repaired with steel sutures or, in smaller fish, with the same suture that closes coelomic cavity [37]. A lateral incision is an alternative approach [6] which can be used for liver, swim bladder, gonad and kidney surgical procedures. These lateral incisions are also useful in pleuronectiform fish when accessing the coelomic cavity. In rays, a dorsal lateral incision can be chosen for ovariectomies [84] or caesarean sections.

Visualization and surgical handling of the coelomic cavity can be facilitated by self-retaining retractors (see Fig. **10**). Because organs are not as mobile as in mammals, most procedures are performed inside the cavity. Rarely, when intracoelomic access is insufficient, the midline incision can be prolonged by a lateral incision. Some species have normal visceral adhesions that can hamper organ handling and can be dissected bluntly [37].

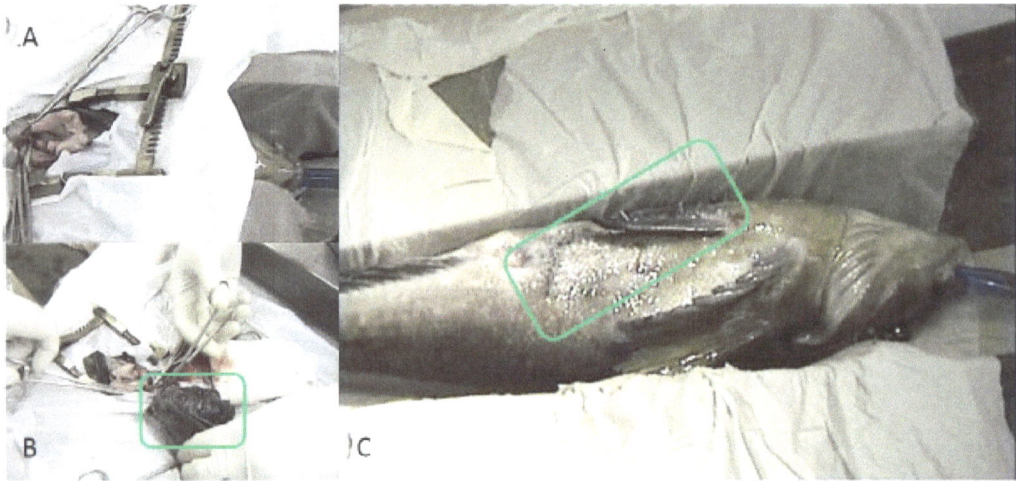

Fig. (10). The access of the coelomic cavity to perform a gonadectomy is facilitated by this small animal abdominal retractor (left images). The extracted gonads are visible inside the green square. It was necessary to prolong laterally the initial incision and the suture in L is visible in the right image, inside the green rectangle. (Courtesy of Oceanário de Lisboa)

Haemostasis (control of bleeding) can be achieved with a bipolar electrocautery or a thermocautery. Both methods involve risks of damaging the tissues so cooling of adjacent structures with saline solution is a prudent measure. Fish tissues are generally more friable than those of mammals or birds, so haemostasis with a clamp or vessel ligament is not as common in fish as in mammal surgery, although larger vessels may be ligated and cut [37].

Suture materials used in fish are fairly well studied regarding healing time and tissue reactivity [37, 85]. A recent review by Wagner focused on telemetry related surgery is available and recommended for a detailed description [86].

Generally, monofilament sutures are more appropriate since the risk of wound contamination is smaller compared to the risk involved with multifilament sutures and also because tissues offer less resistance to its passage [40]. As multifilament sutures have a higher capillarity with a higher risk of contaminated water being introduced in the surgical wound or in the coelomic cavity [37]. Absorbable sutures are absorbed or eventually expelled as a foreign body and synthetic material (*e.g.* nylon or polydioxanone) is preferable than organic suture material (*e.g.* silk, catgut or chromic gut) which can cause a strong tissue inflammatory response [37, 85].

However, absorbable sutures can fail to be absorbed due to inhibition of hydrolysis by low temperature (delayed at lower temperature). There are reports of suture material like Polyglactin 910 (Vicryl) and polydioxanone remaining intact in the body for more than one year [37, 85].

In summary, polydioxanone (PDS) and polygliconate (Maxon), both absorbable monofilaments, are the most recommended sutures for fish. Multifilament sutures like silk, chromic gut and polyglactin 910 can cause moderate to strong inflammatory reaction in fish tissues and higher healing times [6, 37, 85, 87] (Table **13**).

Furthermore, a recent study with subyearling Chinook salmon, in which seven suture materials were compared, selected poliglecaprone 25 (Monocryl) as the best option [89].

There are few studies available on the evaluation of skin suture patterns in fish. In teleost, simple continuous, simple interrupted, horizontal mattress and continuous Ford interlocking, have been described in the literature, thus the choice depends mostly on the surgeon's experience and preferences [6, 37, 66].

In general, continuous patterns reduce drag, knot surface (less bacteria, fungus and algae contamination), risk of liquid entrance in the coelomic cavity and surgery time, however if there is an integrity problem in one point, all sutures can be jeopardized. Interrupted sutures can take more time to perform and have more knot surface but are safer regarding eventual suture material damage. Needles with cutting tip are generally better because fish skin offers some resistance [37].

Table 13. Biomaterials for wound closure or sutures in fish [Adapted from 37, 66, 78, 85, 87 - 89].

Cyanoacrylate	Nile Tilapia (*Oreochromis niloticus*)	More 4-7 days for healing and higher inflammation compared with sutures. Inflammation worse with commercial product compared to medical one (Commercial/Loctite®*versus* Medical/Histoacryl®).
	Rainbow trout	100 % open wounds in 24 hours).
	Zebrafish (*Danio zebra*)	Used to close the coelomic cavity incision without surgical complications.
Polyglactin 910	Bluefin tuna(*Thunnus thynnus thynnus*)	25% shed before healing.
	Doitsu (Scaless) koi carp (*Cyprinus carpio*)	Higher tissue reaction compared with polyglyconate, risk of bacterial contamination due to be a mulfilament.
	Largemouth bass	7 week absorption time in largemouth bass skin and with high tissue reaction compared with polydioxanone. Seven weeks of healing time (three with polydioxanone).
	Sutures can remain intact for more than a year in intracoelomic tissues or in coelomic muscle wall for more than a year.	
	In red pacu at 10 days after suturing, skin inflammation with polyglactin 910 is higher than with polydioxanone and monofilament nylon.	
Silk	African catfish	Can promote transintestinal expulsion.
	Blue tilapia	Fouling by algae and grazing by other fish. Retarded healing.
	Doitsu (Scaless) koi carp(*Cyprinus carpio*)	High tissue reaction (worse than catgut, polyamide and polyglyconate).
Polyamide/ Nylon	Blue tilapia	Faster healing than catgut or braided silk.
	Doitsu (Scaless) koi carp *Cyprinus carpio*	Significant superficial reaction.
	Sea lamprey *Petromyzon marinus*	
Polydioxanone	Bluefin tuna *Thunnus thynnus thynnus*	Remain 2-3 weeks.
	Largemouth bass	7 week absorption time in skin and with lower tissue reaction compared with polyglactin 910. Three weeks of healing time (seven with polyglactin 910).
	Sutures can remain intact for more than a year in intracoelomic tissues or in coelomic muscle wall for more than a year.	
Polyglyconate	Doitsu (Scaless) koi carp *Cyprinus carpio*	Small tissue reaction, less bacterial contamination due to be a monofilament. Ideal for this species.

(Table 13) contd.....

Poliglecaprone 25	Subyearling chinook salmon	The best suture material in 7 studied, for closing incisions created during surgical implantation of acoustic microtransmitters.

To close a coelomic incision, a single or a two-layer suture may be carried out, depending on the fish size and the muscle wall thickness. In smaller fish, a single layer closure, incorporating both skin and muscle layer, is enough because the muscle wall is very thin and firmly adhered to the skin. In larger animals however, the skin and the muscle wall should be sutured separately, with the skin being the strength layer of the closure [6, 37]. For muscle layer suturing, simple continuous or continuous Ford interlocking patterns are generally adequate. All intra coelomic air should be properly evacuated after closing the incision to prevent buoyancy problems [39].

A recent study compared three suture patterns (simple interrupted, interrupted horizontal mattress and subcuticular) to close coelomic incisions of goldfish (*Carassius auratus*). The subcuticular pattern showed the mildest inflammatory reaction [90].

Several reports described different successful approaches for closing a coelomic incision in elasmobranchs: i) a three layers suture in a c. 40kg sandbar shark (*Carcharhinus plum*beus), interrupted cruciate mattress pattern in the muscle wall and peritoneum, a continuous pattern in the superficial fascia and finally an interrupted cruciate pattern in the skin (polydioxanone in all layers) [24]; ii) a continuous horizontal mattress suture pattern (braided glycolide/lactide co-polymer) for the muscle wall and a simple interrupted suture pattern (monofilament nylon) for the skin in a 30-kg male juvenile/subadult sand tiger shark (*Carcharias taurus*) [79]; iii)a left paralumbar incision for a gonadectomie in southern stingrays (*Daysiatis americana*) closed with simple interrupted pattern (polydioxanone), by a two layer suture (peritoneum and skin) [84].

The author favors, in general, an appositional simple interrupted suture with a monofilament suture to close the skin incision (see Fig. **11**).

Skin sutures should be removed within 2-4 weeks in temperate and tropical fish (higher metabolic rate due to temperature) and 6 to 10 weeks after surgery in cold water species [37, 39] (see Fig. **12**). In some cases, sutures can be left without removal because handling and capture stress surpass the benefits of suture removal. Also, it is rarely possible to keep animals in captivity that are to be released in the wild.

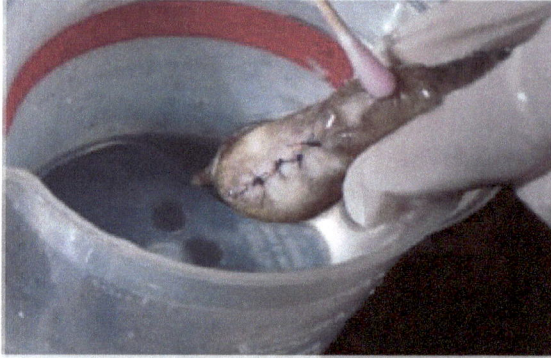

Fig. (11). Closed incision of an infected brood pouch of an big-belly seahorse (*Hippocampus abdominalis*) with a simple interrupted pattern and with a monofilament suture (Courtesy of Oceanário de Lisboa).

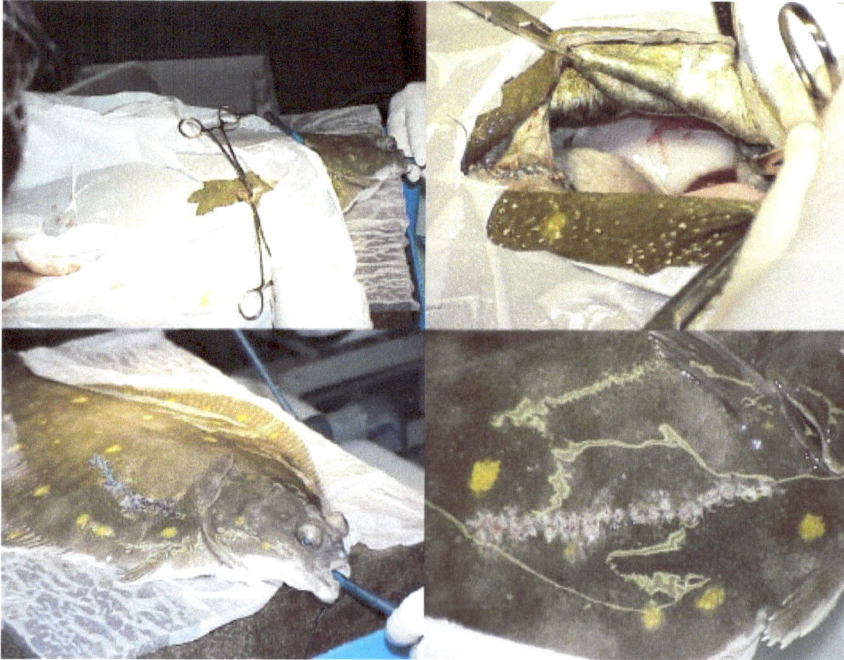

Fig. (12). After a laparatomy in an european plaice (*Pleuronectes platessa*) the incision was closed with a simple interrupted pattern and the sutures were removed 6 weeks after the surgery, which is indicated for cold water species (Courtesy of Oceanário de Lisboa).

Cyanoacrylate can increase tissue inflammation, causing severe dermatitis in some species and delaying healing with a consequent risk of dehiscence (opening of the surgical wound). Closing the incision solely with cyanoacrylate, often leads to reopening of the surgical wound because mucous produced by the skin quickly sheds the layer of adhesive tissue. Used in conjugation with sutures, cyanoacrylate can increase suture drag [37]. Recently the use of only cyanoacrylate was described to close coelomic cavity incisions in zebrafish, without surgical complications [78]. Nevertheless, despite the small size this species can be sutured to close cutaneous incisions [91].

Normally, the suture of the incision is enough to prevent water entrance in the coelomic cavity. Also re-epithelitazion is quite fast in fish due to rapid epithelial cell migration [35, 73, 92, 93].

Nevertheless, the author never detected any visible dermatitis or healing complication following the use of cyanoacrylate (Vetbond®) in marine fish, with the exception of scaleless fish like moray eels (see Fig. **13**). In these species, healing delay should be expected due to cyanoacrylate related dermatitis. In research scenarios, scientists often use commercial cyanoacrylate (*e.g.* Loctite) instead of medical preparations (*e.g.* Vetbond® or Histoacryl®) which should be avoided as they cause a higher inflammatory reaction [87] (see Fig. **14**).

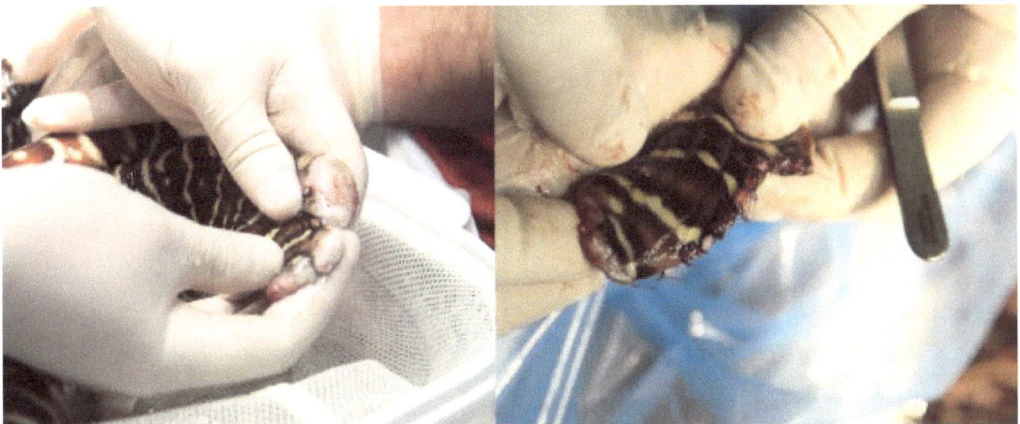

Fig. (13). In scaless fish like this zebra moray eel (*Gymnomuraena zebra*) (excision of caudal fin nodules) the use of cyanoacrylate can provoke an intense dermatitis (Courtesy of Oceanário de Lisboa).

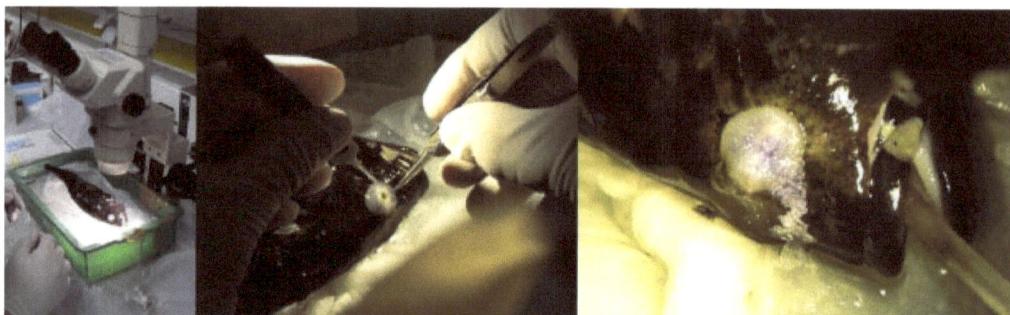

Fig. (14). In this case of a corneal ulcer surgery the use of cyanoacrylate as a final patch is a good solution to protect the corneal suture (Surgery done by Rui Oliveira, DVM, CertVOphthal MRCVS) (Courtesy of Oceanário de Lisboa).

Surgical staples are not easily applied in the fish skin and can be associated with some morbidity (*e.g.* deficient healing) and mortality, so they are rarely used to close surgical incisions [37, 66, 86].

Post-surgical Procedures

Stress mitigation strategies should be implemented also after the surgery to optimize healing and recovery (*e.g.* use of water from the origin tank during anaesthesia, surgery and recovery, stabilization of water parameters, providing hiding places to avoid exposure, salinity manipulation whenever possible and avoidance of intense illumination) [4, 6, 55].

All aspects mentioned in the anaesthesia recovery section of this text are valid for the post-surgery period.

Due to diminutive fish skin plasticity, when comparing with mammals, surgical wounds resulting from masse excisions or from injuries cannot often be closed by direct suturing and need to have second intention healing. Particularly in these cases, minimizing osmotic imbalances with water salinity manipulation can be useful. In fresh water species, salt (1-3g/l) can be added to the system during the healing period [37]. Likewise, in salt water fish, the salinity can be slightly reduced in euryhaline species (decrease gradually 2-4 ppt). This salinity manipulation can also decrease the incidence of external infections (*e.g.* fungal infections of the sutured incision in freshwater species) [37].

Application of iodopovidone ointment in closed incisions can decrease the risk of fungal infection and, in post-surgical skin defects, a silver sulphadiazine cream can be topically applied [37]. The author uses successfully a *Centella asiatica* extract cream (Madecassol) associated with zinc oxide and cod liver oil to enhance second healing of open wounds. These unsuturable skin defects can also be sealed and repaired with Orobase gel or extracellular matrix protein [35] (see Fig. **15**).

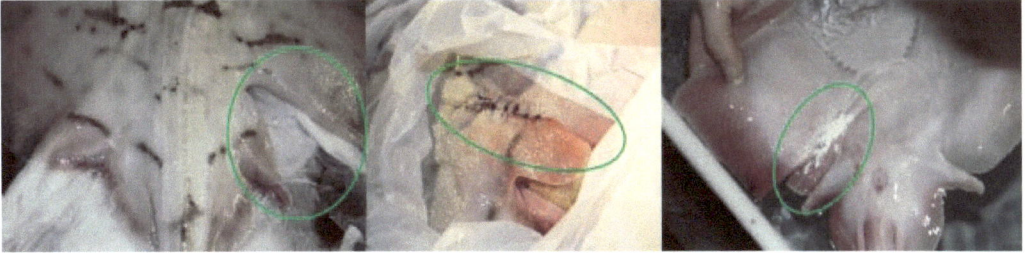

Fig. (15). The final suture of the reconstruction of a birth defect in the lateral fin of a thornback ray (*Raja clavata*) was covered by *Centella asiatica* extract cream (Madecassol) associated with zinc oxide and cod liver oil and Denture adhesive (Corega powder) as a final patch (Courtesy of Oceanário de Lisboa).

In sharks, covering the surgical wound with a lanolin-based cream is an option referred in the literature [79].

Surgical Complications

Surgery complications that are somewhat related with the anaesthesia, were already addressed in this chapter.

Not often, dehiscence of sutured surgical wounds (see Fig. **16**) can occur in fish post-surgery due to lack of aseptic procedures. Other factors can also cause this problem, like choice of an improper suture material, inefficient suturing technique, water with improper microbial load or pressure from an intracoelomic implant [6, 35, 63]. Basically, if aseptic surgical techniques and procedures are not followed, post- surgical infection incidence risk increases.

If the suture tension is to loose, water with microorganisms can contaminate the surgical wound or even the coelomic cavity. However, if suture placement is too tight, it may compromise the healing process [94].

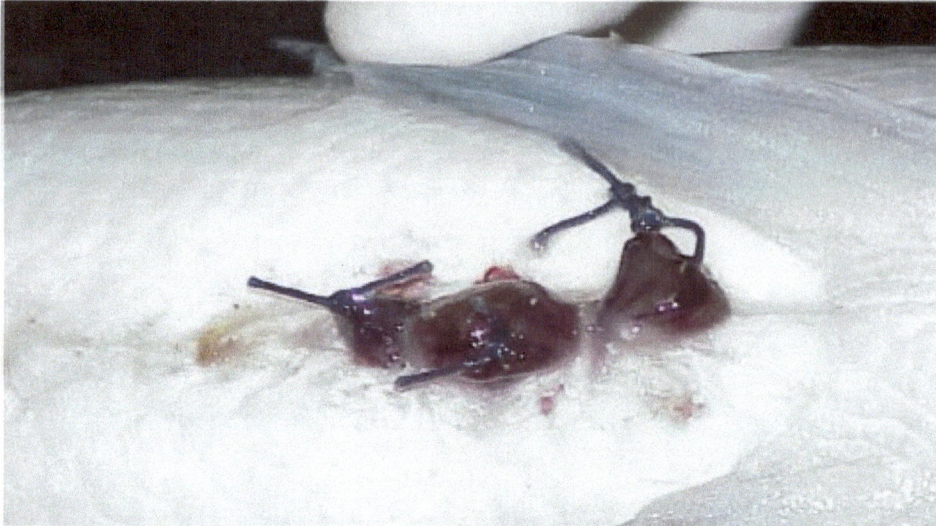

Fig. (16). Infection of a coelomic suture with partial dehiscence of the surgical wound (Courtesy of Oceanário de Lisboa).

The breakdown of the coelomic wall wound is not necessarily a fatal complication. In some reported cases, intracoelomic structures like the omental adipose tissue of rainbow trout or the liver in elasmobranchs (sandbar shark/*Carcharhinus plumbeus*), can be sufficient to protect the interior of the coelomic cavity from the entrance of contaminated water until a second intention healing or a new suture leads to the definite closure of the incision [24, 94].

As previously mentioned, post-surgical buoyancy disturbances can occur, if intra coelomic air is not properly removed before closing the incision. Although a single puncture is not expected to cause major problems, inadvertent perforation of the swim-bladder during surgery can lead to air leaking to the coelomic cavity and buoyancy instability [73].

Intracoelomic implants (*e.g.* telemetry transmitters) can be evacuated by the gonadal pores or by the intestines, after being "handled" as a foreign body ("intestinalization" of the implant). With the use of more inert materials covering the implants, these foreign body reactions are now more rare [94].

In vascular catheterizations, a thrombosis due to mismatch between catheter size and dorsal aorta diameter can occur in certain species and cases (*e.g.* higher risk in hybrid striped bass due to smaller aorta diameter) [37].

A careful haemostasis is necessary in fish because due to low blood pressure, discreet haemorrhaging is not easily detected. Although rarely, post intra coelomic bleeding can occur after extensive visceral surgical manipulation and the use of Spongostan®, a gelatinous haemostatic sponge, can be useful preventing haemorrhages.

Particular Aspects of Surgery Procedures in Fish

Since 1993 [57], the number of described surgical procedures in fish has had a steep increase [6, 24, 35, 37, 39, 40, 73, 79, 84, 94]. Descriptions of fish research surgery procedures are also abundant but are scattered in an immense number of scientific papers, some of which are briefly referenced in this chapter.

Some details of these surgical procedures will be addressed in the following text.

Gonadectomies in teleost is a procedure mainly used in research scenarios. The primary blood supply is located at the cranial pole of the gonads and in smaller fish, haemostasis is done by simple traction (in the author's experience) of the vessels and ligament. In larger animals, it should be ligated and cut. In this surgery, caution should be exercised when dissecting the caudal extremity of the gonads to avoid damaging the excretory adjacent structures [37] (see Fig. **17**). Surgical approach can be by a midline ventral incision [95] or by bilateral incisions [96].

The excision of the caudal part of the gonads can leave some gonadal tissue in place. Some authors cauterize (thermocautery or warm forceps) the remaining tissues [96]. Free eggs at the coelomic cavity (breakdown of the ovarian wall at ovulation or in cases of ruptured ovaries due to egg binding) can be removed by lavage and aspiration [35].

In a reported southern stingray (*Daysiatis americana*) gonadectomy, the surgical approach, was a left paralumbar incision, parallel to the spinal column (2cm lateral from the lumbar muscles) and special care was taken to carefully dissect the caudal pole of the gonad from the adjacent epigonal gland [84].

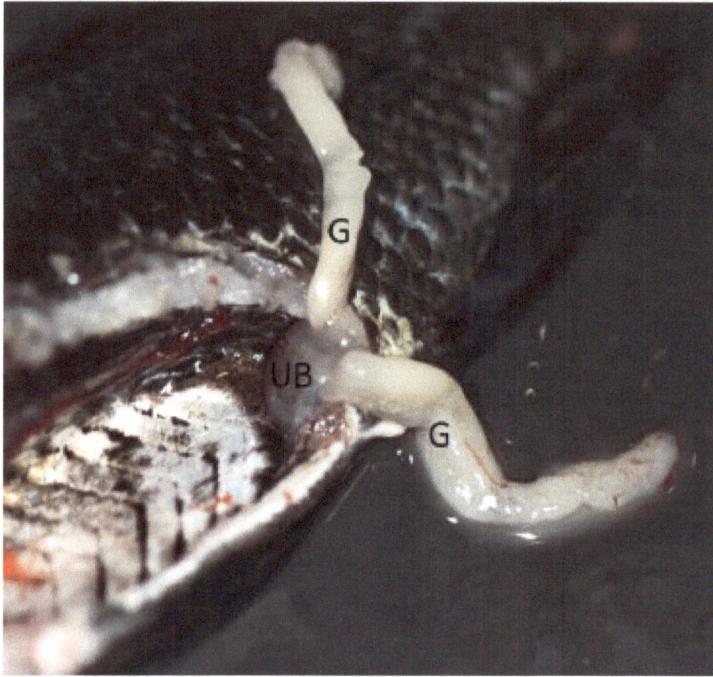

Fig. (17). Mozambique tilapia (*Oreochromis mossambicus*) is a good example of the close proximity between the caudal extremity of the gonads (G) and the urinary bladder (UB). These excretory adjacent structures should be spared when cutting the gonads (Courtesy of Instituto Superior de Psicologia Aplicada - ISPA).

Gaseous exophthalmia can be surgically handled by a pseudobranchectomy. This procedure consists in the excision of the pseudobranch, which is present in most of the teleost and the haemostasis can be achieved by direct pressure, bipolar electrocautery, and ligation of the efferent vessels [37] or by applying haemostatic sponge (Spongostan) glued with Cyanoacrylate, the same method that can be used in haemostasis in eye enucleation.

Due to medical reasons or more rarely in research scenarios, eye enucleation is a common and simple procedure that is well described [37, 73]. There are two surgical approaches to perform a research hypophysectomy, the transorbital and the opercular approaches. For the last (hypophysectomy with a transorbital approach), an eye enucleation is necessary [35].

Haemostasis can be done by direct pressure, by topical application of 2.5% phenylephrine hydrochloride [37, 73] or by applying haemostatic sponge

(Spongostan®) glued with Cyanoacrylate (see Fig. **18**).

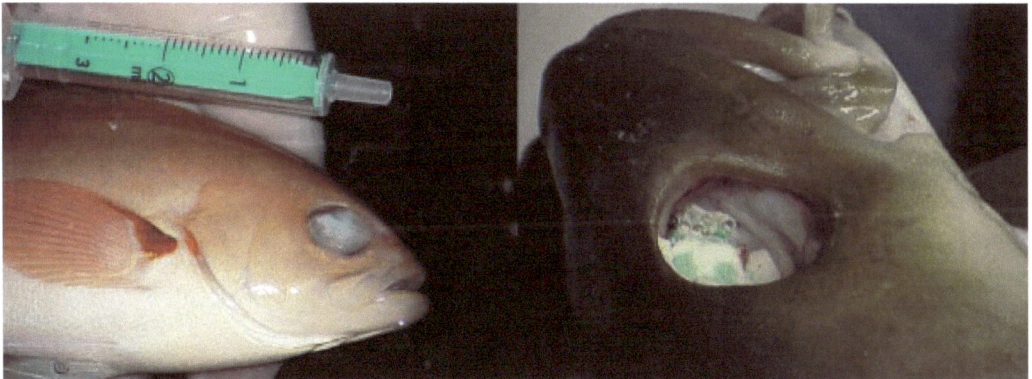

Fig. (18). Eye enucleation hemostasis done by haemostatic sponge (Spongostan®) glued with cyanoacrylate (Courtesy of Oceanário de Lisboa).

In neuroscience research brain surgeries, there is often the need to repair skull fenestrations. This can be done with the use of material like 2% agar gel in physiological saline, vaseline-paraffin oil capped with anchored vinyl polysiloxane impression material [35]. Other possible materials are bone wax [97], fast drying dental cement (Acrílico Auto-PolimerizanteClássico, JET, Brazil, and Líquido Acrílico Auto-Polimerizável, Dental VIPI Ltda., Brazil) [98].

Blood vessel cannulations done in fish research are also extensively referred in the literature, both in teleost and elasmobranchs. Vessels like dorsal aorta, ventral aorta, afferent branchial artery, ductus Cuvier, bulbus arteriosis, conus arteriosos, caudal vein celiac or mesenteric artery [35, 99, 100, 101, 102].

Besides vessels, other catheterizations are described in structures like gall bladder, urinary bladder and intracoelomic cavity [35].

Endoscopies and laparoscopies in fish are described in detail in the literature and are used for gender identification and maturation assessment, examination of coelomic organs and collection of diagnostic specimens, foreign bodies detection and removal, and for minimally invasive surgery [6, 103] (see Fig. **19**). To allow organ visualization, either insufflation of carbon dioxide or eventually filtered room air or either saline-infusion can be used. This last technique in sturgeon, infusion of sterile 0.9% saline, allowing good organ visualization and avoiding the risk of post-surgical buoyancy disturbances due to air that remains after insufflation [6, 77, 103].

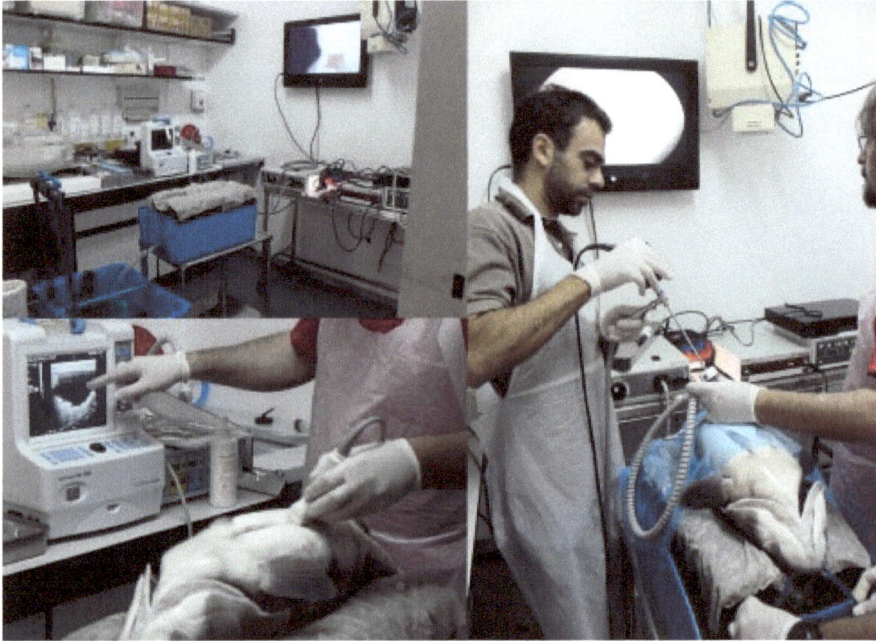

Fig. (19). Although infusion of sterile 0,9% saline in coelomic laparoscopies decreases the risk of post-surgical buoyancy disturbances, this kelp bass (*Paralabrax clathratus*) didn´t showed any post-surgical problem with the use of carbon dioxide (Laparoscopy performed by Rui Bernardino, DVM, Lisbon Zoo) (Courtesy of Oceanário de Lisboa).

Fig. (20). Surgical implantation of a radiotelemetry transmitter with an external antenna in iberian barbel (Luciobarbus bocagei). Note that the antenna exits not at the main incision to minimize the risk of dehiscence and wound contamination (Courtesy of Pedro R. Almeida, Departamento de Biologia, Universidade de Évora and Centro de Oceanografia). "Centro de Oceanografia" by "MARE - Marine and Environmental Sciences Centre.

Implantation of telemetry devices (see Fig. **20**) is extensively described in the literature and there is an excellent special number on fish telemetry in a 2011 publication [104, 105].

Underwater surgery to implant telemetry transmitters can be considered to minimize temperature shock and decompression barotrauma in fish captured in depths fewer than 50 meters. If decompression stops or swim bladder deflation is not a viable alternative, the potential damages related to the aforementioned temperature shock and decompression barotrauma can justify the risks of microbiological contamination in a immersed surgery. Surgeries should be done at around 20 meters and if possible, below the thermocline (see Fig. **21**). Even there are few cases with this dilemma, further research on the efficiency and risks of this immersed surgical approach should be done [57, 66, 105, 106].

Fig. (21). Underwater surgery for intracoelomic implantation of telemetry transmitters in red porgy fish, Pagrus pagrus. (Courtesy of Pedro Afonso, Departamento de Oceanografia e Pescas (DOP) da Universidade dos Açores (UAç). Images taken by Marco Santos.

In the last years, zebrafish emerged as an important model for fish research. Despite its small size, there are several surgical procedures, besides minor surgeries like caudal fin biopsy, described in this species [61].

Several fish models are described for research on spinal cord regeneration but zebrafish is now the main fish model for this research [80, 107 - 109].

Spinal cord transection surgery is briefly described in the literature and there is a recent study with a detailed description of the procedure in zebrafish. Fish are anesthetized with 0.03% MS222 (300ppm or mg/l) and placed in a filter paper groove with crushed ice below. Three or four scales are removed and the skin and muscle are incised with a spring scissor. The spinal cord is exposed by blunt dissection with a microforceps without injuring the spinal cord and the adjacent blood vessels and vertebrae. After this exposure, the spinal cord is completely transected, in a single cut, with a small spring scissor. In this phase, special care should be exercised to avoid damaging the bone structures located just ventrally to the spinal cord. In the bone vertebrae, body spinal cord regeneration could be compromised. After transaction, the wound is quickly sealed with cyanoacrylate adhesive tissue (Histoacryl®) (see Fig. **22**). This procedure is done easily in older fish (c. 6 months) but younger fish (smaller) do regenerate better. In this surgery recovery period administration of an antifungal drug (according to the text) in the water diminishes mycology infections and fish are fed 3 days after the surgery. Prognosis to this procedure is influenced, as in general fish surgery, by fish health status and post-operative stress. Also low water temperatures (< 15°C) increase the mortality rate. In uncomplicated cases, fish should be expected to swim normally within 6 weeks after surgery [80]. The author of this chapter would advise to also add an antibiotic to the water if there is any anticipated bacterial complication.

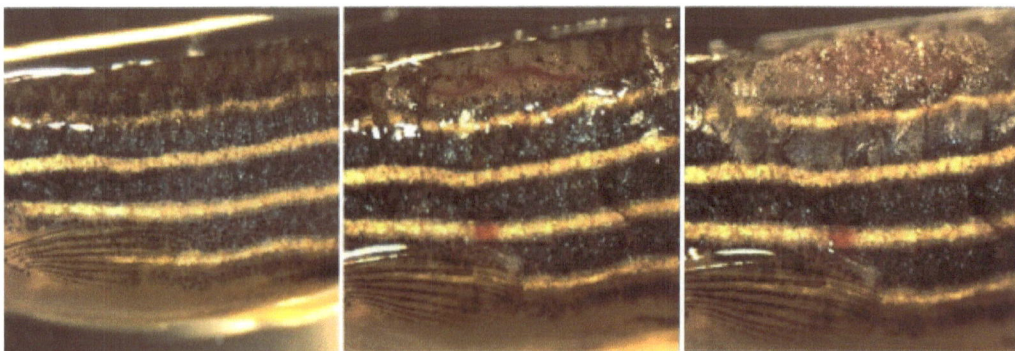

Fig. (22). Spinal cord transection surgery images. To perform the injury an incision is made on the side of anesthetized zebrafish to expose the vertebral column. The spinal cord is then crushed dorsoventrally with a forceps, at a level halfway between the dorsal fin and the operculum. The wound is sealed with Vetbond® (3M) and the fish allowed to recover in individual tanks for up to 6 weeks following injury. (Ribeiro Personal communication). (Images courtesy of Ana Ribeiro. Instituto de Medicina Molecular. Lisboa. Portugal).

Zebrafish is also a fish model for liver regeneration studies. A partial hepatectomy (1/3) in adult fish, with a 90% survival rate, is described in a recent publication. Fish fasting for 24 hours are anesthetized with MS222 at 0.015% (150ppm or mg/l) and placed in dorsal recumbency. After removing a row of scales with forceps, a ventral median incision of 3-4-mm is performed to assess the coelomic cavity. The ventral liver lobe is exteriorized and resected at the lobe base. After reintroducing the remaining liver back to the coelomic cavity, the body wall incision is closed with cyanoacrylate (Nexaband®) [78].

Heart regeneration studies in zebrafish started more than 10 years ago and the surgical procedure involved in this animal model is relatively well documented [74, 110].

Adult (1-2 years old) zebrafish have sufficient body dimensions to be candidates to non-terminal heart resection surgeries. Fish are anesthetized with 0.4% MS222, placed in dorsal recumbency and a row of scales are removed from the incision site, which is done with an iridectomy scissor. The median ventral small incision starts at the skin and continues through the muscle and pericardial sac to assess the heart. The ventricle is exposed by gentle coelomic pressure or pulled by a forceps and 20% of the ventricle at the apex resected by iridectomy scissors. Heart lesion haemostasis is done by mere pressure (15-45 seconds) and the incision is left unsutured, healing in 1-2 days. During anaesthesia recovery phase, fish assisted ventilation is performed. This surgery has a 90% survival rate [74, 110].

CONFLICT OF INTEREST

The author confirms that author has no conflict of interest to declare for this publication.

ACKNOWLEDGEMENTS

The author would like to acknowledge: the Biology staff of the Oceanário de Lisboa and especially those who built our inhalation anaesthesia systems and helped in all of the anaesthesia and surgery procedures (Gonçalo David Nunes, Nuno Antunes, Elsa Santos, Hugo Batista and Núria Baylina); Instituto Superior de Psicologia Aplicada (ISPA) fish research facility technical staff and researchers; Instituto Gulbenkian de Ciência (IGC) Zebrafish fish research facility staff; Boga/Ciimar fish research facility staff; Pedro Afonso, Departamento de Oceanografia e Pescas (DOP) da Universidade dos Açores (UAç); Pedro R. Almeida, Departamento de Biologia, Universidade de Évora and Centro de Oceanografia; Rui Bernardino, Lisbon Zoo; Daniel Garcia-Párraga, Oceanogràfic of the City of Arts and Sciences of Valencia; Rui Oliveira, veterinary ophthalmologist; Francisco Assis Costa and Ana Costa, Clínica Veterinária João XXI; Margarida Duarte, Ana Cassamo and Ana Duarte for their assistance with this chapter preparation.

REFERENCES

[1]   Sneddon LU. Clinical anaesthesia and analgesia in fish. J Exot Pet Med 2012; 21: 32-43.
       [http://dx.doi.org/10.1053/j.jepm.2011.11.009]

[2]   Ward JL, McCartney SP, Chinnadurai SK, Posner LP. Development of a minimum-anestheti-
       -concentration depression model to study the effects of various analgesics in goldfish (*Carassius
       auratus*). J Zoo Wildl Med 2012; 43(2): 214-22.
       [http://dx.doi.org/10.1638/2010-0088.1] [PMID: 22779222]

[3]   Davis MR, Mylniczenko N, Storms T, Raymond F, Dunn JL. Evaluation of intramuscular ketoprofen
       and butorphanol as analgesics in chain dogfish (*Scyliorhinus retifer*). Zoo Biol 2006; 25: 491-500.
       [http://dx.doi.org/10.1002/zoo.20105].
       [http://dx.doi.org/10.1002/zoo.20105]

[4]   Ross LG, Ross B. Anaesthetic and sedative techniques for aquatic animals. 3rd ed. Oxford: Wiley-
       Blackwell 2008; p. 240.
       [http://dx.doi.org/10.1002/9781444302264]

[5]   Machin KL. Zoo animal and wildlife immobilization and anaesthesia. Iowa. 43-59.

[6]   Murray MJ. Fish Surgery. J Exot Pet Med 2002; 11: 246-57.
       [http://dx.doi.org/10.1053/saep.2002.126571]

[7]   Rose J, Arlinghaus R, Cooke S, *et al.* Can fish really feel pain? Fish Fish 2014; 15: 97-133.
       [http://dx.doi.org/10.1111/faf.12010]

[8]   Barton BA. Stress in fishes: a diversity of responses with particular reference to changes in circulating
       corticosteroids. Integr Comp Biol 2002; 42(3): 517-25.
       [http://dx.doi.org/10.1093/icb/42.3.517] [PMID: 21708747]

[9]   Neiffer D. Boney fish (lungfish, sturgeon and teleosts). In: West G, Heard GW, Caulkett N, Eds. Zoo
       animal and wildlife immobilization and anaesthesia. Iowa, USA: Blackwell Publishing Ltd. 2007; pp.
       159-96.
       [http://dx.doi.org/10.1002/9780470376478.ch14]

[10]   Zahl IH, Samuelsen O, Kiessling A. Anaesthesia of farmed fish: implications for welfare. Fish Physiol
       Biochem 2012; 38(1): 201-18.

[http://dx.doi.org/10.1007/s10695-011-9565-1] [PMID: 22160749]

[11]   Thurmon JC, Short CE. History and overview of veterinary anaesthesia. In: At Tranquilli WJ, Thurmon JC, Grimm KA, Eds. Lumb & Jones, veterinary anaesthesia and analgesia. Iowa, USA: Blackwell Publishing 2007; pp. 3-6.

[12]   Flecknell PD. Laboratory animal anaesthesia. San Diego, CA: Academic Press, Inc. 2009.

[13]   Ross LG. Restraint, anaesthesia and euthanasia. In: Wildgoose WH, Ed. BSAVA Manual of Ornamental Fish. London, UK: British Small Animal Veterinary Association 2001; pp. 75-83.

[14]   Brown LA. Anaesthesia and restraint. In: Stoskopf M, Ed. Fish Medicine. W.B.Saunders 1993; pp. 79-90.

[15]   Stamper MA. Immobilizations of elasmobranchs. In: Stamper MA, Ed. Elasmobranch husbandry manual: captive care of sharks, rays and their relatives. Ohio, USA: Ohio Biological Survey 2004; pp. 447-66.

[16]   Fiddes M. Fish anaesthesia. In: Longley L, Ed. Anaesthesia of exotic pets. Edinburgh, UK: ElsevierSaunders 2008; pp. 261-78.

[17]   Matthews M, Varga ZM. Anesthesia and euthanasia in zebrafish. ILAR J 2012; 53(2): 192-204. [http://dx.doi.org/10.1093/ilar.53.2.192] [PMID: 23382350]

[18]   Wilson JM, Bunte RM, Carty AJ. Evaluation of rapid cooling and tricaine methanesulfonate (MS222) as methods of euthanasia in zebrafish (*Danio rerio*). J Am Assoc Lab Anim Sci 2009; 48(6): 785-9. [PMID:19930828].
[PMID: 19930828]

[19]   Snyder DE. Electrofishing and its harmful effects on fish US Geological Survey Eds Information and Technology Report 2003-0002. USA: Denver 2003; p. 149.

[20]   Harms C. Anaesthesiain fish. In: Fowler M, Miller R, Eds. Zoo and wild animal medicine: current therapy. Philadelphia, USA: WB Saunders Co 1999; pp. 158-63.

[21]   Lewbart GA, Harms C. Building a fish anaesthesia delivery system. Exotic DVM 1999; 1: 25-8.

[22]   Zahl IH. Anaesthesia of farmed fish with special emphasis on Atlantic cod (*Gadus morhua*) and Atlantic halibut (Hippoglossus hippoglossus) Doctoral thesis. University of Bergen 2011.

[23]   Popovic NT, Strunjak-Perovic I, Coz-Rakovac R, *et al.* Tricaine methane-sulfonate (MS-222)application in fish anaesthesia. J Appl Ichthyology 2012; 28: 553-64. [http://dx.doi.org/10.1111/j.1439-0426.2012.01950.x]

[24]   Lécu A, Herbert R, Coulier L, Murray MJ. Removal of anintracoelomic hook *via* laparotomy in a sandbar shark (*Carcharhinus plumbeus*). J Zoo Wildl Med 2011; 42(2): 256-62. [PMID: 22946403]

[25]   Brown LA. Recirculation anaesthesia for laboratory fish. Lab Anim 1987; 21(3): 210-5. [http://dx.doi.org/10.1258/002367787781268846] [PMID: 3626467]

[26]   Neiffer DL, Stamper MA. Fish sedation, analgesia, anesthesia, and euthanasia: considerations, methods, and types of drugs. ILAR J 2009; 50(4): 343-60. [http://dx.doi.org/10.1093/ilar.50.4.343] [PMID: 19949251]

[27]   Rotili DA, Devens MA, Diemer O, Lorenz EK, Lazzari R, Boscolo WR. Uso de eugenol como anestésico em pacu. Pesquisa Agropecuária Tropical 2012; 42: 288-94.
[http://dx.doi.org/10.1590/S1983-40632012000300013]

[28]   Alves da Cunha M. Anestesia em jundiás (*Rhamdia quelen*) expostos a substâncias isoladas de plantas. Santa Maria. Universidade Federal de Santa Maria, Centro de Ciências Rurais Master Thesis, Universidade Federal de Santa Maria 2007.

[29]   Simões L, Gomes L. Eficácia do mentol como anestésico para juvenis de tilápia-do-nilo (*Oreochromis niloticus*). Arq Bras Med Vet Zootec 2009; 61: 613-20.
[http://dx.doi.org/10.1590/S0102-09352009000300014]

[30]   Alves da Cunha M, Corrêa de Barros FM, Garcia LO, *et al.* Essential oil of Lippia alba:a new anaesthetic for silver catfish, *Rhamdia quelen.* Aquaculture 2010; 306: 403-6.
[http://dx.doi.org/10.1016/j.aquaculture.2010.06.014]

[31]   Simões LN, Paiva G, Gomes LD. Óleo de cravo como anestésico em adultos de tilápia-do-nilo. Pesquisa Agropecu Bras 2010; 45: 1472-7.
[http://dx.doi.org/10.1590/S0100-204X2010001200019]

[32]   Alves da Cunha M, Ferreira da Silva B, Delunardo FA, *et al.* Anaesthetic induction and recovery of Hippocampus reidi exposed to the essential oil of Lippia alba. Neotrop Ichthyol 2011; 9: 683-8.
[http://dx.doi.org/10.1590/S1679-62252011000300022]

[33]   Stamper MA. Elasmobranchs (Sharks, Rays, and Skates). In: West G, Heard D, Caulkett N, Eds. Zooanimal & wildlife, immobilization and anaesthesia. Iowa, USA: Blackwell 2007; pp. 197-204.
[http://dx.doi.org/10.1002/9780470376478.ch15]

[34]   Lewbart GA. Fish. In: Carpenter J, Ed. Exotic Animal Formulary. St. Louis, Missouri, USA: Elsevier Saunders 2012; pp. 18-52.

[35]   Harms CA. Surgery in fish research: common procedures and postoperative care. Lab Anim (NY) 2005; 34(1): 28-34.
[http://dx.doi.org/10.1038/laban0105-28] [PMID: 19795589]

[36]   Honczaryk A, Inoue LA. Anestesia do piracucu por aspersão da benzocaína directamente nas brânquias. Cienc Rural 2010; 40: 1-4.
[http://dx.doi.org/10.1590/S0103-84782009005000235]

[37]   Harms CA, Lewbart GA. Surgery in fish. Vet Clin North Am Exot Anim Pract 2000; 3(3): 759-74.
[PMID:11228930].
[PMID: 11228930]

[38]   Coyle SD, Durborow RM, Tidwell JH. Anaesthetics in aquaculture. Southern Regional Aquaculture Center 2004; 3900: 1-4.

[39]   Weber EP III, Weisse C, Schwarz T, Innis C, Klide AM. Anesthesia, diagnostic imaging, and surgery of fish. Compend Contin Educ Vet 2009; 31(2): E11.
[PMID: 19288435]

[40]   Roberts HE. Fundamentals of ornamental fish health. Iowa, USA: Wiley-Blackwell 2010.

[41]   Noga EJ. Fish disease, diagnosis and treatment. Raleigh: Willey-Blackwell 2010.

[http://dx.doi.org/10.1002/9781118786758]

[42]   Harper C, Lawrence C. The laboratory zebrafish. Boston, Massachusetts, USA: CRC Press,Taylor & Francis Group 2011.

[43]   Bittencourt F, Souza B, Boscolo W, Rorato R, Feiden A, Neu D. Benzocaína e eugenol como anestésicos para quinguio (*Carassius auratus*). Arq Bras Med Vet Zootec 2012; 64: 1597-602.
[http://dx.doi.org/10.1590/S0102-09352012000600028]

[44]   García-Gomez A, de la Gándara F, Raja T. Utilización del aceite de clavo, *Syzygium aromaticum* L. (Merr. & Perry), como anestésico eficaz y económico para labores rutinarias de manipulación de peces marinos cultivados. Bol Inst Esp Oceanogr 2002; 18: 21-3.

[45]   Boyer S, White J, Stier A, Osenberg C. Effects of the fish aneshetic, clove oil (eugenol), on coral heatlh and growth. J Exp Mar Biol Ecol 2009; 369: 53-7.
[http://dx.doi.org/10.1016/j.jembe.2008.10.020]

[46]   Javahery S, Moradlu AH. AQUI-S, a new anaesthetic for use in fish propagation. Global Veterinaria 2012; 9: 205-10.

[47]   Javahery S, Nekoubin H, Moradlu AH. Effect of anaesthesia with clove oil in fish (review). Fish Physiol Biochem 2012; 38(6): 1545-52.
[http://dx.doi.org/10.1007/s10695-012-9682-5] [PMID: 22752268]

[48]   Di Marco P, Petochi T, Longobardi A, *et al.* Efficacy of tricaine methanosulphonate, clove oil and medetomidine-ketamine and their side effects on the physiology of sturgeon hybrid *Acipenser naccarii* X Acipenser baerri. J Appl Ichthyology 2011; 27: 611-7.
[http://dx.doi.org/10.1111/j.1439-0426.2011.01701.x]

[49]   Fleming GJ, Heard DJ, Francis Floyd R, Riggs A. Evaluation of propofol and medetomidine-ketamine for short-term immobilization of Gulf of Mexico sturgeon (*Acipenser oxyrinchus de soti*). J Zoo Wildl Med 2003; 34(2): 153-8.

[50]   Williams TD, Rollins M, Block BA. Intramuscular anesthesia of bonito and Pacific mackerel with ketamine and medetomidine and reversal of anesthesia with atipamezole. J Am Vet Med Assoc 2004; 225(3): 417-21.
[http://dx.doi.org/10.2460/javma.2004.225.417] [PMID: 15328719]

[51]   Borin S, Crivelenti LZ, De Paula Lima CA. Uso intramuscular da associação de tiletamina e zolazepam na anestesia de tilápias-do-nilo (*Oreochromis niloticus*). Ciência Animal Brasileira 2010; 11: 429-35.

[52]   Gressler LT, Parodi TV, Riffel AP, DaCosta ST, Baldisserotto B. Immersion anaesthesia with tricaine methanesulphonate or propofol on different sizes and strains of silver catfish *Rhamdia quelen*. J Fish Biol 2012; 81(4): 1436-45.
[http://dx.doi.org/10.1111/j.1095-8649.2012.03409.x] [PMID: 22957883]

[53]   Gholipourkanani H, Ahadizadeh S. Use of propofol as an anaesthetic and its efficacy on some hematological values of ornamental fish *Carassius auratus*. Springerplus 2013; 2(1): 76.
[http://dx.doi.org/10.1186/2193-1801-2-76] [PMID: 23539492]

[54]   Garcia-Párraga D, Álvaro D, Valls M, Malabia A. Tiletamine-zolacepam injectable anaesthesia: an easy technique to handle fish in large volume aquaria. IAAAM 38th Annual Conference Proceedings.

Lake Buena Vista, Florida, USA. 2007; pp. 25-7.

[55] Kreiberg H. Stress and Anaesthesia. In: Ostrander GK, Ed. The Laboratory Fish. San Diego, USA: Academic Press 2000; pp. 503-11.
[http://dx.doi.org/10.1016/B978-012529650-2/50038-X]

[56] Sherrill J, Weber ES III, Marty GD, Hernandez-Divers S. Fish cardiovascular physiology and disease. Vet Clin North Am Exot Anim Pract 2009; 12(1): 11-38, v. [v.].
[http://dx.doi.org/10.1016/j.cvex.2008.08.002] [PMID: 19131028]

[57] Stoskopf MK. Surgery. In: Stoskopf M, Ed. Fish Medicine. California, USA: W.B.Saunders 1993; pp. 91-7.

[58] Hadfield CA, Whitaker BR, Clayton LA. Emergency and critical care of fish. Vet Clin North Am Exot Anim Pract 2007; 10(2): 647-75.
[http://dx.doi.org/10.1016/j.cvex.2007.01.002] [PMID: 17577566]

[59] Camus A, Berliner A, Clauss T, Hatcher N, Marancik D. Serratia marcescens associated ampullary system infection and septicaemia in a bonnethead shark, *Sphyrna tiburo* (L.). J Fish Dis 2013; 36(10): 891-5.
[PMID: 23534484]

[60] Stamper MA. Pharmacology in elasmobranchs. In: Stamper MA, Ed. Elasmobranch husbandry manual: captive care of sharks, rays and their relatives Ohio, Ohio Biological Survey. USA 2004; pp. 281-94.

[61] Stoskopf MK. Appendix V: chemotherapeutics. In: Stoskopf MK, Ed. Fish Medicine. California, USA: W.B. Saunders 1993; pp. 832-9.

[62] Harms CA, Lewbart GA, Swanson CR, Kishimori JM, Boylan SM. Behavioral and clinical pathology changes in koi carp (*Cyprinus carpio*) subjected to anesthesia and surgery with and without intra-operative analgesics. Comp Med 2005; 55(3): 221-6.
[PMID: 16089168]

[63] Sneddon LU. Pain perception in fish: indicators and endpoints. ILAR J 2009; 50(4): 338-42.
[http://dx.doi.org/10.1093/ilar.50.4.338] [PMID: 19949250]

[64] Chervova L, Lapshin D. Behavioral control of the efficiency of pharmacological anaesthesia in fish. J Ichthyol 2011; 51: 1126-32.
[http://dx.doi.org/10.1134/S0032945211110026]

[65] Harms CA, Lewbart GA. The veterinarian's role in surgical implantation of electronic tags in fish.rev. Fish Biol Fisheries 2011; 21: 25-33.
[http://dx.doi.org/10.1007/s11160-010-9185-3]

[66] Mulcahy DM. Surgical implantation of transmitters into fish. ILAR J 2003; 44(4): 295-306.
[http://dx.doi.org/10.1093/ilar.44.4.295] [PMID: 13130160]

[67] Leclère FM, Germain MA, Lewbart GA, Unglaub F, Mordon S, Louis D. Microsurgery in liver research: end-to-side portocaval microanastomoses in dogfish. Clin Res Hepatol Gastroenterol 2011; 35(10): 650-4.
[http://dx.doi.org/10.1016/j.clinre.2011.06.016] [PMID: 21821480]

[68] Palay SL. Fixation of neural tissues for electron microscopy by perfusion with solutions of osmium tetroxide. J Cell Biol 1962; 12: 385-410.
[http://dx.doi.org/10.1083/jcb.12.2.385] [PMID: 14483299]

[69] Levy A, Baker BI. Penetration of phosphorothioate oligodeoxynucleotides into rainbow trout brain. A technique for chronic infusion of substances into the brain of free-living adult fish. J Fish Biol 1997; 50: 691-702.

[70] Wollmuth LP, Crawshaw LI, Rausch RN. Adrenoceptors and temperature regulation in goldfish. Am J Physiol 1988; 255(4 Pt 2): R600-4.
[PMID: 3177691]

[71] Davie PS, Franklin CE, Grigg GC. Blood pressure and heart rate during tonic immobility in the black tipped reef shark, *Carcharhinus melanoptera*. Fish Physiol Biochem 1993; 12(2): 95-100.
[http://dx.doi.org/10.1007/BF00004374] [PMID: 24202688]

[72] Wildgoose W. Fish surgery: an overview. Fish Vet J 2000; 5: 22-36.

[73] Harms CA. WH.Surgery. In: Wildgoose WH, Ed. BSAVA Manual of Ornamental Fish. Gloucester, UK: British Small Animal Veterinary Association 2001; pp. 259-66.

[74] Poss KD, Wilson LG, Keating MT. Heart regeneration in zebrafish. Science 2002; 298(5601): 2188-90.
[http://dx.doi.org/10.1126/science.1077857] [PMID: 12481136]

[75] Weisse C, Weber ES, Matzkin Z, Klide A. Surgical removal of a seminoma from a black sea bass. J Am Vet Med Assoc 2002; 221(2): 280-3.

[76] Bakal RS, Hickson BH, Gilger BC, Levy MG, Flowers JR, Khoo L. Surgical removal of cataracts due to Diplostomum species in Gulf sturgeon (*Acipenser oxyrinchus de soti*). J Zoo Wildl Med 2005; 36(3): 504-8.
[http://dx.doi.org/10.1638/04-044.1] [PMID: 17312772]

[77] Divers S, Boone SS, Hoover J, *et al.* Field endoscopy for identifying gender, reproductive stage and gonadal anomalies in free-ranging sturgeon (*Scaphirhynchus*) from the lower Mississippi River. J Appl Ichthyology 2009; 25: 68-74.
[http://dx.doi.org/10.1111/j.1439-0426.2009.01337.x]

[78] Kan NG, Junghans D, Izpisua Belmonte JC. Compensatory growth mechanisms regulated by BMP and FGF signaling mediate liver regeneration in zebrafish after partial hepatectomy. FASEB J 2009; 23(10): 3516-25.
[http://dx.doi.org/10.1096/fj.09-131730] [PMID: 19546304]

[79] Lloyd R, Lloyd C. Surgical removal of a gastric foreign body in a sand tiger shark, *Carcharias taurus* Rafinesque. J Fish Dis 2011; 34(12): 951-3.
[http://dx.doi.org/10.1111/j.1365-2761.2011.01313.x] [PMID: 22074022]

[80] Fang P, Lin J-F, Pan H-C, Shen Y-Q, Schachner M. A surgery protocol for adult zebrafish spinal cord injury. J Genet Genomics 2012; 39(9): 481-7.
[http://dx.doi.org/10.1016/j.jgg.2012.07.010] [PMID: 23021548]

[81] Júnior AB, Maximino C, Pereira A, *et al.* Rapid method for acute intracerebroventricular injection in adult zebrafish. zebrafish protocols for neurobehavioral research. Neuromethods 2012; 66: 323-30.

[http://dx.doi.org/10.1007/978-1-61779-597-8_25]

[82]   Meegan J, Sidor IF, Field C, Roddy N, Sirpenski G, Dunn JL. Endoscopic evaluation and biopsy collection of the gastrointestinal tract in the green moray eel (*Gymnothorax funebris*): application in a case of chronic regurgitation with gastric mucus gland hyperplasia. J Zoo Wildl Med 2012; 43(3): 615-20.
         [http://dx.doi.org/10.1638/2009-0234R1.1] [PMID: 23082527]

[83]   Uribe C, Folch H, Enriquez R, Moran G. Innate and adaptive immunity in teleost fish: a review. Vet Med (Praha) 2011; 56: 486-503.

[84]   George R, Steeil J, Baine K. Ovariectomy of sub-adult Southern stingrays (Dasyatis americana) to prevent future reproductive problems. In: IAAAM 44th Anual Conference Proceedings. 70-1.

[85]   Hurty CA. Evaluation of the tissue reactions in the skin and body wall of koi (*Cyprinus carpio*) to five suture materials. Vet Rec 2002; 151(11): 324-8.
         [http://dx.doi.org/10.1136/vr.151.11.324] [PMID: 12356236]

[86]   Wagner GN. Surgical implantation tecniques for electronic tags in fish. Rev Fish Biol Fish 2011; 21: 71-81.
         [http://dx.doi.org/10.1007/s11160-010-9191-5]

[87]   Jepsen N, Koed A, Thorstad EB, Baras E. Surgical implantation of telemetry transmitters in fish: how much have we learn? Hydrobiologia 2002; 483: 239-48.
         [http://dx.doi.org/10.1023/A:1021356302311]

[88]   Gilliland E. Comparison of absorb able sutures used in large mouth bass liver biopsy surgery. Prog Fish-Cult 1994; 56: 60-1.
         [http://dx.doi.org/10.1577/1548-8640(1994)056<0060:COASUI>2.3.CO;2]

[89]   Deters KA, Brown RS, Carter KM, Boyd JW. Performance assessment of suture type, water temperature, and surgeon skill in juvenile chinook salmon surgically implanted with acoustic transmitters. Trans Am Fish Soc 2010; 139: 888-99.
         [http://dx.doi.org/10.1577/T09-043.1]

[90]   Nematollahi A, Bigham AS, Karimi I, Abbasi. F. Reactions of gold fish (*Carassius auratus*) to three suture patterns following full thickness skin incisions. Res Vet Sci 2010; 89(3): 451-4.
         [http://dx.doi.org/10.1016/j.rvsc.2010.03.022] [PMID: 20434740]

[91]   Diep CQ, Davidson AJ. Transplantation of cells directly into the kidney of adult zebrafish. J Vis Exp 2011; 51(51): 2725.
         [PMID: 21633330]

[92]   Quilhac A, Sire J-Y. Restoration of the sub epidermal tissues and scale regeneration after woundinga cichlid fish (*Hemichromis binaculatus*). J Exp Zool 1998; 281: 305-27.
         [http://dx.doi.org/10.1002/(SICI)1097-010X(19980701)281:4<305::AID-JEZ6>3.0.CO;2-S]

[93]   Fontenot DK, Neiffer DL. Wound management in teleost fish: biology of the healing process,evaluation, and treatment. Vet Clin North Am Exot Anim Pract 2004; 7(1): 57-86.
         [http://dx.doi.org/10.1016/j.cvex.2003.08.007] [PMID: 14768380]

[94]   Johnson GR. Surgical Techniques. In: Ostrander GK, Ed. The Laboratory Fis. SanDiego, USA: Academic Press 2000; pp. 577-67.

[95]   Sakharkar AJ, Singru PS, Sarkar K, Subhedar NK. Neuropeptide Y in the forebrain of the adult male cichlid fish Oreochromis mossambicus: distribution, effects of castration and testosterone replacement. J Comp Neurol 2005; 489(2): 148-65.
[http://dx.doi.org/10.1002/cne.20614] [PMID: 15984003]

[96]   Páll MK, Mayer I, Borg B. Androgen and behavior in the male three-spined stickleback, Gasterosteus aculeatus. II. Castration and 11-ketoandrostenedione effects on courtship and parental care during the nesting cycle. Horm Behav 2002; 42(3): 337-44.
[http://dx.doi.org/10.1006/hbeh.2002.1820] [PMID: 12460593]

[97]   Tsukada T, Nobata S, Hyodo S, Takei Y. Area postrema, a brain circumventricular organ, is the site of antidipsogenic action of circulating atrial natriuretic peptide in eels. J Exp Biol 2007; 210(Pt 22): 3970-8.
[http://dx.doi.org/10.1242/jeb.010645] [PMID: 17981865]

[98]   Garção DC, Canto-de-Souza L, Romaguera F, Mattioli R. Chlorpheniramine impairs functional recovery in *Carassius auratus* after telencephalic ablation. Braz J Med Biol Res 2009; 42(4): 375-9.
[http://dx.doi.org/10.1590/S0100-879X2009000400010] [PMID: 19330266]

[99]   Black MC. Collection of body fluids. In: Ostrander GK, Ed. The Laboratory Fish. San Diego, USA: Academic Press 2000; pp. 513-27.
[http://dx.doi.org/10.1016/B978-012529650-2/50039-1]

[100]   Sandblom E, Axelsson M, Farrell AP. Central venous pressure and mean circulatory filling pressure in the dogfish Squalus acanthias: adrenergic control and role of the pericardium. Am J Physiol Regul Integr Comp Physiol 2006; 291(5): R1465-73.
[http://dx.doi.org/10.1152/ajpregu.00282.2006] [PMID: 16825417]

[101]   Speers-Roesch B, Sandblom E, Lau GY, Farrell AP, Richards JG. Effects of environmental hypoxia on cardiac energy metabolism and performance in tilapia. Am J Physiol Regul Integr Comp Physiol 2010; 298(1): R104-19.
[http://dx.doi.org/10.1152/ajpregu.00418.2009] [PMID: 19864337]

[102]   Johnson KR, Hoagland TM, Olson KR. Endogenous vascular synthesis of B-type and C-type natriuretic peptides in the rainbow trout. J Exp Biol 2011; 214(Pt 16): 2709-17.
[http://dx.doi.org/10.1242/jeb.052415] [PMID: 21795567]

[103]   Murray MJ. Endoscopy in fish. In: Murray MJ, Schildger B, Taylor M, Eds. Endoscopy in Birds, Reptiles, Amphibians and Fish. Tuttlingen, Germany: Endo-Press 1998; pp. 59-75.

[104]   Brown RS, Eppard MB, Murchie KJ, Nielsen JL, Cooke SJ. An introduction to the pratical and ethical perspectives on the need to advance and standardize the intracoelomic surgical implantation of electronic tags in fish. Rev Fish Biol Fish 2011; 21: 1-9.
[http://dx.doi.org/10.1007/s11160-010-9183-5]

[105]   Rosa RM. Desenvolvimento e aplicação de técnicas de telemetria acústica para o estudo dos movimentos de peixes demersais nos Açores Internship 2006.

[106]   Lino PB, Oliveira M, Erzini K, Santos M. The african hind's (Cephalopholis taeniops, serranidae) useof artificial reefs off Sal island (Cape Verde): a preliminary study based on acoustic telemetry. Braz J Oceanogr 2011; 59: 69-76.

[http://dx.doi.org/10.1590/S1679-87592011000500009]

[107] Barreiro-Iglesias A, Shifman M. Use of fluorochrome-labeled inhibitors of caspases to detect neuronal apoptosis in the whole-mounted lamprey brain after spinal cord injury 2012.
[http://dx.doi.org/10.1155/2012/835731]

[108] Allen AR, Smith GT. Spinal transection induces widespread proliferation of cells along the length of the spinal cord in a weakly electric fish. Brain Behav Evol 2012; 80(4): 269-80.
[http://dx.doi.org/10.1159/000342485] [PMID: 23147638]

[109] Zupanc GK, Sîrbulescu RF. Teleost fish as a model system to study successful regeneration of the central nervous system. Curr Top Microbiol Immunol 2013; 367: 193-233.
[http://dx.doi.org/10.1007/82_2012_297] [PMID: 23239273]

[110] Jopling C, Sleep E, Raya M, Martí M, Raya A, Izpisúa Belmonte JC. Zebrafish heart regeneration occurs by cardiomyocyte dedifferentiation and proliferation. Nature 2010; 464(7288): 606-9.
[http://dx.doi.org/10.1038/nature08899] [PMID: 20336145]

[111] Brooks E, Sloman KA, Liss S, Hassan-Hassanein L, Danylchuk AJ, Cooke SJ, *et al.* The stress physiology of extended duration tonic immobility in the juvenile lemon shark, *Neaprion brevirostris* (Poey 1868). Exp Mar Biol Ecol 2011; 409(1-2): 351-60.
[http://dx.doi.org/10.1016/j.jembe.2011.09.017]

CHAPTER 6

Fish Production, Catch, Trade and Processing on Board

António Pedro Correia Margarido[1], Maria Gabriela Lopes Veloso[2,*] and **Miguel José Sardinha de Oliveira Cardo[1,2]**

[1] Direção Geral de Alimentação e Veterinária, Largo da Academia Nacional de Belas Artes 2, 1249 Lisboa, Portugal

[2] Faculdade de Medicina Veterinária / Universidade de Lisboa, Avenida da Universidade Técnica, 1300-477 Lisboa, Portugal

Abstract: World fish production is growing due to an economic growth of developing countries, where the population has greater access to expensive animal protein. The livelihoods and income provided to those involved in the fisheries production and subsidiary activities is bigger than that provided by the agriculture. The substantial demand for fish and fish-based products and overcapacity of fishing fleets are, in turn, responsible for the overexploitation of marine resources and for the negative impact on the economy and wealth of the communities living on fisheries. A proper management of marine resources is important in order to avoid multilevel problems, such as ecological, economic, food security and people's wealth. *The international trade in fish and fishery products* must comply the WTO agreements and with the guidance documents developed by the Codex *Alimentarius* Commission. Maintaining the cold chain is one of the largest contributors to the improvement of the international fish trade. Some fishing vessels are adapted to a single method of catch or fishing gear, in spite of many of them being versatile and equipped for polyvalent captures. Regarding the hygiene requirements and the operations carried out on board to the fishery products, the fishing vessels are classified as primary production vessels, freezer vessels and factory vessels. These categories are independent of the size of the vessel or fishing gear and methods used. The most important is to carry out all the operations on board hygienically to maintain the sanitary quality of the product, avoid contaminations and minimize spoilage by temperature abuse.

* Address correspondence to Maria Gabriela Veloso: Faculdade de Medicina Veterinária, Universidade de Lisboa, Avenida da Universidade Técnica, 1300-477 Lisboa, Portugal; E-mail: gveloso@fmv.ulisboa.pt

Manuela Oliveira, Fernando Bernardo, Joana I. Robalo (Eds.)

Keywords: Certification, Fishing areas, Fleet, Hygiene, Inspection, Trade, World production.

WORLD PRODUCTION

During the last century, there has been a growing demand on food needs caused by a growing population which is expected to reach 8 billion by 2025. Additionally, some developing countries are currently experiencing a strong economic growth, with a growing middle class with a greater access to expensive animal protein. In this context, fish has contributed for 16.6 percent of the consumption of animal protein at the level of the world population in 2009.

In 2008, a population of over 6 billion people were fed by food of animal origin. By 2020, it is expected that this demand undergoes an increase of about 50% [1].

Over the past 50 years the world's supply of fish and fish-based food has grown dramatically in the period 1961-2009 in which the average growth rate was 3.2% per year, outpacing the increase in world population that was 1.7% per year. According to the data provided by FAO on fisheries and aquaculture at a global level [2], fish catch and aquaculture were responsible for the global supply of the world with about 154 million tonnes, of which 131 million tonnes were utilized as food.

The supply of fish *per capita* at world level increased by an average of 9.9 kg in the 1960s to 18.9kg in 2009 and preliminary studies estimate a further increase in fish consumption to 18.6 kg [2]. The figures for fish consumption *per capita* by region are 9.1 kg for Africa, 20.7 kg for Asia, 24.6 kg for Oceania, 24.1 kg for North America, 22.0 kg for Europe, 22.0 kg for Latin America and 9.9 kg for the Caribbean [2].

Between developed and developing countries there are obvious differences on the contribution of fish as a source of animal protein for human consumption.

A substantial part of the fish that is consumed in developed countries is imported, due to a constant demand and a decline in domestic fishing (in the last decade the reduction was 10%). In developed countries, the reliance on imports is projected to grow in the forthcoming years.

In several regions of the world, it became very soon apparent that the natural resources in the fisheries were finite. Although the world captures have maintained a fixed pattern since 1990, which is about 85-90 million tonnes, some changes are noted in trends of captures by country, fishing area and fished species.

Besides this fixed pattern of captures, the world demand on fish has increased and it is clear that the gap is being fulfilled by aquaculture production (Table **1**).

Table 1. World fisheries and aquaculture production and utilization (million tonnes/year) (FAO, 2012).

PRODUCTION	2006	2007	2008	2009	2010	2011
Capture						
Inland	9.8	10.0	10.2	10.4	11.2	11.5
Marine	80.2	80.4	79.5	79.2	77.4	78.9
Total capture	90.0	90.3	89.7	89.6	88.6	90.4
Aquaculture						
Inland	31.3	33.4	36.0	38.1	41.7	44.3
Marine	16.0	16.6	16.9	17.6	18.1	19.3
Total aquaculture	47.3	49.9	52.9	55.7	59.9	63.6
TOTAL WORLD FISHERIES	137.3	140.2	142.6	145.3	148.5	154.0

China is one of the countries where fish production suffered a major increase, mainly from aquaculture, given that their contribution to the global fish production increased from 7% in 1961 to 35% in 2010.

The 2012 FAO figures for aquaculture show that the inland world production suffered an increase of 2.1 to 3.6 Million tonnes each year from 2006 to current days, while the marine aquaculture only had an increase of 0.3 to 1.2 Million tonnes each year, in the same period. The Asian countries are largely responsible for the catch growth in inland waters. In 2010, production figures attributed to Asia a share of global production approaching 89%. In recent years the expansion of aquaculture in North America ceased, but in South America, there has been some growth. At Europe, several important producers have recently ceased expanding. Over the past decade the contribution of Africa to the world production increased from 1.2% to 2.2%. Oceania accounts for minor share of the global aquaculture production.

The least developed countries (LDCs) in 2010 remain in the minority regarding their global fish production quota in aquaculture, 4.1% by quantity, with the main producers included in sub-Saharan Africa and in Asia. However, in some developing countries in Asia, Pacific, sub-Saharan Africa and South America a rapid progress has been made to become major producers in their regions.

In 2010 the world production of farmed food fish by industrialized countries together was 6.9% by quantity, compared with 21.9% in 1990.

In spite of this general growth it is likely that catches in inland waters are seriously underestimated in some regions. In many parts of the world, inland waters are considered as being overfished, moreover important organisms of freshwater are being seriously degraded by the human pressure and by the changes in the environmental conditions.

SOCIAL IMPACT OF FISHERIES

It is estimated that fisheries and aquaculture have provided in 2010 livelihoods and income to 54.8 million people involved in primary fish production sector.

Employment in the primary sector of fisheries and aquaculture has grown faster than employment in agriculture; therefore, in 2010 fishers and fish farmers represent 4.2 percent of the 1.3 billion people working on the agriculture sector worldwide, compared with 2.7 percent in 1990.

The number of people engaged in fish farming has increased by 5.5 percent per year in the last five years, comparing with only 0.8 percent per year for those working in capture fisheries, even though capture fisheries still account for 70 percent of the combined total in 2010.

Fishing and aquaculture are the source of many jobs in subsidiary activities, such as processing of fishery products, packaging, marketing and distribution, manufacture of fish processing equipment, manufacture of fish networks and fishing gear, ice production and supply, building boats and their maintenance, research and administration. It is estimated that all these jobs together with dependent allow the livelihoods of 660-820 million people, which reflects 10-12% of the world's population.

WORLD'S FLEET

In the 2012 FAO report, it is estimated that the total number of fishing vessels in the world in the 2010 was around 4.36 million, from which 74 percent operating in marine waters and the remaining in inland waters.

Asia has the largest fleet in the world which corresponds to 73 % of the world total, the following is Africa with 11%, Latin America and the Caribbean with 8%, North America and Europe with 3% each.

The number of engine powered fishing vessels reaches 60 percent of the world fleet, but this proportion is higher (69%) for marine waters than for inland waters (35%).

This proportion also varies between regions and countries, for instance the non-motorized vessels in marine waters are less than 7 percent in Europe and more than 61 percent in Africa.

More than 85% of the motorized fishing vessels from around the world have less than 12 m in length overall (LOA) and 2% of fishing vessels engaged in industrial fisheries have more than 24 m.

FISHING AREAS

According to FAO figures (2012) [2] in 2010 the area with the highest production rate was the Northwest Pacific (Area 61) with 27% of the global marine catches, followed by the Western Central Pacific (Area 71) with 15%, the Northeast Atlantic (Area 27) with 11% and the Southeast Pacific (Area 87) with 10% (Fig. 1). The number of overexploited stocks is increasing and the possibility to increase the production is through an appropriate management. Meanwhile, some measures in fisheries management are being implemented as precaution or with the aim to rebuild stocks.

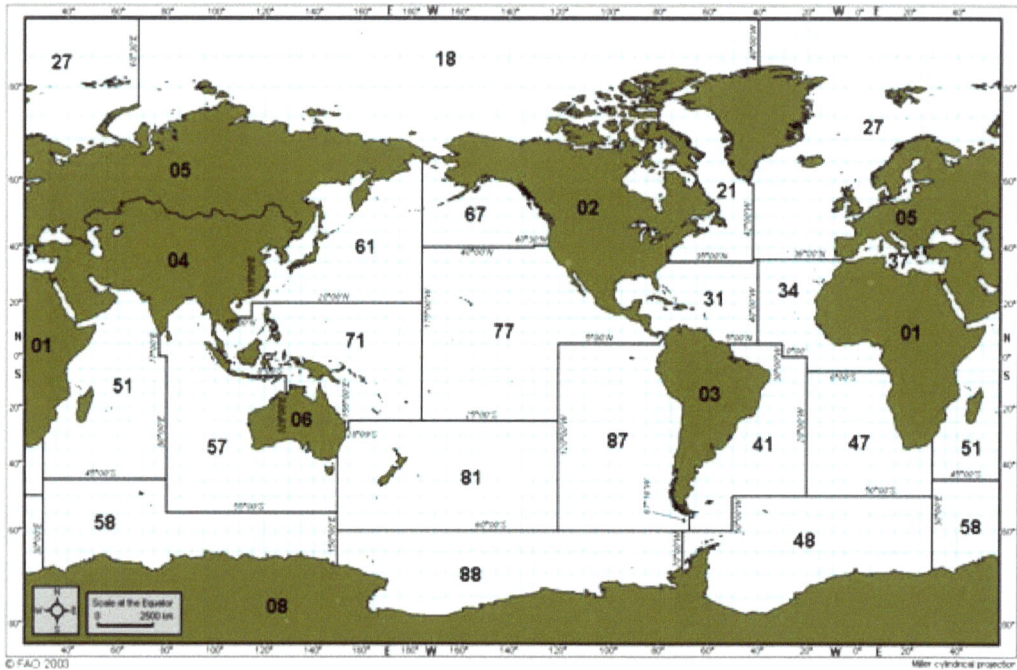

Fig. (1). Fishing areas (ftp://ftp.fao.org/fi/maps/world_2003.gif).

The increasing number of overexploited stocks is responsible for a decrease in production and consequently in catches, with yielding lower than their biological and ecological potential [3]. This is a consequence of substantial demand of fish and fish products and of the significant overcapacity of fishing fleets [4]. For instance at least one-third of the tuna species are estimated to be overexploited [5]. This situation may further deteriorate if a proper management and an effective rebuilding plan in place is not implemented.

Some areas like Northeast Atlantic (Area 27), Northwest Atlantic (Area 21), Western Central Atlantic (Area 31), Southwest Pacific (Area 81), Southeast Atlantic (Area 47), Mediterranean and Black Sea (Area 37) have had higher production rates in the past than now. In some of these areas the decreasing trend in catches can be a reflex of fisheries management measures.

The consequences of overexploitation of marine resources without a proper management plan are very negative not only with regard to ecological aspects and to the volume of fish caught but also with regard to food security, economy and well being of the coastal communities. Many efforts have been made to reduce the overexploitation of fish stocks, in restoring them as well as the marine ecosystems [6].

INTERNATIONAL TRADE AND AGREEMENTS

The Trade of fish results from a balance between the availability of fish in some countries, which is globally very uneven, and the need to meet personal needs, in others countries.

To support the transparency of trade activities there are several international organisms working on standardization of requirements, not only on those related to food safety and animal health principles, but also in those that can prevent the creation of artificial barriers to trade.

The World Trade Organization (WTO) deals with the rules of trade between nations at a global level or near-global (157 member countries). According to the General Agreement on Tariffs and Trade (GATT), WTO assures that international trade rules are developed and approved by the member countries.

Article 20 of the GATT respects to general exceptions to the agreement, and foresees the possibility of governments adopt or enforce measures on imported products in order to protect the life or health of human, animal or plant, provided they do not use it as a form of disguised protectionism or as a form of discrimination.

The WTO has two specific agreements - the Sanitary and Phytosanitary Measures Agreement (SPS) and the Technical Barriers to Trade Agreement (TBT) - to establish the food safety and animal and plant health and safety conditions to use the prerogative. Both agreements identify the compliance of standards and at the same time avoid disguised protectionism.

The SPS agreement tries to guarantee that sanitary requirements are scientifically substantiated. Only when necessary they should be applied in order to protect the life or health of humans, animals or plants and should not be used arbitrarily or unjustifiably as an excuse for protecting domestic producers. The agreement refers to international standards, guidelines and recommendations, when they exist, namely those provided by the FAO/ WHO/ *Codex Alimentarius*, World animal Health Organization (OIE) and International Plant Protection Convention (IPPC) [7, 8].

The TBT agreement tries to guarantee that technical regulations, product standards, testing and certification procedures that may vary from country to country, do not create unnecessary obstacles to producers and exporters or are used as an excuse for protectionism.

The issues covered by the mentioned agreements are becoming more important as tariff barriers in fall. In the light of both specific agreements, if a country applies international standards, it is less likely to be legally challenged in the WTO than if it sets its own standards.

When countries disregard the principles protected by the international agreements, they are likely to be challenged legally, by the injured country, in a WTO dispute.

INSPECTION AND CERTIFICATION SYSTEMS

The official and officially recognized inspection and certification systems are an important basis of the worldwide trade in fish and fishery products.

Fish and fisheries products inspection should be designed to cover all stages from primary production to distribution. For the fish sector, it is recommended the inspection of harvesting areas, first landing of fish in auction halls, the processing, the storage, the transport and the other handling of the product in the retail sale [9].

The inspection systems should be focused on the fish itself to assess their sanitary condition in what respects to diseases caused by bacteria, virus and parasites and also in respect to residues of veterinary drugs, environment contaminants and on their freshness, but also on the process and premises where production, processing and distribution occur along the chain. It cannot be disregarded the inspection of the substances and materials that can be incorporated into or contaminate fish products [10].

In this way, fish inspection and certification systems should be used by governments in the appropriate stages of the commercial chain to ensure that fish and their production systems are according to the requirements in order to protect consumers against food borne hazards and deceptive marketing practices and to facilitate trade.

The Codex *Alimentarius* Commission has developed guidance documents for governments and other interested parties on inspection and certification systems for food import and export. These documents can be very helpful to understand the principles underlying the international trade in food, such as:

- Principles for food import and export inspection and certification (CAC/GL 20-1995).
- Guidelines for food import control systems (CAC/GL 47-2003)
- Guidelines for the design, operation, assessment and accreditation of food import and export inspection and certification systems (CAC/GL 26-1997)
- Guidelines for the development of equivalence agreements regarding food import and export inspection and certification systems (CAC/GL 34-1999)
- Guidelines on the judgment of equivalence of sanitary measures associated with food inspection and certification systems (CAC/GL 53-2003)
- Guidelines for design, production, issuance and use of generic official certificates (CAC/GL 38-2001)
- Principles and guidelines for the exchange of information in food safety emergency situations (CAC/GL 19-1995)
- Guidelines for the exchange of information between countries on rejections of imported food (CAC/GL 25-1997)
- Principles for traceability/product tracing as a tool within a food inspection and certification system (CAC/GL 60-2006)

TECHNICAL EVOLUTION AND TRADE

International fish trade has been growing very rapidly in the last decades. The largest contributor to this improvement was the widespread use of refrigeration, the evolution of the transport capacity and globalization of communications.

As a consequence of the development in the fish preservation and their transportation the share of fish production that enters international trade has increased dramatically [11]. Fish can be traded alive, fresh, frozen, cured or canned and be transported by sea, air or ground. Special care is required for alive, fresh and frozen fish in comparison with cured or canned fish.

The demand for fresh fish in developed countries has had a positive impact on the fisheries production of developing countries, that haven't processing conditions but are exporting high valued fresh fish species to the most demanding markets.

This trading activity is supported mainly by air cargo which is now responsible for the transportation of more than 5% of the world annual catch.

On the other hand, fish transportation by sea or by road is less expensive but is more challenging in terms of the cold chain maintenance, for fresh, chilled and frozen products. It is usually an option for lesser valued species and requires an optimisation of the packing and storage density.

To maintain the cold chain is required the use of proper containers which are usually equipped with mechanical refrigeration and during transportation the air temperature of the containers must be continuously monitored to provide evidence that the cold chain has not been broken [9].

GENERAL AND SPECIFIC HYGIENE REQUIREMENTS

There are different types of fishing vessels that are used throughout the world. They have been adapted and evolved over time taking into account the economic interests, operational conditions, species of fish and shellfish caught or harvested, always aiming to improve the efficiency in all processes, including those related to preserving the quality of the catches. Improvements in selectivity also become a target to achieve.

Some fishing vessels are particularly adapted to a single method of catch or fishing gear, but many of them are versatile and equipped for polyvalent captures.

Nédélec and Prado [12] define and classify the main categories of fishing gear as follows:

 - Surrounding nets (including purse seines);

- Seine nets (including beach seines and boat seines);
- Trawl nets (including bottom trawls, midwater trawls and otter and pair trawls);
- Dredges (including boat dredges and hand dredges);
- Lift nets (portable, boat operated and shore operated lift nets);
- Falling gears (including cast nets);
- Gillnets and entangling nets (including set and drifting gillnets, trammel nets);
- Traps (including pots, stow or bag nets, fixed traps);
- Hooks and lines (including handlines, pole and lines, set or drifting longlines, trolling lines);
- Grappling and wounding gears (including harpoons, spears, arrows, etc.);
- Harvesting gear (pumps, mechanized dredges);
- Miscellaneous.

Due to variability factors within each method, it is difficult to establish correlations between the fish quality and the method used in its capture. However, higher quality is expectable if the fish arrives alive aboard and when excessive fatigue, injuries or pressures are avoided or minimized - frequently, people prefer and value hooked fish than catches using trawl and gillnets.

Regarding the hygiene requirements and the operations carried out on board to the fishery products, the fishing vessels are commonly classified as primary production vessels, freezer vessels and factory vessels. These categories are independent of the size of the vessels or fishing gears and methods used and due to that it is not surprising to observe primary production vessels with 70-80 meters length and much smaller vessels as factory vessels.

PRIMARY PRODUCTION VESSELS

This group is naturally the most representative of each fleet and a wide variety of boats can be observed, regarding fishing methods used and their size, ranging from small boats with 2-3 meters to large vessels with dozens of meters length.

Fishery products, during primary production, are often subjected to different processes after the capture.Sorting, icing/refrigeration, wrapping and storage are usual. However, some complementary preparing operations are frequent, namely gutting/evisceration, bleeding, heading or removing fins. Sometimes "slaughter" takes place on board, by mechanical means or by thermal chock in cooled water.

All the operations must be carried out hygienically to maintain the sanitary quality of the product, avoid contamination and to minimize spoilage by temperature abuse. Summarizing the hygiene requirements applicable to such operations listed below, based on EU legislation, makes perfect sense to say "BE CLEAN, COOL and GENTLE, IF NOT COOL BE FAST" [13].

Sanitary requirements in relation with structures and equipment include:

- Bilge water, fuel, smoke, oil, grease, etc. cannot contaminate products;
- Surfaces in contact with fish must be smooth, easy to clean and non-corrosive;
- Coating surfaces must be durable and non-toxic;
- Equipment to work with fish must be smooth, easy to clean and non-corrosive;
- Water intake must be placed so to avoid contaminated water is taken in the vessel;
- The holds, tanks or containers must be separated from the engine compartments and from the crew quarters;
- The holds, tanks and containers must be adequate to store products at temperatures approaching that of melting ice;
- Containers for ice storage must allow melted water to drain out from products;
- If fishery products are chilled in cooled clean sea water systems they must allow a uniform temperature throughout the tank;
- Temperature in CSW and RSW systems can be monitored;
- The facilities and containers must be clean and in good condition.
Specific hygiene provisions include:
  ○ Products

- Must be protected from contamination and from the sun effect or from other heat sources;
- Clean water is used for washing;
- Handling does not cause bruising on flesh of the products;
- Chilling starts as soon as possible after fish arrival into the boat, or
- Landing takes place as soon as possible after the fish arrival to the boat;
- On CSW and RSW the mixture of fish and clean sea water reach 3 °C in 6 hours after loading and 0 °C after 16 hours;

- In ice cooling systems, the temperatures approaching that of the melting ice is complied with in all parts of the containers;
  ○ Ice
- Must be made of potable water or clean water;
- Must be handled and stored under hygienic conditions;
  ○ Handling on board
- Preparations like heading and gutting are carried out hygienically;
- Only clean water is used for washing the product;
- By-products not intended for human consumption are kept separated from the product;
- The livers and roes for human consumption must be preserved on ice, at a temperature approaching that of the melting ice, or be frozen;
- Crew-staff handling foodstuffs must be healthy and undergone a training on health risks.
- Animal and pest control as far as possible.
- Record keeping (records of the measures put in place to control hazards, small scale fishing can be exempted).
- Recommendations for Guides of Good Hygiene Practice.

FREEZER VESSELS

Due to the modification that the fishery products suffer through the freezing operation such vessels go beyond the primary production and in general, an official approval is required.

In terms of specific structure and equipment, these vessels must have freezing equipment with capacity to rapidly lower the core temperature of the product to a temperature not more than -18 °C and the facilities for the storage of the products must be equipped with a temperature recording device that must be placed where it can be easily read.

The hygiene rules for vessels of primary production must be respected and also those for establishments on land performing similar activity, obviously applying the necessary flexibility.

FACTORY VESSELS

The factory vessels must be equipped and be approved in accordance with the operations of preparation or processing performed on board. In terms of hygiene as well as being fishing vessel is considered a "factory at sea". All the usual hygiene provisions for fishery plants must be followed, including those respecting premises, equipment, production, storage, packaging and labelling, health standards for fishery products, personal hygiene and training. Permanent and suitable procedures based on HACCP principles shall be in place.

Some specific sanitary requirements in relation with structures and equipment should also be met:

- Have a receiving area designed to allow the separation of each successive catch;
- Have a hygienic way to transfer fishery products from the reception area to the working places;
- Working places must be large enough to allow an hygienic processing and preparing (easy to clean, disinfect and avoid contamination);
- Warehouses for finished products must be large enough and conceived to be easily cleaned;
- If a waste-processing unit operates on board it is necessary a separate hold for the waste storage;
- A packaging material warehouse which must be separated from the preparation and processing areas;
- An area and the equipment for the hygienic disposal and/or management of waste/products unfit for human consumption;
- Water collection system located in a position that avoids contamination of the water supply;
- Hand-washing equipment to be used by the staff with taps conceived to prevent spreading of contamination;
- Factory vessels which freeze fishery products must have equipment suitable for freezing.

The following Table **2** sets a comprehensive structure of activities that are usually performed by the different type of vessels.

Table 2. Main operations carried out on board in the different type of vessels; PPV - Primary production vessels, ZV - Freezer vessels, FV - Factory vessels.

Main Operations Carried Out on Board		
Preparing	Bleeding Gutting Heading Fins removal	PPV, ZV, FV
Preparing	Slicing Filleting Skinning Mincing Shucking	FV
Chilling		PPV, ZV, FV
Freezing		ZV, FV
Processing		FV
Wrapping		PPV, ZV, FV
Packaging		ZV, FV
Storage		PPV, ZV, FV

CONFLICT OF INTEREST

The authors confirm that they have no conflict of interest to declare for this publication.

ACKNOWLEDGEMENTS

Declared none.

REFERENCES

[1]   FAO/OIE/WHO (Food and Agriculture Organization/World Organisation for Animal Health/World Health Organisation). One World, One Health: Summary of the FAO/OIE/WHO document 2009 , [2014 Jun 30];2009 Available from: http://www.oie.int/doc/ged/D6296.pdf

[2]   FAO (Food and Agriculture Organization). State of World Fisheries and Aquaculture . Rome, Italy: FAO publication 2012.

[3]   MRAG (Marine Resources Assessment Group). Towards Sustainable Fisheries Management: International Examples of Innovation . London: MRAG Ltd. 2010.

[4]   Hauge KH, Cleeland B, Wilson DC. Fisheries Depletion and Collapse . Geneva: International Risk Governance Council 2009.

[5]   Gilman E, Lundin C. Minimizing Bycatch of Sensitive Species Groups in Marine Capture Fisheries: Lessons from Commercial Tuna Fisheries. In: Q Grafton, R Hillborn, D Squires, M Tait, M Williams, Eds. Handbook of Marine Fisheries Conservation and Management , 2008 [2014 Jun 30]; Available from: http://cmsdata.iucn.org/downloads/minimizing_bycatch_of_sensitive_species.pdf.

[6]   UNEP (United Nations Environment Programme). Marine and Coastal Ecosystems and Human Well-being: A Synthesis Report Based on the Findings of the Millennium Ecosystem Assessment . UNEP 2006; p. 76.

[7]   WTO (World Trade Organization). Agreement on the Application of Sanitary and Phytosanitary Measures . 1995.

[8]   WTO (World Trade Organization). Agreement on Technical Barriers to Trade . 1995.

[9]   FAO (Food and Agriculture Organization). Technical Guidelines for Responsible Fisheries . , Rome: FAO 1998; p. [2014 Jun 30];33. Available from: http://www.fao.org/docrep/003/w9634e/w9634e00.htm#toc_1

[10]   FAO Food and Agriculture Organization). Guidelines for risk-based fish inspection. Rome: FAO 2009; p. 90.

[11]   Johnson T. Fishermen's Direct Marketing Manual Alaska Sea Grant Marine Advisory Program, Oregon, Washington , 2007 [2014 Jun 30];96pp. Available from: http://wsg.washington.edu/communications/online/FishDirectMarMan.pdf

[12]   Nédélec C, Prado J. Definition and classification of fishing gear categories . FAO Fisheries 1990; p. 92. Technical Paper No. 222, Revision 1

[13]   ITC (International Trade Center). Exporting seafood to the UE Bulletin n° 84/2008/Rev , 2008 [2014 Jun 30];35. Available from: http://legacy.intracen.org/tdc/Export%20Quality%20Bulletins/EQM84eng_Rev.1.pdf.

Official Veterinary Inspection of Fish

Fernando Bernardo[*]

CIISA/Faculdade de Medicina Veterinária, Universidade de Lisboa, Avenida da Universidade Técnica, 1300-477 Lisboa, Portugal

Abstract: Fish and shellfish products have been used as food supply since immemorial times. Earliest evidences show that for more than one million years ago, manual capture of gastropods and bivalves bordering the sea shore waterfront have been crucial for survival, development and increasing of human communities. Fish products are a unique source of rich nutrients, being easy to digested, representing a very relevant source of indispensable amino acids, fatty acids, minerals and vitamins for many millions of people worldwide. However, only products obtained from healthy fish and maintained in hygienic environments are suitable for consumption and can present those nutritional advantages. If such products are not handled and or correctly processed, consumer health may be put at risk.

Keywords: Fish products health, Nutrition, Official control, Sea food safety.

INTRODUCTION

Inspection of Fish has been performed for many centuries. Since medieval times in Europe, there were individuals from burgs that were responsible for checking if fish sold in popular markets or streets was fitted for human consumption or were "injured by the sunshine warming". The safety concept applied to fish products came from the Ancient Egypt, where priests declared that some particular fish were unfitted for human consumption, considering that fish without scales were "unclean". In Ancient times, technologies to preserve fish were developed to assure populations fish supply. Phoenicians, along their trips through the

---

[*] **Address correspondence to Fernando Bernardo:** CIISA/Faculdade de Medicina Veterinária, Universidade de Lisboa, Avenida da Universidade Técnica, 1300-477 Lisboa, Portugal; E-mail: fbernardo@fmv.ulisboa.pt

Manuela Oliveira, Fernando Bernardo, Joana I. Robalo (Eds.)

Mediterranean Sea, sold many fish products, including "Oenogarum", "Allec", "Garum", "Muria", "Lymphatum" and "Liquamen", previously processed by "salsamentum". Romans expanded Phoenicians heritage. They would frequently ate fish and transmitted their habits to Western cultures. For many centuries, Europeans ate fish products dried and salted, as the fundamental base of their daily diet. Only after the Industrial Revolution, with the triumph of the Physiocratic doctrine, the red meat substitute fish in the consumption behaviour of occidental populations. In recent decades, the nutritional value of fish based diet has been acknowledged once again. However, modern consumers are more sensitive regarding the diet role in their survival and personal development. Today's consumers are extremely aware of all diet factors that may affect their health, and fish and fishery products are included in such scrutiny [1]. The first principle to consider is the one that states that "it is not possible to obtain fish products fitted for human consumption when the originating fish have some kind of illness". To assure fish health is crucial that all fish should be subjected to inspection by a veterinarian before being placed into market. Globalization of fish trading is a risk factor for microbial and parasitic diseases worldwide dissemination. Therefore, veterinary certification of the fish and fishery products health status is a key issue for public health safeguard. This inspection is especially important when fish is eaten raw, dry, partially salted or smoked. There are many hazards that can be present in fish meat, including: biological hazards (pathogenic bacteria, virus, parasites); chemical hazards (environmental contaminants, biotoxins, biogenic amines, drug residues from aquaculture, additives in processed fish), and physical hazards (spines, hooks, foreign bodies such as sand, mud or soil)

ASSESSING BIOLOGICAL HAZARDS

In some food products, potentially pathogenic bacteria, viruses and parasites may be present. These hazards may be associated with fish and fishery products and its consumption [2].

BACTERIAL HAZARDS

Pathogenic bacteria may cause illness in animals, including fish, humans and plants. They can be found in all environments including the aquatic one, and so, can be present on fish before and after capture [3]. Pathogenic bacteria carried by humans and terrestrial animals can also contaminate the aquatic environment *via* sanitary waters and contaminate fish and shellfish due to deficient hygiene practices before and after capture. Some of the potential pathogenic bacteria found in fish may cause food poisoning, either through toxins release or as a result of bacterial colonization or invasion. Exposure to these food poisoning agents can cause illness, being vomiting and diarrhoea the most common signals. Clinical expression depends on several epidemiological determinants, such as microorganism virulence, infective doses, and the consumers' health status, including the immunity competence, individual predispositions, age and previous contacts with the agent. To trigger infection, the presence and multiplication of pathogenic bacteria in fish is determinant, being necessary to ingest live pathogenic bacteria which have to multiply inside the host. Under these circumstances, fever will also appear. In food poisoning, bacteria do not necessarily need to be ingested alive, being sufficient the presence of previous excreted toxins in food before ingestion. In fish, specific pathogenic bacteria from water can contaminate fish and shellfish, being a potential food safety hazard: *Clostridium botulinum, Listeria monocytogenes, Plesiomonas shigelloides, Aeromonas hydrophila* and *Vibrio* ssp. *Clostridium botulinum,* specifically types B, E and F, can be regularly found in sludges. This specie is anaerobic and psychrotrophic (> +3.3 °C), particularly the E type, which can be found in cold waters. It can produce toxins that are stable to salt or acidic treatments [4]. Toxins are responsible for a major disease, known as botulism, in humans, ruminants, wild ducks and carnivores. Botulism mortality rate is very high and hospitalization cares are crucial for recovering. Poor handling and deficient hygiene practices are risk factors for bacterial contamination and growth [5]; toxin release is increased by thermal shock that occurs during sterilization of canned and vacuum packed fish [6]. Different fishery products have been incriminated in botulism outbreaks. Clinical signals vary with toxin doses and individual vulnerability, from a mild illness with head hake and nausea 30 minutes after ingestion, to a severe comatose disease, often fatal. Typical poisoning clinical signs include visual impairment, mouth and throat sensibility loss, muscular incoordination and breathing difficulties. *Clostridium botulinum* proteolytic types

A, B, C and D are widely distributed in soil and vegetables but are also found in the intestinal tracts of herbivores, carnivores and fish. These biotypes only multiply in anaerobic conditions and are thermotropic. Once again, botulism is caused by the toxins previously released in fish, being a particular problem in inadequate processed canned fish and vacuum- packed fish sliced or smoked products [7, 8]. Toxins affect the nervous system, causing double vision, lachrymal drooping, speech impairment, vomiting, diarrhoea and slowed respiratory movements. Toxins effects have a fatal evolution in 10 to 20 % of the cases, especially if affected people are not treated. Pathogenic strains of halophylic *Vibrio* spp. are found in sea and estuarine waters and, consequently, can be found in fisheries, farmed fish and shellfish captured from those ecological environments [9]. Pathogenic strains are more frequent in warmer sea waters, being found in temperate zones in summer [10]. Consumption of heavily contaminated raw products causes severe diarrhoea leading to rapid dehydration [11]. Sometimes disease becomes severe, particularly in individuals affected by hypersideremia, thalassemia and hepatic disorders. Some "Kanagawa" positive pathogenic strains of *Vibrio parahaemolyticus* occur naturally in sea foods [12]. Humans are contaminated *via* ingestion of raw fish (ceviche, sushi, carpaccio) and also raw crustaceans, gastropods and bivalves (barnacle, oysters, abalones) [13]. Consumption of these contaminated products may cause an intestinal infection, expressing diarrhoea, nausea, vomiting, headache, fever and chills [14]. *Vibrio vulnificus* is very common in warm sea or estuarine waters and has also been incriminated in human illness subsequent to ingestion of raw fish products. *V. vulnificus* food poisoning typical signals include fever, diarrhoea, chills and nausea, especially in individuals belonging to the risk group (hypersideremia). It is also able to cause septicaemia and wound infections [15]. The severe disease caused by *Vibrio cholerae* serotypes O1 and O139 is completely different. This agent is responsible for many pandemic cholera outbreaks [16]. It is an inhabitant of human intestine and low salinity environments, being more common in freshwater and estuarine environments [17, 18]. Infected humans are the main source of sanitary waters contaminations. Urban effluents treatments are the basic key procedure to avoid waters contaminations. Deficient hygiene practices, use of contaminated water and ice, and cross- contamination during processing can increase the risk of fish, fish products and crustacean contamination [19, 20].

Clinically, cholera is responsible for intensive watery diarrhoea that can lead to rapid dehydration and a very high mortality rates. *Plesiomonas shigelloides* may be found in natural temperate freshwater and marine environments [21, 22]. Although it is a Gram-negative bacterium, it can survive in frozen fish and shellfish. Deficient hygienic procedures during food manipulations and cross-contamination from waters or raw products can lead to consumers' contamination [23]. Some *P. shigelloides* food poisoning outbreaks have been described, and related to raw shellfish and fish products. Typical symptoms of the illness caused by *P. shigelloides* are chills, fever and watery diarrhoea. *Aeromonas hydrophila* is a Gram-negative mobile bacterium, present in freshwater, estuarine and marine environments [21]. Although it is an opportunist fish pathogen, it can multiply in fish products with modified atmosphere, and also at low temperatures [24]. Consumption of raw, undercooked or contaminated ready to eat fish products, allowing its multiplication, may lead to food borne disease, promoting headache, fever, abdominal pain and diarrhoea. *Bacillus cereus* is a Gram-positive bacterium, able to produce spores, present in soil, vegetation and natural waters, and associated with food poisoning outbreaks [25]. The anemophilic spores, resistant to drying and warming, are more abundant during summer. It produces toxins that cause sickness, abdominal aching and diarrhea, 30 minutes afterwards the consumption of contaminated undercooked fish and fishery products (dry, smoked, salted). *Listeria monocytogenes* is a Gram-positive small rod, present in the soil, decaying vegetation and vegetables, as well as in humans and other animals' faeces. It contaminates fish and fish products *via* water effluents [26]. During fish processing, deficient handling conditions and poor hygiene practices may lead to cross-contaminations of final products. Smoked fish, sliced fish and ready to eat food are the most relevant *L. monocytogenes* vehicles [27, 28]. For illness to develop, ingestion of viable bacteria must occur. Listeriosis symptoms typically include fever, chills, headache, back ache, abdominal pain and diarrhoea, and subsequent complications include septicaemia, meningitis and encephalitis. Incubation period may be as long as 3 weeks [29]. Symptoms may be most severe in several risk groups, including pregnant women, newborns, elderly and immunocompromised individuals. Mortality rates can range from 20 to 40%. There are many other pathogenic bacteria that may be transmitted by fish and fish products that are not transmitted through the aquatic environments [30]. These include bacteria present in the outer (skin and scalp) and inner surfaces (intestinal and respiratory tracts) of infected humans and animals. These foreigner bacteria colonise fish and fishery products through poor hygienic habits, including deficient hygiene of the staff, processing systems or water [31]. Sometimes, these pathogens may attain aquatic ecosystems as a result of effluents contamination with urban sewage. Nevertheless, they are usually found in fish at low levels of contamination. Highest levels are usually found in bivalves (bio concentration) and in the intestines of predatory fish. Some of these bacteria can seriously impair human health, being responsible for food poisoning (*Staphylococcus aureus* and *Clostridium perfringens* toxins) and intestinal fevers (*Salmonella,* thermophilic *Campylobacter,* some of *Escherichia coli* pathotypes and *Shigella* spp.). *Staphylococcus aureus* is one of the most halotolerant bacteria, inhabitant of the human skin, upper respiratory tract, infected wounds and abscesses. Deficient professional or personal hygiene practices can lead to fish and fish products surfaces contamination, *via* unprotected sneezing, coughing, spit and wounds excretions [32], and also of salted fish [33]. This toxigenic bacterium grows and releases its toxin rapidly at temperatures superior to 8 °C and inferior to 40 °C; once *S. aureus* populations reaches 10^5 cells/g of food matrix, the amount of enterotoxin A is sufficient to cause illness in humans. This toxin is a heat-stable protein, which is not inactivated by heating at 100 °C for minutes. Considering these characteristics, foods at risk are those prepared for periods more than 12 hours, ready to eat or eaten without further cooking, subjected to intensive human manipulation (salads, pastries), not refrigerated in an appropriate time. People exposed to *S. aureus* contaminated food, including products from low concentrated bines (cod, anchovies and smoked fish), rapidly manifest severe nausea, abdominal cramps, vomiting and diarrhoea and exhaustion (without fever). *Salmonella sp.* is an intestinal resident bacterium of homoeothermic animals and some reptiles. Some serotypes, such as *Salmonella* Typhi and S. Paratyphi A, are strictly adapted to human intestine. It is frequently responsible for human food borne diseases: annually, about 100 thousand cases occur in the European Union [34]. Disseminated from the intestine, Salmonella is released in water environments through sewage or sanitary water, *via* faecal-oral route [35]. Wild birds may also play a role in fish contamination, especially when fish are discharged from vessels at fishery ports. Inappropriate hygienic handling and bad practices lead to fish, shellfish and fish products containers contamination [36, 37]. Contaminated raw materials and their manipulation during food processing or cooking, may allow direct or cross-contamination of final products. Consumption of contaminated raw shellfish and fish may conduce to Salmonellosis, after 8 to 24 hours. First signs of disease are abdominal pain, fever, headache nausea, chills and mucous diarrhoea. Clinical evolution depends on many risk factors, such as strain virulence, serotype (*Salmonella* Typhi is responsible for severe typhoid fever) and individual vulnerability, being observed that infants, elderly and immunocompromised are more susceptible. Another disease that may be associated to fish products and shellfish consumption is "aestival diarrhoea". This foodborne illness is an intestinal infection caused by different *Shigella* spp. Biotypes, including *Shigella dysenteriae, S. sonney, S. boydii* and *S. flexneri* [38]. These pathogenic bacteria are found in primates' intestines and are transmitted to fish and fish products through the faecal-oral cycle, sanitary waters or deficient personal hygiene practices of food manipulators. Ingestion of raw or undercooked fish products may be responsible for Shigellosis development, characterised by abdominal pain, chills, fever, vomiting, mucous diarrhoea, sometimes bloody. Infants are particularly vulnerable. Shigellosis is a typical intestinal fever in countries where the human sewages are not conveniently treated. Direct or cross-contaminations of fish products involving water environments and aquatic animals used for food, are also responsible by human intestinal infections identified as colibacillosis (*Escherichia coli*) and campylobacterioses (*Campylobacter* spp.), although less frequently [39]. *Escherichia coli* is a commensal bacteria of animal intestines, including human. Effluents of human or animal sewages disseminate this agent to aquatic environments. Some strains can cause human illness with different levels of severity. It accumulates in bivalve living in contaminated aquatic environments [40]. Foodborne disease arises as a result of raw or undercooked products consumption, including oysters, barnacle and raw fish. Most common clinical signs are abdominal pain, nausea and watery, white pasty and, sometimes, bloody diarrhoea. Particular verotoxigenic (VTEC) and enterotoxigenic (ETEC) strains are especially aggressive to infants and, sometimes, fatal, inducing haemolytic and uremic syndromes.

Campylobacter jejuni, C. coli, C. lari and *C. upsaliensis* are vibrio shaped bacteria, inhabitant of homoeothermic animals' intestinal tracts, reaching aquatic environments through sewages. They contaminate shellfish and fish products *via* cross-contamination and the faecal-oral route. Foodborne disease may occur as a consequence of contaminated raw or undercooked fish and shellfish ingestion. First disease signs arise 12 to 36 hours after contaminated food ingestion and include headache, abdominal pain, nausea, fever, muscle pain (flu-like syndrome) and diarrhoea. Campylobacteriose is a mild foodborne disease, being children the most vulnerable group. At present, this foodborne disease is the most frequently reported of all bacterial foodborne diseases in developed societies, although fish products are not reported as a relevant vehicle for this agent [34].

VIRAL HAZARDS

Viruses are not able to replicate in food matrices, but they can survive for some time in fish and fishery products, being concentrated in bivalves [41]. Most of the viruses having nosologic relevance in the fish or shellfish chains come from humans [42]. Fish, but specially shellfish and bivalves, captured from contaminated aquatic environments or contaminated post harvesting, and ingested raw or undercooked, are hazardous products [43, 44, 45, 46]. Faecal contamination of aquatic environments with urban effluents is mainly responsible for fish products contamination with enteric human viruses. Survival of those microorganisms in waters depends on many ecological factors, including water transparency, UV radiations, salinity and organic pollution levels. They can enter the food chain through the ice used as preservative of fish, or *via* faecal-oral route. Most common viral illnesses transmitted by raw fish products or shellfish are hepatitis A, Norwalk virus (Norovirus), Calicivirus, Astrovirus and Rotavirus. Symptoms of hepatitis A arise develop days after contaminated food ingestion, and include abdominal discomfort, fever, malaise, nausea, acholic faeces and jaundice. Clinical evolution is very slow, with sequels [47]. In 1988, nearly 300 000 people developed hepatitis A in China, after ingestion of contaminated clams collected from a sewage polluted area. This is the largest recorded incident of food-borne viral infection reported [48]. Norovirus, Rotavirus, Calicivirus and Astrovirus infections evolve as a typical gastroenteritis expressing clinical signs such as abdominal cramps, fever, nausea, vomiting and yellowish and pasty diarrhoea. Infants are most at risk; the most relevant determinants for those infections are poverty, lack of basic health cares, including waters and sewage treatments. Human cases of water-borne virus infections are dramatically increasing worldwide, and millions of cases are reported each year. Probably, this is a consequence of many ecological transformations, including urban concentration, aquatic environment degradation, decrease of potable water supply and new consumption behaviours. The risk management of seafood-borne viruses demands high technological capabilities.

PARASITIC HAZARDS

As previously described, there are many parasites that can affect fish health. Most of them have high economic impacts, but about 50 species also represent a threat to human health [49]. Human cases of illness due to fish parasites are regularly reported worldwide [50]. Foodborne parasitic diseases arise from the ingestion of raw, minimally processed, brine salted, or inadequately cooked fish and fish products, containing cyst, larvae or any other infectious stage of zoonotic parasites [51]. Some human pathogenic Protozoa, including *Criptosporidium, Giardia* and *Entamoeba,* may also be transmitted by fish products, especially shellfish eaten raw or without proper thermal treatment [52, 53]. Pathogenic Protozoa are typical transmitted *via* faecal-oral cycle [54]. There are three distinct types of zoonotic parasites that may be found in fish: Nematodes, Cestodes and Trematodes [55]. Most of them have very complex life cycles, passing through a large number of intermediate hosts, such as aquatic mammals or wild birds, crustaceous and gastropods [56]. Not all fish zoonotic nematodes have a worldwide distribution; however, global trade is a risk factor for disseminating fish parasites all over the world. Among the most common nematodes related to food safety are *Anisakis* spp., *Pseudoteranova* spp. and *Contracaecum* spp., that can be present in the abdominal cavity, liver capsule, ova and muscular tissues.

Anisakiadae are the most frequent zoonotic fish parasites, having a worldwide distribution. Human illness is caused through the ingestion of raw or undercooked fish or cephalopods hosting viable larvae (L3) of *Anisakis simplex, A. pegreffii* or *Pseudoterranova decipiens*. Every year, approximately 20 thousand cases of human anisakiasis are reported worldwide. Over 90% of the cases occur in Japan, and the others in Spain, Germany and the Netherlands. Occurrence depends on fish consuming habits [57]. Viable larvae can invade the digestive tract or, occasionally, other organs, causing local inflammatory lesions, haemorrhage, peritonitis and local granulomatous masses. Infection persistency induces allergic reactions, such as skin rash, angioedema and anaphylaxis. Like other parasitic infestations, Anisakidae larva induces an immune adaptive response increasing IgE production, expressing eosinophilia and mastocytosis. It is generally assumed that a previous infection is necessary to induce allergic sensitivity to *Anisakidae*. Effective treatment requires the removal of live larvae by endoscopy or surgery. The best way to prevent anisakiasis is to avoid the ingestion of raw or inadequately thermally treated marine fish or cephalopods [58]. Official authorities may also impose to freeze fish products before raw consumption, promoting larvae inactivation. Gnathostoma nematodes are found in fish from Africa, Asia and Latin America. Reaching humans, parasites may migrate to the skin, causing creeping eruption; they can also migrate to the eye or other internal organs causing serious injuries. *Capillaria* spp. are also nematodes, being a major public health concern in some countries, such as Thailand. Infection causes severe diarrhoea and sometimes death due to dehydration, especially in children. *Angiostrongylus* spp. are common nematodes in fish native from Southeast Asia. Some species are able to migrate to the human brain, causing meningitis, or can remain in the abdomen where they induce serious intestinal disorders. Few tapeworms (Cestoda) are associated with fish consumption. The most problematic specie is Diphyllobothrium latum, which occurs worldwide. Fresh and marine fish act as intermediate hosts. Symptoms are mild but may include abdominal distension and cramps, flatulence and intermittent diarrhoea [59].

Trematode infestations are a main public health issue, occurring frequently in some regions [60], being a frequent problem in Asia. Most important flatworm species belong to *Clonorchis, Ophisthorchis, Paragonimus, Heterophyes* and *Echinochasmus* [61]. *Clonorchis sinensis* is described as having infected 20 million people in Asia, being endemic in Southeast Asia: Korea, China, Taiwan and Vietnam [62]. About 80 species of freshwater fish are referred as carriers of this parasite [63]. To avoid risks transmitted by fish and shellfish consumption, adequate controls measures must be adopted. During fish products production, capture, processing preservation, it is absolutely crucial that food operators are aware of these risks, as they have a social responsibility to control risk factors associated to fish products. Biological hazards from the faecal-oral route may be easily reduced, by: preventing contamination of waters used for fish production with sewage; monitoring the microbiological characteristics of waters used for aquaculture or aquatic farms; monitoring capture fish or shellfish pre and post-harvesting operations; applying Good Hygienic Practices (GHP) to prevent cross-contamination along the food chain, including processing, storage, transportation and cooking; and ensuring good sanitation and professional hygiene of fish manipulators. Proper cooking (thermal processing at 85-90 °C for few minutes) will inactive viruses, bacteria, protozoa and parasites in fish and shellfish.

CHEMICAL HAZARDS

The fish food chain may also be affected by the presence of exogenous harmful chemical compounds (xenobiotics), able to originate adverse effects in humans [64]. There are different categories of xenobiotics, considering their origin: drug residues used in aquaculture; banned substances; contaminants; not desirable substances and compounds generated or added by processing [65]. Some natural fish intrinsic compounds may also be harmful, if able to induce abnormal immune-responses in fish consumers (allergens). In the last four decades, aquaculture has been one of most successful animal production technologies, reaching systematically a growth rate of 1%/year. To achieve this rate, technical and scientific work has been developed, concerning aspects related with the health management of farmed fish or shellfish. Due to the very high concentration of individuals present in aquaculture systems, health problems are dramatically amplified, with major sanitary and economic impacts. To prevent and combat these fish health problems, there is an inexorable need of veterinary medicines use for fish therapy and prevention [66], which can generate residues related problems [67]. In fact, the application of biocides and veterinary drugs in aquaculture may generate residues of medicines, comprising a large number of substances, being relevant chemical hazards [68]. Residues in fish above the authorized limits may enhance adverse effects in humans, including sensitizations, anaemia and antimicrobial resistances cross-transmission [69, 70]. Antimicrobial drugs are generally applied in aquaculture to cure fish disease due to a wide range of pathogenic bacteria, previously mentioned. They are usually administrated orally, being available a broad range of antimicrobial drugs for aquaculture [71] (Table 1). However, it is important to follow manufacturers guidelines regarding dosage and withdrawal periods. The use of antimicrobials in aquaculture is largely regulated worldwide, for global trade proposes [72]. All antimicrobial used as medicines need to be previously approved, its maximum residues levels (MRL) must be determined and all fish administration needs to be registered [73]. Antibiotics prudent use policies are a key issue. Antimicrobials use requires a veterinarian's prescription, and hence, their use is therapeutic [74].

Table 1. Antibiotics used in aquaculture and maximum residues level (MRL) in fish muscle (Adapted from FAO and UE).

"Families"	Compound	Use / matrices	MRL
Sulphonamides	Sulphamerazine	Furunculosis treatment in *Salmonidae*	100 pg/ kg
	Sulphaimidine	Furunculosis treatment in *Salmonidae*	100 pg/ kg
	Sulfadimethoxine	Furunculosis treatment in *Salmonidae*	100 pg/ kg
Potentiated Sulphonamides	Trimethoprim+ sulfadiazine	Furunculosis, vibriosis and Enteric Red Mouth treatment	50 pg/ kg
Tetracyclines	Chlortetracycline	Against fish, shrimp and lobsters pathogens	100 pg/ kg
	Oxytetracycline	Against fish, shrimp and lobsters pathogens	100 pg/kg
Penicillins (Beta-lactams)	Ampicillin	Furunculosis in salmon and rainbow trout fry syndrome (RTFS)	50 pg/kg
	Amoxycillin	All fish and shellfish bacterial diseases	50 pg/kg
	Oxacillin	All fish and shellfish bacterial diseases	300 pg/kg
	Benzyl penicillin	In yellowtail and sea bream in Japan	50 pg/kg
Quinolones	Ciprofloxacin	In shrimp farms in Asia	100 pg/kg
	Enrofloxacin	In shrimp farms in Asia	100 pg/kg
	Norfloxacin	In shrimp farms in Asia	100 pg/kg
	Oxolinic acid	Worldwide for fish diseases	100 pg/kg
	Cloxacillin	Worldwide for fish bacterial diseases	300 pg/kg
	Difloxacin	Worldwide for fish bacterial diseases	300 pg/kg
	Flumequine	Worldwide for fish diseases	500 µg/kg
	Nalidixic acid	To treat furunculosis	300 pg/kg
	Sarafloxacin	Furunculosis treatment in *Salmonidae*	30 pg / kg

(Table 1) contd.....

"Families"	Compound	Use / matrices	MRL
Macrolides	Erythromycin	All fish and shellfish bacterial diseases	200 pg/kg
	Spiramycin	All fish and shellfish bacterial diseases	300 pg/kg
	Tylosine	All fish and shellfish bacterial diseases	100 pg/kg
	Tilmicosine	All fish and shellfish bacterial diseases	50 pg/kg
Aminoglycosides	Gentamicin	All fish and shellfish bacterial diseases	500 pg/kg
	Paromomycin	All fish and shellfish bacterial diseases	500 pg/kg
	Neomycine	All fish and shellfish bacterial diseases	500 pg/kg
Other	Florfenicol	To treat RTFS and furunculosis	1000 pg/kg
	Thiamephenicol	All fish and shellfish bacterial diseases	50 pg/kg
	Tiamulin	All fish and shellfish bacterial diseases	50 pg/kg

Fish producers must respect dosage levels and strictly adhere to withdrawal periods for the approved antibiotics. Antimicrobial drugs use to promote fish growth rate is not allowed, since data indicate that antibiotics may actually slow fish growth. Before approval, drugs are assessed for the definition of their maximum residue limits (MRL) (Table **1**), and their environmental impact and efficacy [75, 76]. Aquaculture products having residues above legal limits are not allowed to be placed in market or be exported. MRL are generated by several regulatory bodies, such as the EU and the FAO/WHO Codex Alimentarius Commission, scientifically advised by JECFA (Joint FAO/WHO Expert Committee on Food Additives). The use of antibacterial agents in food animal species, including fish, is controlled, particularly in developed societies [77]. Hazards concerning antibiotic deposits in comestible aquaculture fish and crustaceans comprise allergies, toxicity, and deviations in human intestine microbiota and environmental disbiosis [78], besides the emergent problem of antimicrobial resistance increase in fish and water bacteria [79]. Other problems are related with the use of drugs that are banned from animal farming systems, including aquaculture [80], such as nitrofurans (AOZ- furazolidone; AMOZ - furaltadone; SEM-nitrofurazone), malachite green [81, 82], chloramphenicol and hormones [83]. Presence of these banned drugs in aquaculture fish and crustacean is frequently reported [84]. There are no admissible limits in aquatic animals for these banned drugs; however, some metabolites may occur in nature [85, 86]. Drugs are not the only chemical hazards present in fish; there are chemical molecules that are released in the environment and may be concentrated in the trophic chains, from which aquatic animals are relevant components. These environmental pollutants (contaminants) are frequently reported in wild fish and shellfish and can be introduced in aquaculture through feed and water [87]. Most of these contaminants are generated by industrial activities, especially if adequate management programs of dangerous wastes are not applied. During the last decades, some hazardous contaminants, like heavy metals, pesticides, polychlorinated biphenyls (PCBs), poly-chlorinated dibenzo-p-dioxins (PCDD), polychlorinated dibenzofurans (PCDF) and polycyclic aromatic hydrocarbon (PAH) have been found in wild and farmed fish and crustacean, representing a major concern to food risk managers [80].

Only a limited number of these compounds can be found at high concentrations in aquatic animals used for food, enough to pose a threat to human health [89]. Reported human poisoning has been related with accidental exposure, but the real probability of intoxication occurrence is extremely small. However, chronic contact with low concentrations of some of these chemicals has been related with severe illness, including neurological and digestive tract injuries, nephro and teratogenic effects and several cancer types. Excluding spontaneous fires, almost all problems related to environment chemical pollution are generated by humans.

Aquatic ecosystems contamination may originate from industry effluents; sludges from water treatment plants; draining of agricultural pesticides; and city sewages [90]. Once discharged in water, chemicals reach all organisms that live in aquatic ecosystems, including fishes. Cumulative quantities of these contaminants can be found in tissues of predators due to biomagnification, responsible for the accumulation of these molecules in the food chains superior trophic levels. Bioaccumulation may also occur, once higher quantities of these compounds accumulate in the body of individual during their lifetime. In this case, the largest and older fish will have a higher chemical content.

The occurrence of chemicals in fish depends upon their geographical position, feeding patterns, species, and chemicals characteristics, including solubility and environmental persistence.

Heavy metals related to fish and shellfish products and having health impacts include methyl mercury, lead, cadmium, cooper, alumina, tin, arsenic and selenium. Most international regulations established maximum admissible levels based on scientific risk assessment, and considering human patterns of fish products consumption. Maximum admissible limits are quite variable, from 0.02 mg/kg to 200 mg/kg (tin). Low concentrations of metals are omnipresent in these ecosystems. Some metals, including copper, zinc, iron and selenium, remain fundamental for aquatic animals' nutrition. To be considered as contaminants, a statistically significant increase must occur in the average physiological levels. Many compounds groups, with a wide range of industrial use, have been reported in aquatic products and correlated with human health problems: PCBs, PCDDs, PCDFs, pesticides (chlorinated hydrocarbons), PAHs, in a total of several hundreds of chemical species (congeners), having a potential cumulative effect. One of the reasons for higher concern arises from the fact that these compounds persist long time in the environment, allowing accumulation. Generally, levels of organochloride compounds found in fish and shellfish are low and, most times, are present below levels that may be a risk for humans. Nevertheless, they can be considered a hazard for societies having seafood as the base of their nutrition regimen, especially to risk groups that ingest high amounts of oily fish, including children. The high frequency of human exposure, even to low levels of contaminants, is not comprehensively assessed. Some significant risks comprise reproductive impairment due to PCBs and methyl mercury and increased cancer risk due to PCB related compounds, dioxins and chlorinated hydrocarbon pesticides.

Quantitative risk assessment procedures must be upgraded to include non-carcinogenic consequences, such as developmental problems, acne and immune system damage. Control and declining of human exposition can be achieved through actions aiming at the font, including reducing industrial smoke emissions and environmental release of dioxins. Another strategic exposure control is the application of systematic surveillance plans, monitoring the presence of contaminants in edible seafood. International, regional and national regulations establish contaminants maximum admissible levels in multiple food matrices, including fish products (Table **2**). In the absence of harmonized international regulations, contaminants presence in fish products may block the global market, unless the exports operator is able to assure that products comply with levels allowed in the destination market.

Table 2. Critical limits of environmental chemical contaminants in fish and fish products (Adapted from 51, 91); Legend: (TEF) - Toxic Equivalency Factors; PAHs - Polycyclic Aromatic Hydrocarbons.

Substances	Maximum levels (mg /Kg)		Matrices
	US	EU (mg/kg wet weight)	
Arsenic	76-86	-	molluscs, crustaceans
Cadmium	3-4	0.05 - 1.0	fish, molluscs
Lead	1.5-1.7	0.2 - 1.0	fish, molluscs
Methyl mercury	1.0	0,5 -1.0	all fish
PCB and PCDF (TEF)	2.0	0.0000001	all fish
Diedrin	0.0	0.00001	all fish
Dioxin (TEF)	1.75 - 3.5	0.000000004	all fish
PAHs	-	0.002-0.005	all fish and shellfish

There are other classes of chemical hazards that may be found in fish and shellfish products, known as "undesirable substances". Among this group, the most relevant for Public Health are natural biotoxines and biogenic amines (histamine). One of the most relevant health problems associated with the presence of biotoxins in fish is "Ciguatera". Intoxication is endemic in some geographical areas, especially in the Pacific and Indian Oceans and the Caribbean [92]. The hazardous toxins, ciguatoxins (CTX), accumulate through the food chain, from small phytoplankton including microalgae and the dinoflagellate *Gambierdiscus toxicus,* to bigger predatory fish, feeding on them [93, 94, 95]. At the moment, ciguatera poisoning is the most frequent sea food intoxication, having an annual occurrence of 50000 people worldwide; it constitutes a global health problem [96]. Ciguatoxins are a complex of lipophilic chemical compounds (CTX-1, CTX-2 and CTX-3) that are rapid and efficiently absorbed from the intestine, although vomiting and diarrhoea onset may occur before toxins are absorbed. After absorption, neurological symptoms may appear in mild intoxications, including headache, dizziness, anxiety, nervousness, tremors, inverse temperature perception and sensation of "electric currents", nightmares, hallucinations, agitation, muscle weakness and paraesthesia. More severe intoxications symptoms include myosis, mydriasis, seizures, body rigidity, respiratory difficulties, pulmonary edema, cardiovascular failures, bradycardia, gingiva hemorrhages and coma. All exposed people exhibit clinical signals and human mortality rates are also high, depending on exposure doses and medical assistance.

Ciguatoxins may also be transmitted through the skin and mucous membranes, through breast milk and through the placenta [94]; cases of sexual transmission are also referred [96]. Worldwide, many fish, mollusc and crustacean species, belonging to different families, have been incriminated in ciguatera intoxication, including *Acanthuridae*, *Scaridae*, *Muraenidae*, *Lutjanidae*, *Serranidae*, *Epinephelidae*, *Lethrinidae*, *Scombridae*, *Carangidae* and *Sphyraenidae* [94, 96] (Table **3**).

Table 3. Commercial fish species more frequently associated with ciguatera (Adapted from [95]).

Fish species	Distribution
Lined surgeonfish (*Acanthurus linearis*)	Indo-Pacific
Bonefish (*Albula vulpes*)	Worldwide in warm seas
Gray triggerfish (*Balistes carolinensis*)	Atlantic, Gulf of Mexico
Gaucereye porgy (*Calamus calamus*)	Western Atlantic
Horse-eye jack (*Caranx latus*)	Atlantic
Whitetip shark (*Carcharinus longimanus*)	Worldwide
Humphead wrasse (*Cheilinus undulatus*)	Indo-Pacific
Heavybeak parrotfish (*Chlorurs gibbus*)	Indo-Pacific
Red groupper (*Epinephelus morio*)	Western-Atlantic
Giant moray (*Gymnothorax javanicus*)	Indo-Pacific
Hogfish (*Lachnolaimus maximus*)	Western Atlantic
Northern red snapper (*Lutjanus campechanus*)	Western Atlantic, Gulf of Mexico
Tarpon (*Megalops atlanticus*)	Eastern Atlantic
Narrowhead gray mullet (*Mugil capurri*)	East CentralAtlantic
Yellowtail snapper (*Ocyurus chrysurus*)	Western Atlantic
Spotted coralgrouper (*Plectropomus maculatus*)	Western Pacific
Blue parrotfish (*Sparus coeruleus*)	Western Atlantic
Spanish mackerel (*Scomberomorus maculatus*)	Western Atlantic

(Table 3) contd.....

Fish species	Distribution
Lesser amberjack (*Seriola fasciata*)	Western Atlantic
Great barracuda (*Sphyraena barracuda*)	Indo-Pacific, Western Atlantic
Chinamanfish (*Symphorus nematophorus*)	Western Pacific
Swordfish (*Xiphias gladius*)	Atlantic, Indo-Pacific, Mediterranean

Common thermal treatments apply to contaminated fish do not inactivate ciguatoxins, as well as basic or acidic treatments [94, 97]. Reactive strategies based on laboratorial detection and quantification of CTX in fish muscles or internal organs are available, but they are not practicable or efficient, due to the heterogeneity of fish species and fisheries batches. Assuming these limitations, the precautionary principle must be applied, namely to ban from the commerce the fish species that present a higher risk. Some regional authorities, including the European, ordered the interdiction to commercialize fishery products obtained from toxic fishes belonging to the *Tetraodontidae, Molidae, Diodontidae* and *Canthigasteridae* families. However, in other geographical zones, other fish species are usually contaminated with ciguatoxin. Monitoring waters and the detection of *Gambierdiscus toxicus* may be assumed as a significant indicator, associated with ciguatera [96].

Another important intoxication due to the presence of non-desirable substances in fish is the so called "scombroide poisoning" [98]. It is caused by the scombrotoxin (histamine), a biogenic amine formed by bacteria (mesophilic *Vibrionaceae, Lactobacillus* and some *Enterobactereaceae*), which are able to metabolize large amounts of histidine, as they are equipped with specific decarboxylases [99, 100]. In most cases, histamine levels necessary to promote illness have to be above 200 mg/kg or even above 500 mg/kg.

Histamine is not destroyed by cooking [101]. "Scombroid fish poisoning" is always related with the consumption of specific fish, belonging to *Scombridae, Coryphaenidae, Scomberesocidae, Clupeidae* and *Engraulidae* families [100]. Clinical signs of this fish poisoning resemble those of food allergies, including sickness, diarrhea, rash, itching and hypotension. In more severe poisoning, generalized urticaria, facial and body erythema, palpitations, tachycardic, tachypnoeic and anaphylaxis may be expressed. Individual sensitivity is also relevant [102]. Symptoms onset commonly take place after approximatly15 minutes after consumption, and they can last up to 24 hours. Antihistamines may be administered for intoxication treatment [97]. To avoid or fully control this chemical hazard, critical limits should be acknowledged [103]. Fish with histamine concentrations exceeding 100 mg per 100 g are generally considered to be hazardous. Some regional and international legislation state that above that level, fish are considered "not fited" for human consumption, and are not allowed to be commercialized. There are some exceptions, especially for salted anchovies, in which the admitted level is double. This analytical control must be performed by owners and supervised by official health authorities. As referred above, problematic fish are those belonging to five families, with a large lateral "red muscle", such as mackerel, tuna, marlin, swordfish, albacore, bonito, skipjack, amberjack, dolphinfish, herring, sardine, anchovy and bluefish. On these, histamine formation can be prevented by proper handling and refrigerated storage, insuring that histamine-producing bacteria are not present or that they are not able to multiply, as producing bacteria are usually mesophilic [104, 105, 106]. Sensorial examination of fish, previous to commercial distribution and utilization, are quite accurate on detection of fish decomposition. Other chemical compounds that are generated or technological added during fish products transformation, have been referred as having a potential adverse effect to human health. These are the cases of benzopyrenes added through fish smoking procedures, acrylamide in fried fish and shrimps, free radicals generated from fish lipids oxidation, heterocyclic amines, "advanced glycation end products" and most of the food additives, such as sulphites, methabi-sulphites and nitrites, which have incorporation limits in fish products. Most risks associated with these substances are still not fully established. Some regional legislation enforces maximum levels for benzophyrens formed during smoking fish technology applied to eels, salmon, *Scombrideae* and *Clupeideae*. In Europe, critical levels for smoked fish and respective by-products are stated at a maximum of 5 µg/kg [91]. Also, the list of approved food additives is not worldwide unanimous. For example, "Borax" (E285) continues to be used in "caviar" production, but most occidental countries have banned this technological additive. Another problem related with fish chemical hazards include sensitization to specific protein compounds of fish and shellfish.

Allergy episodes after ingestion of a peculiar food are frequent in occidental societies, and fish has been associated with some of these events. Food allergies severity is quite variable, but sudden fatalities have been registered, justifying the need for a very accurate labelling of processed food. Fish from the *Gempylidae* family, including escolar or snake mackerels, have been referred as able to cause intolerances in some individuals, due to gempylotoxin. This metabolite has a laxative effect, able to cause dramatic, short- lived gastrointestinal responses in some people. European legislation establishes the need of unequivocal labelling of food and fish oils that use these fish as ingredients.

PHYSICAL HAZARDS

Foreign bodies found in fish used for consumption may cause injuries in fish manipulators and consumers. Also, specific radiations produced by fish contamination with radioisotopes may be potentially harmful. Major nuclear disasters, like Chernobyl (Ukraine, 1986) or Fukushima (Japan, 2011) may have extremely serious food safety consequences due to environmental contamination with radioisotopes, including aquatic ecosystems contamination. Cesium-137 (137Cs55) is the most relevant radioactive isotope formed by uranium- 235 nuclear fission. This isotope has a half-life of about 30 years and human exposure increase the risk of thyroid cancer and leukaemia. Depending on the extension and volume of radioactive isotopes released in the natural aquatic environments, different consequences may be expected after fishery products consumption. Fish captured in the Japan Sea after Fukushima disaster considerably exceed the Japanese legal limit of 500 Becquerel per kilogram at that time. Japanese authorities had detected a record of 740000 Becquerel per kilogram of radioactive cesium in fish caught near to that plant, 7400 times the government limit established for safe human consumption. To adequately control this physical hazard, wild fish monitoring for radioisotopes is crucial and absolute captures interdiction and prohibition to commerce fish and shellfish coming from contaminated areas are the only efficient tools. Nevertheless, it must be noticed that marine currents and fish migration may spread radioisotopes to a very far away distances, namely from Pacific Ocean to Antarctic, Indic ocean and even to the Atlantic. Therefore, sporadic monitoring of wild fish for radioisotopes is a good practice for fish safety assurance. Other physical hazards may also occur in fish products, although with less relevance, including bones, thorns, small hooks, fin rays, spines, solid fragments of glass and metal and stones, amongst others. These can be responsible for obstructions and injuries in the entire gastrointestinal system. Foreign objects smaller than 7 mm are hardly responsible for severe injuries, with the exception for risk groups, which include children and seniors [51].

Though these hazards seldom promote severe damages, they are frequently object of consumer protests, since damages occur during or immediately after consumption, and the accident cause is promptly identified. Control measures for physical hazards have been implemented in fish plants, and include frequently monitoring equipment for damages or misplaced parts and subjecting fish to metal detection, separation equipment or X-ray detector. In developed societies, official veterinary and environmental authorities are aware of all these hazards (biological, chemical and physical) and associated risks, and developed risk management systems aiming to public health assurance. Without an appropriate knowledge of food safety hazards it will be impossible for fish inspectors to develop and implement an effective preventive risk-based inspection system.

FISH AND FISHERY PRODUCTS

Fish and shellfish are important nutrient sources. Several thousands of oceanic, freshwater and aquaculture species are used for consumption worldwide. Fish and fish trade are important for countries economy. On a global scale, capture fisheries and aquaculture supply approximately 160 million tonnes of fish and shellfish for human consumption, an equivalent of approximately 19 kg per capita. While capture fisheries have remained relatively stable since 1999 (95 million tonnes), aquaculture production of fish and shellfish has consistently grown 6% per year. Aquaculture contribution to the world total fish production is 40.1% [107]. Fish and shellfish are important sources of easily digested protein for many millions of people, being the main or only source of affordable animal protein in some regions. Fish and shellfish protein composition varies from 13% to 25%, and includes essential amino acids. Some fish species are valuable sources of mono and polyunsaturated fatty acids, particularly omega 3, which is known to reduce the risk of cardiovascular disease and to support brain and nervous system development. Fish is also an important source of vitamins A, B12, D and E and of vestigial elements (phosphorus, potassium, iodine, fluoride, molybdenum, selenium). Fish and shellfish fat content varies from 0.1% to 16%, depending on species, season and life cycle stage. Fish flesh chemical composition also varies according to species, season, maturity status, fishing ground and feed availability. Fish products trade is extremely important for developing countries economy. Fish and fish products international trade is predicted to grow along with the expansion of world economy, food trade liberalization, growing consumer demand, food science and technology developments, and transport and communication advances. In 2012, overall value of worldwide fish exports was higher than 100 billion dollars. Globally, workers from fisheries primary production, including aquaculture, are probably around 40 million, 87% of which can be found in Asia. It is also estimated that about 20 million individuals work in parallel activities, including processing, marketing and trading. Considering the fisherman, the rest of the workers and their families, it may be estimated that globally, about 200 million individuals are sustained by small fisheries, 50% of them working on post-harvest. These include fisherman, fish farmers, processors, traders, transporters, fish carriers and janitors, who are involved in fish and fishery products handling and distribution. All resources, benefits and incomes pending on the fishery sector are difficult to sustain, since fish and shellfish are highly perishable, and once harvested and placed on the market, they will rapidly spoil and become unfit, losing their nutritional, economic and social values. Only fishery products obtained from healthy fish and shellfish handled using appropriated hygienic conditions are eligible for placing on the market [108].

GENERAL PRINCIPLES OF FISH AND FISHERY PRODUCTS OFFICIAL CONTROL

In developed societies, when fish consumers' welfare is impaired, the operator that supplied the incriminated product can be prosecuted, suffering administrative sanctions and economic losses. Fish industries reputation may be shattered if illness cases proved to be transmitted by their fish products. The first social obligation of any food chain operator is to assure the safety of its products, since without safety guaranties there is no food, only by-product. On another hand, the first obligation of official authorities and regulatory organisms is to verify if the economic food operators comply all legal requirements concerning procedures hygiene and safety assurance [109]. The absence of an official control system able to attest all food chain operators can impair the principles of fair competition and security assurance levels to consumers. Official fish inspection systems have evolved to be based on direct proactive interventions on the food chain and on safety problems prevention. Nowadays, all methodologies are supported by systematic risk analysis, based on scientific and political assessments. There is an effort to upgrade inspection systems and private sector activities in order to meet the food safety requirements stated by the major fish importing markets and the World Trade Organization Sanitary and Phytosanitary Agreement (WTO SPS). WTO promotes the application of safety control measures and a risk assessment approach to assure food safety at the national level, which are based on international guidelines from international organizations, such as *"Codex Alimentarius"* [110, 111]. Most consensual guidelines regarding fish used as food are provided by FAO's Code of Conduct for Responsible Fisheries (CCRF) [72], which states that "harvesting, handling, processing and distribution of fish and fishery products should be carried out in a manner which will maintain the nutritional value, quality and safety of the products, reduce waste and minimize negative impacts on the environment".

Ababouch *et al.* (2005) have shown a high level of "nonconformities" in fish products originated from developing countries to be placed on major markets [64]. Therefore, there is a need for further improvement of sanitary conditions throughout the fish food chain in these countries, by Good Aquaculture Practices (GAP) implementation and national food control systems development. Legislation and food safety control systems applied to fisheries sector have been permanently evolving, due to changing policies, priorities, technology, research and equipment. Official fish inspection systems must be organized to attend to local or regional specific characteristics, including: harvested species, aquaculture productions, most frequent hazards, local habits of food preparation and consumption, species and volumes intended for external market, specific external market requirements and available technologies. One of the most relevant constrains to the accurate official fish inspection services implementation concerns financial resources. Each society needs to clarify how much of its social resources need to be allocated to the official fish inspection services, taking into consideration the population exposure levels (according to the per capita intake of fish) and specific health risk problems associated to fish consumption. It is obvious that populations that consume frequently multiple fish species, presenting a per capita consumption superior then 50 kg/ habitant/ year, including consumption of raw fish, are more exposed to biological hazards than populations that only eat a restrict number of species, always fried, being more exposed to chemical hazards. Nevertheless, in some societies fishery products safety is considered to be a low priority and resources are diverted to other food products control. Furthermore, official fish inspection services workers may be insufficient, or may not have the required knowledge or skills. It is also observed that sometimes legislation do not focus on these issues, or that data on foodborne illness risk factors associated with fish products is missing, being essential for the development of risk-based inspection plans. The absence of official monitoring programmes, to detect contaminants, residues or non-desirable substances in the fish supply chain, impairs an effective hazards control. Absolutely critical is the control of imports and exports safety. Difficulties may arise due to diverse legislative requirements of different export markets, although WTO agreements may be assumed as a generic legal base. Another question concerns to the fact that most of the food fish chain operators are small-scale or artisanal manufacturers, especially those associated with local fishery activities [112]. These operators may not have the technological know-how required for the successful application of proactive food safety procedures such as the Hazard Analysis Critical Control Point (HACCP) and traceability systems [113]. Specific knowledge and capabilities are required to implement these proactive systems, which are very difficult since local fish processing and distribution activities are frequently unregulated. This sector, however, is responsible for the subsistence of many people who may lack education skills and access to information, services and inputs required for the production of safe fishery products and to meet national legislative requirements. These social aspects must always be taken into consideration. Official controls should be based on documented procedures to ensure that they are carried out uniformly, fairly and independently, with consistently high standards [31].

MEASURES AND ACTIONS

Official fish inspection is responsible for insuring consumers that they will have access to safe and nutritious fish and fish products, independently of its origin. The main objectives of official fish inspection acts are: to confirm if each operator (fishing vessels, processing plant, storage facilities and transport) is registered and approved to produce or distribute fish and derived products; to investigate if these products are being handled and produced hygienically, respecting the principles stated in law; to identify risks associated with fish, including all applied processes, materials, ingredients or substances that may influence food fish safety; to check fish operators' past record regarding food legislation, fish health and animal welfare rules compliance, confirming that fish products were obtained from healthy fish; to verify the liability of self-records; to identify foreseeable incidents of food poisoning or injury; to collect samples for analytical verification of fish products compliance with legislation; also, in specific situations, to officially certify the fish product sanitary status, when demanded by imports or external markets. All procedures executed by the official inspection services must be documented, and official controls results must be reported. These reports should include the official controls purpose, control methods applied, results and the actions that operators must follow as a consequence of "noncompliance" detections. If the noncompliance detected is related to hygiene conditions, inspector should provide a copy of the report to the fish business operator. Follow-up of the corrective measures implemented by fish operators should be carried out and confirmed through analytical procedures. These must comply with reference sampling and analysis methods, so that results can be used to solve a dispute. In this context, it is also necessary to assure that analysis criteria, parameters and validation methods are adequately documented. If a non-compliance is detected or if there are doubts regarding the identity, the destination or the certification of a fish product, the official inspector must carry out controls in order to confirm or reject suspicions. Procedures must be based on official rules, namely: the scenario in which the official certification is required; the harmonized certification forms; the credentials of the certifying officials; and the adopted codes aiming to guarantee trustworthy accreditation.

The consignments should remain under official control until the final decision and the official final documents must follow the consignments.

AUDITING FOOD FISH BUSINESS OPERATORS

According to Huss *et al.* (2003), food inspection comprises all actions performed to guarantee its quality and safety along the food chain, including control activities implemented and applied by the operator, which accuracy must be verified and confirmed by official controls [31]. Every production phase, from primary production, to handling, storage, distribution and preparation, should be considered in a structured programme, with different levels of responsibilities. The overall purpose is to establish a efficient methodology to all regulator actions using a programme founded on scientific knowledge and risk assessment, leading to targeted inspection.

At present, preventive official fish inspection services oversee the safety of fish imports, exports and supplying chain the domestic consumer, being responsible for verifying compliances with national, regional and international legislation. The aims of their interventions are: to evaluate the associated risks regarding different fish and fishery products and establish the risk level for each operation, relevant to determine the frequency of inspections; to inspect facilities and health and hygiene practices in fish production, processing and distribution establishments, including fishing vessels, landing ports, vehicles, premises, aquaculture establishments, ice plants, cold stores and markets; to audit and approve HACCP and traceability systems set up by the operators, as well as the system pre-requisites; to approve production and processing sites, activities and premises; to implement corrective actions, whenever private sector practices do not meet with legislation requirements; to follow up and monitor the implementation of those corrective action; to approve and help to develop good practices codes for the private sector, in compliance with the best practices and food safety legislation; to conceive and promote training courses on best practices and food safety; to confirm and verify health status and hygiene conditions, including sampling and analysis of products, water and other premises, aiming to demonstrate the effective compliance of hygiene standards on handling and processing facilities; to attribute official authorizations (approval numbers) and health certificates, for processing and market activities and for fish products for international market; to ensure that non-compliant products are rapidly withdrawn from the food chain. These generic objectives and procedures should always be present, to assist official fish inspectors to carry out their responsibilities as a risk-based approach. The official fish inspector must be able to understand the food safety hazards associated with particular fish species or products, and what are the practicable measures for controlling associated hazards. The essential knowledge requirements and skills that official fish inspectors need to include are: fish anatomy and pathology; fish health and its management, including fish diseases requiring international notification and therapeutic managements including drug registries; ichthyology, taxonomy, identification and composition, essential for fish authenticity; post-mortem fish alterations due to decomposition and freshness evaluation; fish production systems, including aquaculture, fisheries, processing, distribution and traceability systems; veterinary public health, food hygiene, microbiology, toxicology, parasitology and epidemiology; specific fish inspection procedures, standards and codes of practice, including sampling planning and procedures; comprehensive approaches to risk analysis system; safety assurance and risk management programmes in the fisheries sector, such as good manufacturing and hygiene practices and HACCP for fish and fishery products.

AUDITING "GOOD HYGIENIC PRACTICES" (GHP) AND HACCP

Fish product hazards can occur at different stages of the fish food chain and vary according to the fish or shellfish species, aquatic environment, and especially post-harvest handling, processing, storage and distribution procedures. All fish premises included in the food supply chain need to comply with hygienic requirements mandatory by law or international guidelines [114]. Basically, the operators have to comply with "good hygienic practices" (GHP) generic rules and establish a proactive control system based on HACCP principles. Only after meeting these two requirements, fish operators are allowed to put fish products on the national, regional or international markets. GHP, also designed "prerequisite programmes", aim to ensure that: the fish plant or vessels structure and their equipment's were designed and constructed in a way that do not cause contaminations; the availability and use of safety water and ice; the hygiene and disinfection of surfaces in contact with fish; the prevention of cross-contamination due to operation sequences; the personal hygiene of the food workers, including prevention of problems resulting in employee health issues; the storage of toxic compounds is safe; there are efficient pest control practices for insects, rodents and birds; there are adequate liquid or solid wastes management procedures; the storage and transportation of fish, fish products and other ingredients are adequate; the traceability system is working; and the people in contact with food fish received a proper training. All procedures concerning these GHP components need to be documented according to internal procedures, so that they can be verified by official inspection systems. Primary production includes aquaculture and fishing, while correlated operations comprise all the subsequent processes performed on the fishing boats and vessels, including transport of products with minor changes. Official inspections of these operators aim to verify if they are really able to guarantee that fishing vessels present the required structure and tools, and also if processing is performed according to good hygienic practices rules. Vessels must be designed and constructed in a way that structures do not promote the fishes' contamination with polluted water, sewage, petrol, lubricant and Vessels surfaces and equipment coming in contact with fish should be made of strong non-toxic material resistant to corrosion, easy to wash and disinfect. If a water pipe is present in the vessels, it should be located in order to avoid contamination of the water supply. If fish have to remain on the vessels for more than 24 hours, the boats and vessels must possess a warehouses or containers for the storage of fishery products at temperatures inferior to +3 °C, that should be located far from engine rooms and staff accommodations, to prevent contamination. These warehouses should be preserved in excellent hygiene conditions and it should be guaranteed that melting water does not reach fishes in storage. The same conditions must be followed by industrial boats and factory vessels, to prevent fish contamination and guarantee hygienic conditions in processing and storage areas.

In the fish sector, water is an omnipresent element and a key risk factor. Water is the fish life environment and is used at all stages of the fish chain: to wash fish, equipment and places in contact with it, to produce ice, to glaze frozen products, to make brines for salting, as an ingredient in transformed fish products, some technical procedures such as commercial sterilization and canning, and as vehicle of liquid wastes. Therefore, its characteristics are extremely relevant to avoid or to eliminate biological, chemical and physical hazards. Water used as fish products ingredient must be potable, complying with human consumption standards. Water supply for fish plants may be obtained from different origins, including local captions from ground water, from sea, rivers and lakes, or from the public water supply. Each origin poses specific problems concerning biological or chemical contaminations, and its adequate application depends on treatment. Some water treatments also enhance chemical risks. Water used for ice production, chilling or as fish product ingredient has to be potable. Standards for potable water are established by WHO International Guidelines for Drinking Water Quality. Seawater may be used for cleaning fishing nets or fishing port areas where products are unloaded. To be considered "cleaned", this seawater needs to meet similar microbiological criteria as good quality water, absent of visible solid particles in suspension [115]. Fish operators must establish an accurate control of water characteristics. For that, it is absolutely crucial to analyze all used water regularly, keeping records and adopting the appropriate corrective actions in case of noncompliance. All the documents produced are fundamental for official inspections and verifications. In fact, all documents produced by the GHP program are essential pieces for the safety system and official inspection activities targets. During the auditing process of the GHP established by food fish operators, official inspector will always search for: design peculiarities of spaces, equipment and operation sequences, allowing cross-contaminations or difficult disinfection; fish food obtained from unsafe sources; inadequate control of thermal treatments; improper refrigeration or freezing temperatures; equipment, containers and places cleaning and disinfection evidences; deficient personal or professional hygiene; documentation regarding official approval of biocides and devices certificates; food manipulator' health status; water characteristics control and analytical reports; presence of pests or its testimonial evidences; evidence of programs concerning identification, conditioning and elimination of unfit by-products; and evidences of adequate storage and elimination of dangerous wastes. In the last twenty years, generic and systematic control approaches have been proposed and applied, such as Good Aquaculture Practice (GAP), Good Hygienic Practice (GHP), Standard Sanitation Operating Procedures (SSOP) and traceability, as well as specific technical approaches used to control particular food safety hazards. This is the case of HACCP principles. GHP needs to be established before

HACCP (hazard analysis and critical control points) system development and implementation. HACCP system is a food safety management tool that helps to ensure that food, including fish and fishery products, are safe and will not cause adverse health effects in consumers. It is now internationally recognized and is promoted by FAO/WHO Codex Alimentarius. Application of HACCP in the fisheries sector is a legislative requirement in many regional and international markets. HACCP helps fish producers and processors to centre attention on preventing or eliminating known hazards. Biological, chemical and physical hazards anticipation and identification of control points at which these hazards are prevented or eliminated are key elements of HACCP. Development and application of an HACCP plan is based on seven steps or principles: hazard analysis, identification of critical control points (CCP), setting CCP parameters, monitoring, counteractive measures, certification and record storage.

CAC (2005a) provides useful guidance on HACCP to control specific hazards in fish and fishery products [111]. In developed markets, food fish operators need to implement specific proactive procedures based on HACCP principles, so that they can place their fish products on the markets. Procedures must be audited and validated though official inspections. After this formal approval, the operator may access to an official authorization, the Approval number or code, that allows him to develop its activities. These procedures are inevitable when fish production is intended for international markets or developed urban societies. However, in some small markets at piscatorial local communities, compliance with HACCP principles is not always applied. Moreover, workers within the fish food chain may not have enough knowledge on GHP, neither the resources to invest in training or in equipment and personal hygiene. For example, small-scale fish producers may lack potable water for cleaning and personal hygiene maintenance, or even for ice production. In such scenario, the food safety system should be encouraged to evolve so that facilities and locations could gradually be improved, redesigned and human capacities developed to enable reaching minimum appropriate standards. Simple changes can be obtained through education. These rudimentary small behavioral changes may reveal to be efficient on avoiding contaminations and reducing bacteria multiplication. Examples of these simple processing controls include: to wash fish or shellfish and remove the gills and abdominal cavity organs as soon as possible; to use control methods to kill bacteria and inactivate enzymes, including temperature (chilling, freezing or heat processing), water removing and fish drying, salt addition (additives), pH control (marinades, brines) and to reduce the possibility contact of fish fats and fish oils with oxygen, using appropriate packaging. Low temperatures, such as those associated with chilling, will prevent the growth of bacteria responsible for decomposition and also pathogenic bacteria. Heating treatments, including cooking, frying and grilling, prior to consumption, will kill bacteria and destroy most toxins, eliminating the risk of potential adverse effects to local communities' health. However, toxins and spores of some pathogens such as Clostridium botulinum are relatively heat resistant and require specific processing controls. Promoting very acidic environments for fish and fishery products by lowering marinades pH will also prevent pathogens growth. Removing water by drying or salting will also create unfavourable conditions for bacterial growth. Controlling access to oxygen by appropriate use of packaging can also affect the growth of pathogens, as certain pathogens require oxygen in order to grow, while others require an environment with little or no oxygen [116]. In general, these informal local markets are not targeted by official inspections.

DIRECT OFFICIAL INSPECTION OF FISH AND FISH PRODUCTS

The official veterinary inspection acts may be directed to fish products, especially when the fish lot is imported (BIP), intended for the internal market, and also in aquacultures or at unloading at fishing ports. Before fish direct examination, official inspectors must be aware that operators, accountable for fishery products arrival and reception, are able to: ensure that specific tools in contact with fish are built with non-damaged material that should be easily cleaned and disinfected; and to prevent fish microbial contamination and physical damage in all manipulation steps.

If fishing vessels do not have appropriate refrigeration facilities or equipment, fresh fish products should be refrigerated and stored at melting ice temperature prompt after unloading. Operators responsible for auction and wholesale markets should guarantee the following necessities: have lockable facilities for the refrigerated storage of fish products must be available, as well as isolated lockable accommodations for unfit fish storage products. If required by the official inspection services, an adequately equipped lockable facility should be available, as well as a room for the exclusive use of inspection services. Regarding locations and building, premises used for fish presentation or storing should not have other uses. Polluting vehicles and pets access should also be forbidden. All facilities should be illuminated to help official monitoring.

FRESH FISH

Physical examinations of fresh fish must take into consideration: fish identity, including species identification in accordance with international nomenclature and legal statements, comprising the scientific name and its trade name; origin, such as place of capture or aquaculture); lot dimension, or net weight of the packed fish products; documental dates of grading and dispatch; name and address of consignor or owner to assure traceability; and freshness and size categories. Identification or authenticity of fish species and its origin is quite relevant for two main reasons. Firstly, there are diseases with specific tropism for a particular fish species and secondly, fish species have quite different trade value, depending on size, meat structure and composition, technological aptitudes and sensorial characteristics [117]. Fish identification also impairs forbidden species to be inadvertently put on the market, including toxic fish species of families *Diondontidae, Canthigasteridae, Molidae, Tethraodontidae* and *Gempilidae*. Herring, mackerel, sprat, wild salmons or other species intended to be raw eaten or minimally processed using brine, cold smoking process, marinated and/or salted, should be compulsory cooled to a temperature of at least -20 °C for not less than 24 hours, to guarantee the destruction of contaminating nematodes larvae. Both raw and processed final fish products can be subjected to this treatment.

Documentation stating the processing protocols must accompany these fishery products when placed on the market. Unfortunately, some culinary preparations are not compatible with freezing treatment because the muscle texture is impaired, not allowing the production of Carpaccio and Ceviche. In this case, the only alternative is the performance of a meticulous veterinary inspection of these fish products. Concerning freshness, maximization of nutrient values can be evaluated using two main types of methodologies: subjective or organoleptic and objective, including chemical, physical and microbiological analysis. Subjective techniques are quite practical, allowing obtaining a definitive decision in a given time, compatible with the good hygiene practices. However, they require an experimented technician, with previous training. Subjective or organoleptic examination of fish freshness consists on the evaluation of specific fish aspects, namely the colour and brightness of skin gilts, odour and fish flesh consistency (Tables **4** and **5**).

Table 4. Organoleptic characteristic of Teleost fresh fish.

Parameter	"White fish"	"Blue fish"
Skin aspect	Bright iridescent pigment or opalescent non discolored	Bright pigmentation, shining iridescent colors; distinction between dorsal and ventral surfaces
Skin mucus	Aqueous, transparent or slightly cloudy	Scarce, aqueous, transparent or slightly cloudy
Eye	Convex or slightly sunken; black bright or dull pupil; transparent or slightly opalescent cornea	Convex or slightly sunken; black bright or dull pupil; transparent or slightly opalescent cornea; Transparent eyelid
Gills	Bright or less colored; No or transparent mucus	Uniformly dark red to purple; no or transparent mucus
Cover gills	Bright, without hemorrhage	Silvery or very slightly red or brown
Peritoneum	Smooth or slightly dull; bright; adherent or difficult to detached from flesh	Smooth or slightly dull; bright; adherent or difficult to detached from flesh
Smell	No smell, seaweed or neutral	Fresh seaweed, pungent, iodine
Muscle	Translucent, firm and elastic, smooth surface	Translucent, firm and elastic, smooth surface (when cut)
Blood vessels	Sharp outline and vivid color	Sharp outline, sometime collapsed
Spine	Adherent, difficult to detached	Adherent, difficult to detached

Legend: "White fish"- fish like haddock, cod, saithe, pollack, redfish, withing, ling, breams, anglerfish, pouting, poor cod, bogue, picarel, conger, gurnard, mullet, plaice, megrim, soles, dab, flounder, scabbard fish; "Blue fish"- fish species of *Scombridae, Clupeidae, Carangidae* and *Engraulidae* families.

Table 5. Organoleptic characteristic of Selachii (elasmobranches) fresh fish.

Parameter	Sharks, dog fish and skates
Body aspect	In rigor mortis, partially or beyond rigor stage; stiff fins
Skin mucus	No or aqueous mucus on skin and especially in mouth and gills cameras
Eye	Convex or slightly sunken; bright or loss of brightness and iridescence, oval pupils
Smell	Seaweed or very slight sour but not an ammonia smell
Flesh	Opalescent, firm and elastic, smooth surface

(Table 5) contd.....

Parameter	Sharks, dog fish and skates
Blood vessels	Sharp outline and vivid color

This examination may be adapted to structured valorization scales to minimize subjectivity [2], being the Torry's scheme and the Quality Index Method (QIM) the most universally accepted. QIM methodology is centered on the surveillance of variations that happen in fresh fish through decomposition, allowing evaluating its freshness grade. These changes are observed and calculated using a demerit scale, and explanations and matching grades for individual characteristics are available in a QIM table. The addition of scores determined for individual parameters originates a total sensory grade recognized as the "Quality Index" (QI).

Official inspectors also need to verify the origin and the commercial purpose of the fishery lots. Fish products destined for external market or for export should be sold only if their packages include the following labelling, representing an inherent condition for traceability [118]: scientific name of the fish and trade designation; country of origin; presentation (whole, sliced); freshness and dimension classification; net weight of fish products in package; dates of grading and dispatch; and the name and address of consignor. Commercial shelf life of fresh fish can be considered as the period of days after capture in which fish is allowed to be maintained in storage, before becoming unfit for consumption. It is established considering the examination of the sensorial variations that occur in each fish species along its commercial life. Decomposition inherent to organic materials enhance deep modifications on sensorial characteristics of fresh fish, including decrease in skin brightness, abundance of opaque mucus, softness of flesh, concave eye and malodorous ammoniac and acre odours. Evolution from freshness to a final stage of decomposition (unfitted) is conditioned by different factors, such as: fish species, capture technique, processing circumstances, post-capture cooling velocity, cool chain integrity, season and general hygiene of each procedure. Evaluation of fish freshness or decomposition stage should be performed on an identified batch, and the sample collected for analysis should be illustrative of the entire batch, which is guaranteed by the application of proper sampling guidelines.

Trying to eliminate the subjectivity of sensorial evaluations, many efforts have been made in the last decades to develop accurate objective methods to evaluate fish decomposition, using physical, chemical and microbiological parameters [119]. Most referred physical methods are related with the electrical properties of the fish muscle, pH and fish flesh texture. Electrical variations if fish skin and other tissues can be used for measuring post-mortem changes or decomposition degree, which could be used in processing facilities. However, many difficulties have been found in instrumental development, especially due to: species diversity; variation within the same fish batch; different readings in the same fish as a consequence of skin injuries, freezing, filleting, bleeding; and a low relation between results from measuring devices and sensory analysis.

There is a popular instrument, the GR Torrymeter, that is not able to measure the freshness from only one fish, but it can be applied for the evaluation of fish lots. The use of a potentiometer for pH value measurement could allow evaluate fish meat freshness, and this quantitative evaluation may be performed by putting the electrodes in direct contact with the fish flesh or placing them in a solution of distilled water and fish flesh.

Consistency is a major quality of raw or cooked fish flesh, and it might be altered by freezing and decomposition. It can be evaluated through rheological testing, using several methods, including: the Instron Model TM, the compressive deformability method, the determination of fish flesh shear force, or a portable penetrometer. It is important to refer that sometimes evaluation results can be are hard to interpret.

Most objective parameters to evaluate are chemical based, namely the levels of total volatile nitrogen (TVB-N), trimethylamine (TMA), dimethylamine (DMA), biogenic amine, nucleotide catabolites, etanol and oxidative rancidity. TVB-N has been used for many years as criteria for fish freshness evaluation. This value is known as the most important criteria for raw industrial fish. TVB-N amount in fish flesh is considered as an indicator for bacterial growth and protein attack, since ammonia originates amino acids decomposition (deamination). TVB- N levels above 40mg/100g of fish mass are regarded as decomposition and fish are declared not fitted for human consumption. Some markets do not allow levels superior to 35 mg/100g. Main constituents of TVB-N are trimethylamine (TMA) and ammonia. Trimethylamine originates from bacterial decomposition, increasing with storage period in the unfrozen state. It has been proposed that TMA levels above 10 mg/100g of fish flesh should be considered the maximum allowed levels in international trading.

However, analytical errors may result from amines volatilization from stored samples, especially in those species, such as *Gadidae*, that produce the enzyme TMAO dimethylase (TMAO-ase), able to transform TMAO in of DMA and formaldehyde (FA), that induces proteins strengthening. DMA levels can be determined through colorimetric techniques, using a spectrophotometer, or through more precise chromatographic techniques, such as gas-liquid and high performance liquid chromatography. These techniques have the disadvantage of being destructive and not appropriated for analyzing big samples.

Biogenic amines have been evaluated in canned products as an accurate indicator for previous spoilage of fishes before heating, since they are stable to thermal processing. Fish flesh is decomposed by microbiota growth, inducing the catabolic production of several amines due to amino-acids decarboxylation, and subsequent pH raise through amines production [120]. Since it has already been related with scombroid fish poisoning, the best known biogenic amine is histamine, which results from the decarboxylation of histidine, but putrescine (ornithine), cadaverine (lysine) and tyramine (tyrosine) can also be tested [121]. Protocols for biogenic amine determination include chromatography (high pressure liquid and gas chromatography) and spectrofluorometry [122].

The quantitative analysis of nucleotide metabolites has also been used to evaluate autolysis catabolic changes. Nucleotides breakdown can be due to autolytic enzymes (for example: adenosine triphosphate, ATP → inosine monophosphate, IMP) or to bacterial growth (for example: IMP → inosine, Ino → hypoxanthine, Hx). Freshness assessment can be performed by determining the levels of catabolic intermediates. Especially when fresh fish is intended for industrial processing, the quantitative determination of the whole nucleotide catabolite profile is suggested, by high pressure liquid chromatographic (HPLC). However, some cautions are needed, as these protocols often require an initial deproteinization step using perchloric or trichloracetic acids. In these cases, it is necessary to neutralize the acid extracts quickly using potassium hydroxide, to avoid the degradation of nucleotides and promote the stabilization of the extracts.

Another molecule that can be evaluated as an indicator of fish preservation status is ethanol. Ethanol is commonly produced by several bacteria, through anaerobic fermentation of carbohydrates and deamination and decarboxylation of amino-acids such as alanine. It has already been applied to quantitative measurement of different fish species freshness and spoilage, including tuna, salmon, redfish, pollock, flounder and cod, and commercial enzyme test kits are available for ethanol measurement. As it is heat-stable, it can be applied to the evaluation of canned fish products.

But not only protein degradation and glycolysis happen during fish decomposition. Highly unsaturated fatty acids can also be altered by oxidation, originating several metabolites, such as lipid hydroperoxides, that can be determined using chemical-based methods, spectrophotometry or iodometry. However, results analysis must be performed carefully, as these metabolites cannot be correlated with results from sensory analysis, as they are odor and flavor free. Also, they are not stable over time, so late in oxidation, secondary oxidation products, such as aldehydes, ketones and short chain fatty acid produced *via* beta-oxidation due to cryophyle microbes, should be determined for oxidative rancid evaluation. Some of these secondary metabolites, in particular aldehydes, can be determined by spectrophotometry, using thiobarbituric acid. The results regarding the development of a red product by acid-reactive substances (TBA-RS) are based on the control molecule tested, malonaldehyde, and represented as malonaldehyde (MDA) micromoles present in 1g of fat, as mg in 1g of fat, or as µmol or µg in relation to total of analyzed material. Thus, interpretation of TBA-RS values must be prudent. Acceptable fish products should not have above 20 meq of PV/kg of fish fat, as determined by iodometric titration; fish with TBA-RS values superior to 1-2 µmol MDA/g or superior to 10 µmol MDA/kg are not adequate for consumption.

Microbes quantification in fish may also be used to evaluate the level of fish flesh decomposition. The aim of fish products microbial examination is to evaluate the possible presence of public health significance microorganisms and to evaluate fish hygienic characteristics, including temperature abuse during handling and processing. Microbial data may give information about processes efficiency and freshness [123]. The total quantity of spoilage bacteria (saprophyte) is associated to shelf life and its quantification should be used to predict it in fresh fish [124]. Conventional bacteriological methods are time-consuming, laborious and costly, and require professional skills in execution and results interpretation.

Several rapid protocols have been established, some of which are mechanized, allowing the analysis of a large number of samples [125]. Total aerobic count is a standard plate count (SPC) procedure that represents the total number of viable bacteria [126], and can be used as a general idea of bacterial contamination and of good hygienic practices. In fish kept in ice, when the total count reaches $10^{7.5}$ cfu (colony forming units) /g of fish flesh, the product has to be recollected from the market. Within levels superior to 10^8 cfu/g of fish flesh, there are sensorial changes.

Common plate count agars (PCA) are widely used for total counts determination. However, when seaborne fish are examined, performances using seawater or more nutrient rich agars give considerably higher bacteria totals. Also, iron agar or other supplemented media should be used for determining the quantity of bacteria able to produce hydrogen sulphide, since these include specific spoilage bacteria for some fish products, such as *Shewanella putrefaciens* and is often due to *Vibrionaceae*. There also available selective media for specific bacteria genera, including *Pseudomonas* spp., bacteria that spoils tropical and freshwater fish, and for Photobacterium phosphoreum, which spoils packed fresh fish.

Also, incubation temperature should be adapted to the microbiological examination of these products. Seafood samples stores at refrigeration temperatures should be incubated for 3 to 4 days at 25°C for the isolation of relevant psychotrophic microbes. For products where Photobacterium phosphoreum is expected, incubation temperature should not exceed 15 °C.

Rapid procedures have been developed to simplify these methods, namely through miniaturization. The main advantage of these rapid methods in comparison with standard plate counts are the prices and simplicity. So, pathogenic bacteria presence may be determined using conventional bacteriological methods [127, 128] or using immunological or molecular-based techniques, such as PCR. These methods can be used for specific spoilage bacteria, including indicators of fecal contaminations, comprising phages of human enteric bacteria as virus indicators [129]. Many rapid techniques have been developed for estimating bacterial levels in fish, adapting physico-chemical based technologies (microscopy, densitometry, impedance, microcalorimetry, conductance, APTmetry), although most of them do not have direct equivalence. Some of these methods are applied for the rapid determination of bacteria counts and for making a decision regarding fish products withdrawal.

PROCESSED FISH

Official inspections of manufactured fish must take into consideration the results of previous audits of plant processing conditions and the risk level associated to the product. Approaches must be systematic and based on structured documentation and check lists and, when necessary, verifications have to be performed. To be placed on the market, processed food fish must comply with all legal frames concerning labelling, analytical requirements (chemical, physical and biological), including authenticity.

Also, operators must not process or place in the market fish products that are obviously contaminated with parasites, so the previous visual observation of fish for detection of visible parasites should be mandatory. Official verification can be made by observation under transilumination according to a representative sampling plan. Examinations records have to be kept for documental check by official controls, which should also verify fish storage conditions. These must agree with several guidelines, as follows: fresh and defrosted fishes should be kept at melting ice temperature [130]; frozen fishes should be storage at a temperature inferior to -18 °C; frozen salted fishes to be used for cans' manufacture should be maintained at a temperature inferior to - 9 °C; and live fishes should be maintained at a temperature and in environmental conditions adequate for the safeguarding of food safety and fishes viability.

Official inspectors need to verify if the operations are in accordance with legal frames or good hygiene practices, in particular: fresh fish products, thawed unprocessed fish products, should be kept at melting ice temperature; frozen fishes, except for salted products to be used in canned food production, should be upheld at a stable temperature below -18 °C during transport, allowing short variations of less than 3 °C; canned fishes need documentation proving the thermal treatments efficiency, such as time and temperature of commercial sterilization procedures, including devices calibration. These requirements can be officially confirmed using appropriate devices for thermal measurements.

CONCLUDING REMARKS

Fish products are special sources of rich nutrients having high digestibility rates and general acceptance. These food products are sources of essential amino acids, fatty acids, minerals and vitamins. To achieve all these benefits some previous conditions need to be attended, including guaranteeing the health status of each fish. Fish obtained from aquaculture or the wild must not suffer from any unhealthy condition. Official veterinary inspection of fish is needed to detect these problems and proceed to diagnosis and, if required, to the official international notification, which will stop international commerce of fish and resulting products. Globalization of such products trading is a risk factor for microbial and parasitic disease worldwide spreading. Under this perspective, veterinary certification of fish and fishery products health status is a key piece for the general health safeguard, as stated by multiple international organizations, including WTO and OIE. It is absolutely fundamental that products that do not comply with health standards or that have been handled and incorrectly processed be declared unfit and never allowed to be placed in the market. Also, knowledge of therapies applied in aquacultures is relevant for residues assessments. In conclusion, there are many hazards that can be carried by fish products: biological (pathogenic bacteria, virus and parasites); chemical (environmental contaminants, biotoxines, biogenic amines, drug residues from aquaculture and additives) and physical hazards (spines, hooks and other foreign bodies such as sand, mud or soil). All these hazards must be appropriately controlled and risks have to be assumed and shared by all food chain stakeholders, and supervised by competent and independent official services.

CONFLICT OF INTEREST

The author confirms that author has no conflict of interest to declare for this publication.

ACKNOWLEDGEMENTS

Declared none.

REFERENCES

[1]   Kurien J. Responsible fish trade and food security. FAO Fisheries and Aquaculture Technical 2005; 456: 102.

[2]   Huss HH. Quality and quality changes in fresh fish. FAO Fisheries and Aquaculture Technical 1995; 348: 195.

[3]   Feldhusen F. The role of seafood in bacterial foodborne diseases. Microbes Infect 2000; 2(13): 1651-60.
[http://dx.doi.org/10.1016/S1286-4579(00)01321-6] [PMID: 11113384]

[4]   Huss HH, Petersen ER. The stability of *Clostridium botulinum* type E toxin in salty and/or acid environment. J Food Technol 1980; 15: 619-27.
[http://dx.doi.org/10.1111/j.1365-2621.1980.tb00982.x]

[5]   Lalitha KV, Surendran PK. Occurrence of *Clostridium botulinum* in fresh and cured fish in retail trade in Cochin (India). Int J Food Microbiol 2002; 72(1-2): 169-74.
[http://dx.doi.org/10.1016/S0168-1605(01)00632-8] [PMID: 11843409]

[6]   Dufresne I, Smith JP, Liu JN, Tarte I, Blanchfield B, Austin JW. Effect of films of different oxygen transmission rate on toxin production by *Clostridium botulinum* type E in vacuum packaged cold and hot smoked trout fillets. J Food Saf 2000; 20: 251-68.
[http://dx.doi.org/10.1111/j.1745-4565.2000.tb00303.x]

[7]   Cann DC, Taylor L. The control of botulinum hazard in hot smoked trout and mackerel. J Food Technol 1979; 14: 123-9.
[http://dx.doi.org/10.1111/j.1365-2621.1979.tb00856.x]

[8]   Hauschild AHW. *Clostridium botulinum.* In: Doyle MP, Ed. Foodborne Bacterial Pathogens. New York: Marcel Dekker, Inc. 1989; pp. 111-90.

[9]   *Codex Alimentarius* Commission Food and Agriculture Organization / World Health Organization. Discussion Paper on Risk Management Strategies for *Vibrio* spp. in Seafood. Rome, Italy 2002.

[10]   *Codex Alimentarius* Commission Food and Agriculture Organization / World Health Organization.. Discussion Paper on Risk Management Strategies for Vibrio spp in Seafood. Rome, Italy 2001.

[11]   Hayashi F, Harada K, Mitsuhashi S, Inoue M. Conjugation of drug-resistance plasmids from vibrio anguillarum to Vibrio parahaemolyticus. Microbiol Immunol 1982; 26(6): 479-85.
[http://dx.doi.org/10.1111/j.1348-0421.1982.tb00201.x] [PMID: 6752666]

[12]   Oliver JD, Kaper JB. *Vibrio* species. In: Doyle MP, Beuchat LR, Montville TJ, Eds. Food microbiology fundamentals and frontiers. Washington, DC: ASM Press 1997; pp. 228-64.

[13]   Gooch JA, DePaola A, Bowers J, Marshall DL. Growth and survival of *Vibrio parahaemolyticus* in postharvest American oysters. J Food Prot 2002; 65(6): 970-4.
[PMID: 12092730]

[14]   EC. Opinion of the Scientific Committee on Veterinary Measures relating to Public Health on Vibrio vulnificus and Vibrio parahaemolyticus in raw and undercooked seafood . Brussels, Belgium 2001a.

[15]   Bisharat N, Agmon V, Finkelstein R, *et al.* Israel Vibrio Study Group. Clinical, epidemiological, and microbiological features of Vibrio vulnificus biogroup 3 causing outbreaks of wound infection and bacteraemia in Israel. Lancet 1999; 354(9188): 1421-4.
[http://dx.doi.org/10.1016/S0140-6736(99)02471-X] [PMID: 10543668]

[16]   Stewart-Tull DE. Vaba, Haiza, Kholera, Foklune or Cholera: in any language still the disease of seven pandemics. J Appl Microbiol 2001; 91(4): 580-91.
[http://dx.doi.org/10.1046/j.1365-2672.2001.01493.x] [PMID: 11576291]

[17]   Barbieri E, Falzano L, Fiorentini C, *et al.* Occurrence, diversity, and pathogenicity of halophilic Vibrio spp. and non-O1 *Vibrio cholerae* from estuarine waters along the Italian Adriatic coast. Appl Environ Microbiol 1999; 65(6): 2748-53.
[PMID: 10347072]

[18]   Chiavelli DA, Marsh JW, Taylor RK. The mannose-sensitive hemagglutinin of *Vibrio cholerae* promotes adherence to zooplankton. Appl Environ Microbiol 2001; 67(7): 3220-5.
[http://dx.doi.org/10.1128/AEM.67.7.3220-3225.2001] [PMID: 11425745]

[19]   Torres-Vitela MR, Castillo A, Finne G, Rodriguez-Garcia MO, Martinez-Gonzales NE, Navarro Hidalgo V. Incidence of *Vibrio cholera* in fresh fish and ceviche in Guadalajara, Mexico. J Food Prot 1997; 60: 237-41.

[20]   Weber JT, Mintz ED, CaAizares R, *et al.* Epidemic cholera in Ecuador: multidrug-resistance and transmission by water and seafood. Epidemiol Infect 1994; 112(1): 1-11.
[http://dx.doi.org/10.1017/S0950268800057368] [PMID: 8119348]

[21]   Farmer JJ, Arduino MJ, Hickman-Brenner FW. The genera *Aeromonas* and *Plesiomonas*. In: Balows A, Trüper HG, Dworkin M, Harder W, Schleifer K-H, Eds. The Procaryotes. 2nd ed. Heidelberg: Springer Verlag 1997; pp. 3012-28.

[22]   Krovacek K, Eriksson LM, GonzAlez-Rey C, Rosinsky J, Ciznar I. Isolation, biochemical and serological characterisation of *Plesiomonas shigelloides* from freshwater in Northern Europe. Comp Immunol Microbiol Infect Dis 2000; 23(1): 45-51.
[http://dx.doi.org/10.1016/S0147-9571(99)00058-2] [PMID: 10660257]

[23]   Kirov SM. *Aeromonas* and *Plesiomonas* species. In: Doyle M, Beuchat LR, Montville TJ, Eds. Food Microbiology Fundamentals and Frontiers. Washington, DC: ASM Press 1997; pp. 265-87.

[24]   KnA,chel S. Growth characteristics of motile *Aeromonas* spp. isolated from different environments. Int J Food Microbiol 1990; 10(3-4): 235-44.
[http://dx.doi.org/10.1016/0168-1605(90)90071-C] [PMID: 2397155]

[25]   Granum PE, Baird-Parker TC. *Bacillus* species. In: Lund BM, Baird-Parker TC, Gould GW, Eds. The microbiological safety and quality of foods gaithersberg. Aspen Publishers Inc. 2000; pp. 1029-39.

[26]   Fonnesbech Vogel B, Huss HH, Ojeniyi B, Ahrens P, Gram L, Gram L. Elucidation of *Listeria monocytogenes* contamination routes in cold-smoked salmon processing plants detected by DNA-based typing methods. Appl Environ Microbiol 2001; 67(6): 2586-95.
[http://dx.doi.org/10.1128/AEM.67.6.2586-2595.2001] [PMID: 11375167]

[27]   Autio T, Hielm S, Miettinen M, *et al.* Sources of *Listeria monocytogenes* contamination in a cold-smoked rainbow trout processing plant detected by pulsed-field gel electrophoresis typing. Appl Environ Microbiol 1999; 65(1): 150-5.
[PMID: 9872773]

[28]   Huss HH, JA,rgensen LV, Vogel BF. Control options for *Listeria monocytogenes* in seafoods. Int J Food Microbiol 2000; 62(3): 267-74.
[http://dx.doi.org/10.1016/S0168-1605(00)00347-0] [PMID: 11156271]

[29]   Miettinen MK, Siitonen A, Heiskanen P, Haajanen H, Bjorkroth KJ, Korkeala HJ. Molecular epidemiology of an outbreak of febrile gastroenteritis caused by *Listeria monocytogenes* in cold-smoked rainbow trout. J Clin Microbiol 1999; 37(7): 2358-60.
[PMID: 10364616]

[30]   JimA(c)nez L, MuAiz I, Toranzos GA, Hazen TC. Survival and activity of *Salmonella typhimurium* and *Escherichia coli* in tropical freshwater. J Appl Bacteriol 1989; 67(1): 61-9.
[http://dx.doi.org/10.1111/j.1365-2672.1989.tb04955.x] [PMID: 2674097]

[31]   Huss HH, Ababouch L, Gram L. Assessment and management of seafood safety and quality. FAO Fisheries Technical Paper 2003; 444: 230.

[32]   Jablonski LM, Bohach GA. *Staphylococcus aureus* In: Doyle MP, Beuchat LR, Montville TJ, Eds. Food Microbiology Fundamentals and Frontiers. Washington, DC: ASM Press 1997; pp. 353-75.

[33]   Pedro S, MagalhAes N, Albuquerque MM, Batista I, Nunes M, Bernardo F. Preliminary observation on spoilage potential of flora from desalted Cod (*Gadus morhua*). J Aquat Food Prod Technol 2002; 11: 143-50.
[http://dx.doi.org/10.1300/J030v11n03_11]

[34]   EFSA. The European Union Summary Report on Trends and Sources of Zoonoses, Zoonotic Agents and Food-borne Outbreaks in 2010. The EFSA J 2012; 10: 2597-422.

[35]   Reilly PJ, Twiddy DR. Salmonella and *Vibrio cholerae* in brackishwater cultured tropical prawns. Int J Food Microbiol 1992; 16(4): 293-301.
[http://dx.doi.org/10.1016/0168-1605(92)90031-W] [PMID: 1457289]

[36]   Saheki K, Kobayashi S, Kawanishi T. *Salmonella* contamination of eel culture ponds. Nippon Suisan Gakkai Shi 1989; 55: 675-9.
[http://dx.doi.org/10.2331/suisan.55.675]

[37]   Reilly PJ, Twiddy DR, Fuchs RS. Review of the occurrence of *Salmonella* in cultured tropical shrimp. FAO Fisheries Circular 1992; 851: 23.

[38]   Lampel KA, Madden JM, Wachsmuthk IK. *Shigella* species. In: Lund BM, Baird-Parker TC, Gould GW, Eds. The Microbiological Safety and Quality of Foods Gaithersberg. Aspen Publishers Inc. 2000;

pp. 1200-16.

[39]   Nachamkin I. *Campylobacter jejuni* In: Doyle MP, Beuchat LR, Montville TJ, Eds. Food Microbiology Fundamentals and Frontiers. Washington, DC: ASM Press 1997; pp. 159-70.

[40]   Rhodes MW, Kator H. Survival of *Escherichia coli* and *Salmonella* spp. in estuarine environments. Appl Environ Microbiol 1988; 54(12): 2902-7.
[PMID: 3066291]

[41]   Lees D. Viruses and bivalve shellfish. Int J Food Microbiol 2000; 59(1-2): 81-116.
[http://dx.doi.org/10.1016/S0168-1605(00)00248-8] [PMID: 10946842]

[42]   Caul EO. Foodborne viruses. In: Lund BM, Baird-Parker TC, Gould GW, Eds. The microbiological safety and quality of foods. Gaithersberg: Aspen Publishers Inc. 2000; pp. 1457-89.

[43]   Lees D. Control measures in seafood. In: Advisory Committee on Microbiological Safety of Foods Workshop on Foodborne Viral Infections. London, UK. 1995; pp. 61-71.

[44]   Opinion of the Scientific Committee on Veterinary Measures relating to Public Health on Norwalk-like viruses. Brussels, Belgium 2002.

[45]   Gerba CP, Haas CN. Assessment of risks associated with enteric viruses in contaminated drinking water. American Society for Testing and Materials 1988; p. 976.
[http://dx.doi.org/10.1520/STP26732S]

[46]   Gantzer C, Dubois E, Crance JM, *et al.* Influence of environmental factors on the survival of enteric viruses in seawater. Oceanol Acta 1998; 21: 983-92.
[http://dx.doi.org/10.1016/S0399-1784(99)80020-6]

[47]   Tang YW, Wang JX, Xu ZY, Guo YF, Qian WH, Xu JX. A serologically confirmed, case-control study, of a large outbreak of hepatitis A in China, associated with consumption of clams. Epidemiol Infect 1991; 107(3): 651-7.
[http://dx.doi.org/10.1017/S0950268800049347] [PMID: 1661240]

[48]   Fiore AE. Hepatitis A transmitted by food. Clin Infect Dis 2004; 38(5): 705-15.
[http://dx.doi.org/10.1086/381671] [PMID: 14986256]

[49]   Deardoff TL, Overstreet RM. Seafood transmitted zoonosis in the United States: the fishes, the dishes and the worms In: Ward DR, Hackney CR, Eds. Microbiology of marine food products. New York, USA 1991; pp. 211-65.
[http://dx.doi.org/10.1007/978-1-4615-3926-1_9]

[50]   Olson RE. Marine fish parasites of public health importance In: Kramer DE, Liston J, Eds. Seafood quality determination. The Netherlands: Elsevier Science Publishers 1987; pp. 339-55.

[51]   Food and Drug Administration. Fish and fishery products hazards and controls guidance. In: Department of Health and Human Services, Public Health Service, Food and Drug Administration, Center for Food Safety and Applied Nutrition, Office of Food Safety, Eds. Guidance for the Industry: Fish and Fishery Products Hazards and Controls Guidance, 4th Edition. Gainesville, USA. 2011; p. 468.

[52]   Ives K. *Cryptosporidium* and water supplies: Treatment process and oocyst removal. *Cryptosporidium* in water supplies. London, UK: Her Majesty's Stationery Office 1990; pp. 154-84.

[53]   Fayer R. Waterborne and Foodborne Protozoa. In: Hui YH, Sattar SA, Murell KD, Nip WK, Stanfield PS, Eds. Foodborne Disease Handbook. 2nd ed. New York, US: Marcel Dekker Inc. 2001; pp. 289-321.

[54]   Bristow GA, Berland B. A report of some metazoan parasites of wild marine salmon (*Salmo salar* L.) from the west coast of Norway with comments on their interactions with farmed salmon. Aquaculture 1991; 98: 223-9.
[http://dx.doi.org/10.1016/0044-8486(91)90395-N]

[55]   Adams AM, Murrell KD, Cross JH. Parasites of fish and risks to public health. Rev - Off Int Epizoot 1997; 16(2): 652-60.
[PMID: 9501379]

[56]   Higashi GH. Foodborne parasites transmitted to man from fish and other aquatic foods. Food Technol 1985; 39: 69-111.

[57]   Karl H, Roepstorff A, Huss HH, Bloemsma B. Survival of *Anisakis* larval in marinated herring fillets. Int J Food Sci Technol 1995; 29: 661-70.
[http://dx.doi.org/10.1111/j.1365-2621.1994.tb02107.x]

[58]   Pravettoni V, Primavesi L, Piantanida M. *Anisakis simplex*: current knowledge. Eur Ann Allergy Clin Immunol 2012; 44(4): 150-6.
[PMID: 23092000]

[59]   Kino H, Hori W, Kobayashi H, Nakamura N, Nagasawa K. A mass occurrence of human infection with *Diplogonoporus grandis* (Cestoda: *Diphyllobothriidae*) in Shizuoka Prefecture, central Japan. Parasitol Int 2002; 51(1): 73-9.
[http://dx.doi.org/10.1016/S1383-5769(01)00106-4] [PMID: 11880229]

[60]   Chai J-Y, Lee S-H. Food-borne intestinal trematode infections in the Republic of Korea. Parasitol Int 2002; 51(2): 129-54.
[http://dx.doi.org/10.1016/S1383-5769(02)00008-9] [PMID: 12113752]

[61]   Tesana S, Kaewkes S, Phinlaor S. Infectivity and survivorship of *Opisthorchis viverrini* metacercariae in fermented fish. J Parasitol Trop Med Assoc Thailand 1986; 9: 21-30.

[62]   Fan PC. Viability of metacercariae of *Clonorchis sinensis* in frozen or salted freshwater fish. Int J Parasitol 1998; 28(4): 603-5.
[http://dx.doi.org/10.1016/S0020-7519(97)00215-4] [PMID: 9602382]

[63]   Sithithaworn P, Phinlor S, Tesana S, Keawkes S, Srisawangwonk T. Infectivity of *Opisthorchis viverrini* metacercariae stored at 4A C. J. Trop Med Parasitol 1991; 14: 14-20.

[64]   Ababouch L, Gandini G, Ryder J. Causes of detentions and rejections in international fish trade. FAO Fisheries and Aquaculture Technical Paper. 2005; p. 110.

[65]   Price RJ. Residue concerns in seafoods. Dairy Food Environ Sanit 1992; 12: 139-43.

[66]   Cabello FC. Heavy use of prophylactic antibiotics in aquaculture: a growing problem for human and animal health and for the environment. Environ Microbiol 2006; 8(7): 1137-44.
[http://dx.doi.org/10.1111/j.1462-2920.2006.01054.x] [PMID: 16817922]

[67]   World Health Organization in collaboration, Food and Agriculture Organization of the United Nations, World Organisation for Animal Health. Report of a Joint FAO/OIE/WHO Expert Consultation on

Antimicrobial Use in Aquaculture and Antimicrobial Resistance Seoul, Republic of Korea. 2006; p. 97.

[68] Lupin HM, Subasinghe R, Alderman D. Antibiotic residues in aquaculture products. The state of world fisheries and aquaculture 2002. Rome, Italy: FAO 2003; pp. 74-83.

[69] Tan LK. Chloramphenicol-induced aplastic anaemia--should its topical use be abandoned? Singapore Med J 1999; 40(7): 445-6.
[PMID: 10560268]

[70] Acar J, Rostel B. Antimicrobial resistance: an overview. Rev - Off Int Epizoot 2001; 20(3): 797-810.
[PMID: 11732423]

[71] Alderman DJ, Hastings TS. Antibiotic use in aquaculture. Int J Food Sci Technol 1998; 33: 139-55.
[http://dx.doi.org/10.1046/j.1365-2621.1998.3320139.x]

[72] Food and Agriculture Organization. Code of Conduct for Responsible Fisheries. Rome, Italy: FAO 1995; p. 41.

[73] Howgate P. Review of the public health safety of products from aquaculture. Int J Food Sci Technol 1998; 33: 99-125.
[http://dx.doi.org/10.1046/j.1365-2621.1998.3320099.x]

[74] Grave K, Lillehaug A, Lunestad BT, Horsberg TE. Prudent use of antibacterial drugs in Norwegian aquaculture? Surveillance by the use of prescription data. Acta Vet Scand 1999; 40(3): 185-95.
[PMID: 10605135]

[75] Samuelson OB, Lunestad T, Husevag B, Holleland T, Ervik A. Residues of oxolinic in wild fauna following medication in fish farms. Dis Aquat Organ 1992; 12: 111-9.
[http://dx.doi.org/10.3354/dao012111]

[76] Coyne R, Hiney M, O'Connor B, Kerry J, Cazabon D, Smith P. Concentration and persistence of oxytetracycline in sediments under a marine salmon farm. Aquaculture 1994; 123: 31-42.
[http://dx.doi.org/10.1016/0044-8486(94)90117-1]

[77] Food and Agriculture Organization. Responsible use of antibiotics in aquaculture. Fisheries Technical Paper 2005; 469: 97.

[78] Bjorklund HV, Raberg CM, Bylund G. Residues of oxolinic acid and oxytetratcycline in fish and sediments from fish farms. Aquaculture 1991; 97: 85-96.
[http://dx.doi.org/10.1016/0044-8486(91)90281-B]

[79] Midvedt T, Lingass E. Putative public health risks of antibiotic resistance development in aquatic bacteria. In: Michael C, Alderman DJ, Eds. Chemotherapy in aquaculture: from theory to reality. Paris, France: Office International d?(tm)Epizooties 1992; pp. 302-14.

[80] Serrano PH. Responsible use of antibiotics in aquaculture. FAO Fisheries Technical Paper. 2005; 469: p. 97.

[81] Srivastava S, Sinha R, Roy D. Toxicological effects of malachite green. Aquat Toxicol 2004; 66(3): 319-29.
[http://dx.doi.org/10.1016/j.aquatox.2003.09.008] [PMID: 15129773]

[82] National Toxicology Program. Toxicology and carcinogenesis studies of malachite green chloride and

leucomalachite green. (CAS NOS. 569-64-2 and 129-73-7) in F344/N rats and B6C3F1 mice (feed studies). Natl Toxicol Program Tech Rep Ser 2005; 527(527): 1-312.
[PMID: 15891780]

[83]   Okamoto A. Restrictions on the use of drugs in aquaculture in Japan. In: Michel C, Alderman DJ, Eds. Chemotheraphy in Aquaculture. Paris, France: From Theory to Reality. Office International d?(tm)Epizooties 1992; pp. 109-14.

[84]   Tittlemier SA, Van de Riet J, Burns G, *et al.* Analysis of veterinary drug residues in fish and shrimp composites collected during the Canadian Total Diet Study, 1993-2004. Food Addit Contam 2007; 24(1): 14-20.
[http://dx.doi.org/10.1080/02652030600932937] [PMID: 17164212]

[85]   Capone DG, Weston DP, Miller V, Shoemaker C. Antibacterial residues in marine sediments and invertebrates following chemotherapy in aquaculture. Aquaculture 1996; 145: 55-75.
[http://dx.doi.org/10.1016/S0044-8486(96)01330-0]

[86]   Commission of the European Communities. Commission Decision 2001/699/EC concerning certain protective measures with regard to certain fishery and aquaculture products intended for human consumption and originating in China and Vietnam. Official Journal of the European Communities 2001; L125: 11-2.

[87]   Commission of the European Communities. Commission Regulation (EC) No. 466/2001 setting maximum levels of certain contaminants of foodstuffs. Official Journal of the European Communitites 2001; L77: 1-14.

[88]   Focardi S, Corsi I, Franchi E. Safety issues and sustainable development of European aquaculture: New tools for environmentally sound aquaculture. Aquacult Int 2005; 13: 3-17.
[http://dx.doi.org/10.1007/s10499-004-9036-0]

[89]   Smith AG, Gangolli SD. Organochlorine chemicals in seafood: occurrence and health concerns. Food Chem Toxicol 2002; 40(6): 767-79.
[http://dx.doi.org/10.1016/S0278-6915(02)00046-7] [PMID: 11983271]

[90]   Lin AY, Yu T-H, Lin CF. Pharmaceutical contamination in residential, industrial, and agricultural waste streams: risk to aqueous environments in Taiwan. Chemosphere 2008; 74(1): 131-41.
[http://dx.doi.org/10.1016/j.chemosphere.2008.08.027] [PMID: 18829065]

[91]   Commission of the European Communities. Commission regulation (ec) No 1881/2006 of 19 December 2006 setting maximum levels for certain contaminants in foodstuffs of the European Parliament and of the Council. Official Journal of the European Communitites 2006; L 364: 5-24.

[92]   Legrand AM. Ciguatera toxins: origin, transfer through the food chain and toxicity to humans. In: Reguera B, Blanco J, Fernandez M, Wyatt T, Eds. Harmful Algae, Proceedings of the VIII International Conference on Harmful Algae. 39-43.

[93]   Scoging AC. Marine biotoxins. Symp Ser Soc Appl Microbiol 1998; 27: 41S-50S.
[http://dx.doi.org/10.1046/j.1365-2672.1998.0840s141S.x] [PMID: 9750361]

[94]   Lehane L, Lewis RJ. Ciguatera: recent advances but the risk remains. Int J Food Microbiol 2000; 61(2-3): 91-125.
[http://dx.doi.org/10.1016/S0168-1605(00)00382-2] [PMID: 11078162]

[95]   Food and Agriculture Organization. Marine biotoxins. Food and Nutrition. 2004; 80: p. 233.

[96]   De Fouw JC, Van Egmond HP, Speijers GJ. Ciguatera fish poisoning: A review. RIVM Report No388802021 1999; 66.

[97]   Lehane L, Olley J. Histamine fish poisoning revisited. Int J Food Microbiol 2000; 58(1-2): 1-37.
[http://dx.doi.org/10.1016/S0168-1605(00)00296-8] [PMID: 10898459]

[98]   Bartholomew BA, Berry PR, Rodhouse JC, Gilbert RJ, Murray CK. Scombrotoxic fish poisoning in Britain: features of over 250 suspected incidents from 1976 to 1986. Epidemiol Infect 1987; 99(3): 775-82.
[http://dx.doi.org/10.1017/S0950268800066632] [PMID: 3428380]

[99]   Okuzumi M, Yamanaka H, Kubozuka T. Occurrence of various histamine-forming bacteria on/in fresh fishes. Bulletin of the Japanese Society of Scientific Fisheries 1984; 50: 161-7.
[http://dx.doi.org/10.2331/suisan.50.161]

[100]   Ababouch L. Histamine food poisoning: An update. Fish Tech News 1991; 11: 3-5.

[101]   Shalaby AR. Significance of biogenic amines to food safety and human health. Food Res Int 1996; 29: 675-90.
[http://dx.doi.org/10.1016/S0963-9969(96)00066-X]

[102]   Taylor SL. Histamine food poisoning: toxicology and clinical aspects. Crit Rev Toxicol 1986; 17(2): 91-128.
[http://dx.doi.org/10.3109/10408448609023767] [PMID: 3530640]

[103]   Gessner BD, Hokama Y, Isto S. Scombrotoxicosis-like illness following the ingestion of smoked salmon that demonstrated low histamine levels and high toxicity on mouse bioassay. Clin Infect Dis 1996; 23(6): 1316-8.
[http://dx.doi.org/10.1093/clinids/23.6.1316] [PMID: 8953082]

[104]   Klausen NK, Huss HH. Growth and histamine production by *Morganella morganii* under various temperature conditions. Int J Food Microbiol 1987; 5: 147-56.
[http://dx.doi.org/10.1016/0168-1605(87)90032-8]

[105]   Ijomah P, Clifford MN, Walker R, Wright J, Hardy R, Murray CK. The importance of endogenous histamine relative to dietary histamine in the aetiology of scombrotoxicosis. Food Addit Contam 1991; 8(4): 531-42.
[http://dx.doi.org/10.1080/02652039109374005] [PMID: 1806404]

[106]   JA,rgensen LV, Huss HH, Dalgaard P. The effect of biogenic amine production by single bacterial cultures and metabiosis on cold-smoked salmon. J Appl Microbiol 2000; 89(6): 920-34.
[http://dx.doi.org/10.1046/j.1365-2672.2000.01196.x] [PMID: 11123465]

[107]   Food and Agriculture Organization Fisheries and Aquaculture Department. Global Aquaculture Production Statistics for the year , 2011 [2014 Jul 21]; Available from: ftp://ftp.fao.org/ FI/news/GlobalAquacultureProductionStatistics2011.pdf.

[108]   Commission of the European Communities. Council Directive 91/493/EEC of 22 July 1991 laying down the health conditions for the production and the placing on the market of fishery products. Official Journal of the European Communities 1991; L 268: 0015-34.

[109]   World Health Organization, Food and Agriculture Organization. Working principles for risk analysis for application by Governments Codex Alimentarius. Rome, Italy: Commission 2007; p. 33.

[110]   World Health Organization, Food and Agriculture Organization. General principles of food hygiene Codex Alimentarius. Rome, Italy: Commission 2003; p. 31.

[111]   World Health Organization, Food and Agriculture Organization. Code of practice for fish and fishery products Codex Alimentarius. Rome, Italy: Commission 2005; p. 98.

[112]   Food and Agriculture Organization. Increasing the contribution of small-scale fisheries to poverty alleviation and food security. Technical guidelines for responsible fisheries . Rome, Italy 2005; p. 79.

[113]   Derrick S, Dillon M. Guide to traceability within the fish industry. Eurofish, Humber Institute of Food and Fisheries, Swiss Import Promotion Programme. Denmark: CCopenhagen 2004; p. 78.

[114]   World Health Organization, Food and Agriculture Organization. Code of Practice for Fish and Fishery Products. 2nd ed., Rome, Italy: Codex Alimentarius Commission 2012.

[115]   Huss HH, Dillon M, Derrick S. A guide to seafood hygiene management. Accessing the European and American market Eurofish, Humber Institute of Food and Fisheries, Swiss Import Promotion Programme. Denmark: Copenhagen 2005; p. 76.

[116]   Dalgaard P. Qualitative and quantitative characterization of spoilage bacteria from packed fish. Int J Food Microbiol 1995; 26(3): 319-33.
        [http://dx.doi.org/10.1016/0168-1605(94)00137-U] [PMID: 7488527]

[117]   Goulding I, do Porto O. Manual/handbook for the execution of sanitary inspection of fish as raw material and fish products as food for human consumption Secretariat of ACP Group of States Strengthening Fishery Products Health Conditions in ACP/OCT Countries (SPF) , 2014 [2014 Jul 22]; Available from: http://www.sfp-acp.eu/EN/B15-Handbook.htm

[118]   World Health Organization, Food and Agriculture Organization. Codex general standard for the labelling of prepackaged foods Codex Alimentarius. Rome, Italy: Commission 2005; p. 7.

[119]   Hobbs G. Changes in fish after catching. In: Aitken A, Mackie IM, Merritt JH, Windsor ML, Eds. Fish handling and processing. Edinburgh, UK: Torry Research Station 1982; pp. 20-7.

[120]   Flick GJ, Oria MP, Douglas L. Potential hazards in cold-smoked fish: Biogenic Amines. J Food Sci 2001; 66: S1088-99.
        [http://dx.doi.org/10.1111/j.1365-2621.2001.tb15528.x]

[121]   Clifford MN, Walker R, Ijomah P, Wright J, Murray CK, Hardy R. Is there a role for amines other than histamines in the aetiology of scombrotoxicosis? Food Addit Contam 1991; 8(5): 641-51.
        [http://dx.doi.org/10.1080/02652039109374018] [PMID: 1818838]

[122]   Luten JB, Bouquet W, Seuren LA, *et al.* Biogenic amines in fishery products. Standardization methods within EC. In: Huss HH, Jakobsen M, Liston J, Eds. Quality assurance in the fish industry. The Netherlands: Elsevier Science Publishers 1992; pp. 427-40.

[123]   Huss HH. Control of indigenous pathogenic bacteria in seafood. Food Contr 1997; 8: 91-8.
        [http://dx.doi.org/10.1016/S0956-7135(96)00079-5]

[124]   Middlebrooks BL, Toom PM, Douglas WL, Harrison RE, McDowell S. Effects of storage time and temperature on the microflora and amine development in Spanish mackerel (*Scomberomorus*

maculatus). J Food Sci 1988; 53: 1024-9.
[http://dx.doi.org/10.1111/j.1365-2621.1988.tb13522.x]

[125]   Gram L, Huss HH. Fish and shellfish products. In: Lund BM, Baird-Parker TC, Gould GW, Eds. The microbiological safety and quality of foods. Gaithersberg, USA: Aspen Publishers Inc. 2000; pp. 472-506.

[126]   Gram L, Wedell-Neergaard C, Huss HH. The bacteriology of fresh and spoiling Lake Victorian Nile perch (*Lates niloticus*). Int J Food Microbiol 1990; 10(3-4): 303-16.
[http://dx.doi.org/10.1016/0168-1605(90)90077-I] [PMID: 2397157]

[127]   D'Aoust J-Y, Gelinas R, Maishment C. Presence of indicator organisms and recovery of *Salmonella* in fish and shellfish. J Food Prot 1980; 43: 769-82.

[128]   Gram L. Potential hazard in cold-smoked fish: *Listeria monocytogenes.* J Food Sci 2001; 66: S1072-81.
[http://dx.doi.org/10.1111/j.1365-2621.2001.tb15526.x]

[129]   DorA(c) WJ, Henshilwood K, Lees DN. Evaluation of F-specific RNA bacteriophage as a candidate human enteric virus indicator for bivalve molluscan shellfish. Appl Environ Microbiol 2000; 66(4): 1280-5.
[http://dx.doi.org/10.1128/AEM.66.4.1280-1285.2000] [PMID: 10742200]

[130]   Graham J, Johnston WA, Nicholson FJ. Ice in fisheries FAO Fisheries Technical Paper No 331 . Rome, Italy 1992; p. 75.

CHAPTER 8

Processed Fishery Products

Maria João Fraqueza[1,*] and **Manuel Abreu Dias[2]**

[1] *CIISA/Faculdade de Medicina Veterinária, Universidade de Lisboa, Avenida da Universidade Técnica, 1300-477 Lisboa, Portugal*

[2] *Alicontrol, Tecnologia e Controlo de Alimentos, Lda., Rua Fernando Vaz, lote 26-B, 1750 108 Lisbon, Portugal*

Abstract: Many fish preservation technologies are based on multiple hurdles used to inhibit or reduce biochemical changes and microbial contamination that leads to spoilage. This chapter aims to describe some of the most traditional technologies currently applied to fishery products based on the control of water activity, pH, oxide-reduction potential, temperature, relative humidity and gaseous composition of the atmosphere. Emergent technologies, such as Ohmic heating and microwaving; amongst others, can be seen with particular interest for future application at industrial scale. In fact, the combination of traditional and emergent technologies (thermal or non-thermal processes) in processed fishery products, might be the trend to achieve and supply safer and high quality products.

Keywords: Chemical treatments, Emergent technologies, Non-thermal treatments, Processed fishery products, Thermal treatments.

INTRODUCTION

A major purpose of the processed fish production is to preserve fish, minimising losses in the distribution chain, over time and space. Another important goal for that industry is to add value to raw fish materials, which might be related to the application of emergent technologies, in order to develop new fishery products. Nowadays, there is an increased demand of ready-to-eat food products, or those

* **Address correspondence to Maria João Fraqueza:** Faculdade de Medicina Veterinária, Universidade de Lisboa, 1300-477 Lisboa, Portugal; E-mail: mjoaofraqueza@fmv.ulisboa.pt

Manuela Oliveira, Fernando Bernardo, Joana I. Robalo (Eds.)

requiring little or no preparation before serving. So, the further industrial processing of fish is not only common, but also desirable, to sustain the demand for a wide variety of value-added products. Fishery products processing industry receives raw fish materials (sardines, tuna, cod, mackerel, anchovy, crustaceans, molluscs) directly from fishery and preserves them using traditional preservation technologies. Presently, these technologies are more mechanized and automated in some processing steps. Processed fishery products include preparations, canned, chilled, frozen, smoked and dried fish, crustaceans and molluscs. The technologies for fish preservation aim the inhibition or reduction of the metabolic and biochemical changes that lead to spoilage. These preservation technologies are based on the control of specific parameters such as water activity, pH, oxide-reduction potential, temperature, relative humidity and gaseous composition of the atmosphere surrounding food under packaging. In general, technological treatments are physical (thermal treatments: cooling, freezing, pasteurization, sterilization, Ohmic heating, microwaving, and non-thermal treatments: irradiation, high hydrostatic pressure, ultrafiltration, pulsating electric fields, ultrasound and even dehydration) or chemical (addition of sugar, salt, acids, additives), the last with repercussion on the chemical composition of foods. Some of these technologies are currently applied to fishery products processes (pasteurization, sterilization, dehydration, addition of sugar, salt, acids, additives), others have been tested but without large industrial application (irradiation) while emergent technologies, such as Ohmic heating, microwaving, high hydrostatic pressure, pulsating electric fields, may be seen with particular interest by industry. Many fish preservation technologies are based on Leistner's [1] multiple hurdle theory, such as pasteurization-refrigeration, cook-chill, modified atmosphere packaging - refrigeration, salting-drying, salting-smoking, drying-smoking and salting-marinating .

All these technological processes may be applied to raw fish material aiming to have the best quality; however, it is mandatory to assure food safety. With that purpose, not only unitary preservation technologies should be applied but also safety measures founded on proactive methodologies such as Hazard Analysis Critical Control Point (HACCP) method. Finally, proper waste structures should be included in fish processing operations.

PRELIMINARY PROCESSING FISH OPERATIONS

The type of raw fish material affects its processing, being relevant to distinguish demersal fish (codfish, flatfish), pelagic fish, crustaceans (shrimp, prawn, lobster) and molluscs. Preliminary processing should assure the best market quality and safety of fish products, a suitable presentation of semi- processed final product, labor saving on raw materials processing, and waste reduction. Considering different fish, crustaceans and molluscs raw materials, with different specificities that enter a processing unit and all the obtained final products, proactive methodologies need to be implemented for hazards prevention and for decreasing human health and environmental risks.

Concepts related to HACCP methodologies and "cleaner production" must not be forgotten. The application of a proactive methodology, such as HACCP, involves a precise methodology of preventive measures to be managed throughout the fish products processing steps, according to an established plan which is a document for identifying, evaluating and controlling food safety hazards [2]. "Cleaner production" is a concept that seeks efficiency improvement and reduction of risks to humans and environment by the applying precautionary ecological approaches to practices, produces and services. It intends the preservation of fresh resources and energy, removal of toxic constituents, and decrease in quantity and toxicity of litters and discharges [3]. Typically, a fish processing industry spends considerable amounts of energy and water, and releases substantial amounts of biological effluents and solid wastes. Also, because of all the hand labor and manual operations, the fish industry is dependent on the operator performance, which in turn influences plant performance, mainly in small-scale units and low automated automat operations. The main steps of fish preliminary processing can be illustrated in a diagram as shown in Fig. (**1**). Fish is a perishable raw material, and immediately after death must be beheaded, gutted, washed and chilled to prevent enzymatic degradation and spoilage by microorganisms. Main fish preliminary processing operations are common to different methods of preservation beginning by sorting, grading, washing, flaking, head and tail removal, gutting, de-skinning, filleting, and trimming.

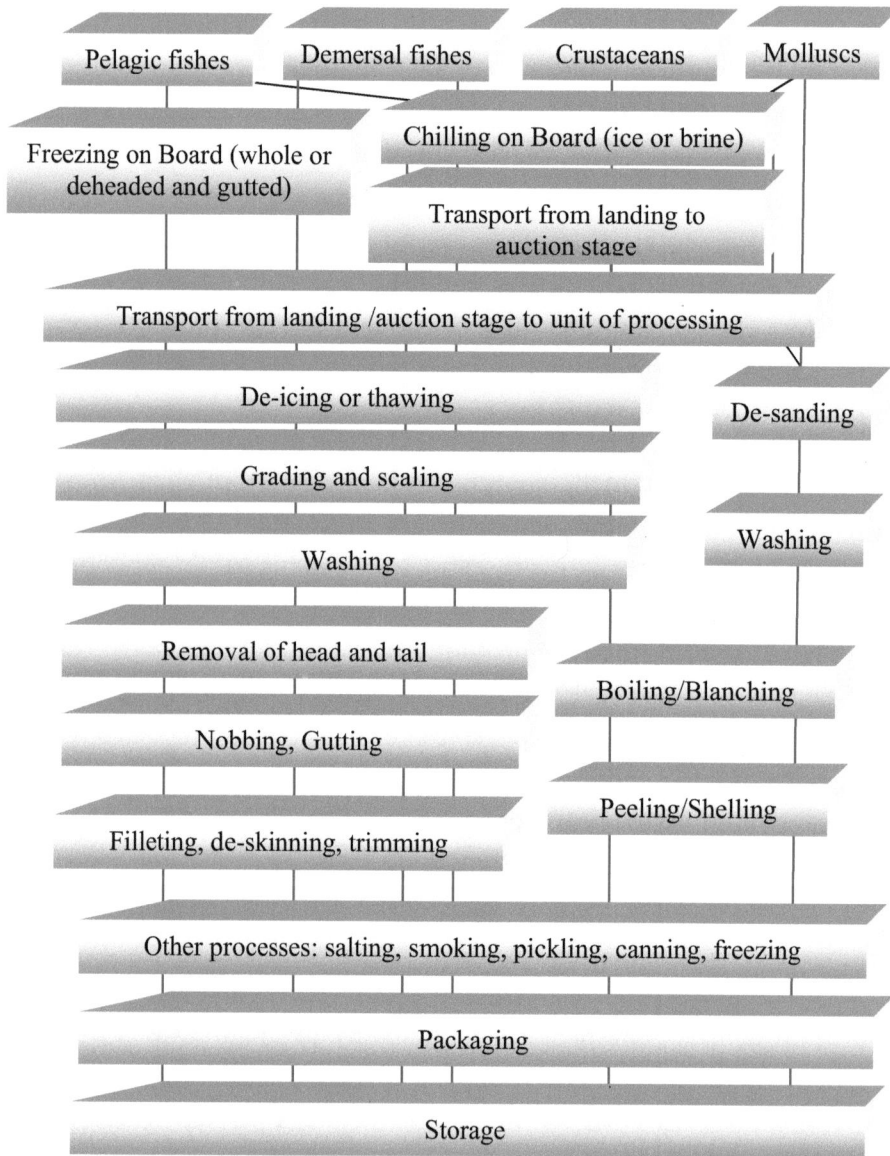

Fig. (1). Preliminary operations in fish processing industry.

Demersal fish processing is usually quite straightforward, but pelagic fish involves further processing for the production of more elaborated foods, including pickled herring and canned mackerel and sardine. Crustaceans and bivalve molluscs can be transformed into several products, being usually boiled. Fish transport inside an industry should be seen as a fundamental operation that contributes to a cleaner production. Dry transport is a better solution to avoid large quantities of water waste and to reduce the content of organic material in wastewater, with a high benefit for the environment [4], but it damages fish more than water transportation. This kind of transport is able to move and position fish in the machinery by the use of specific filter conveyer belts and brushes [4].

Sorting and grading can be done by specie and size, either manually or mechanized. Roller, vibration and belt graders are commercially available depending on the fish products to be graded. Small or large pelagic, demersal fish and crustaceans are usually sorted by grading machines. Fish grader is designed and built for high capacity grading of fish into several size categories according to species, ranging from sardines, herring and mackerel to large cod, salmon and tuna. The machines segregate the fish into different grades using as example rollers tracks (Fig. **2**).

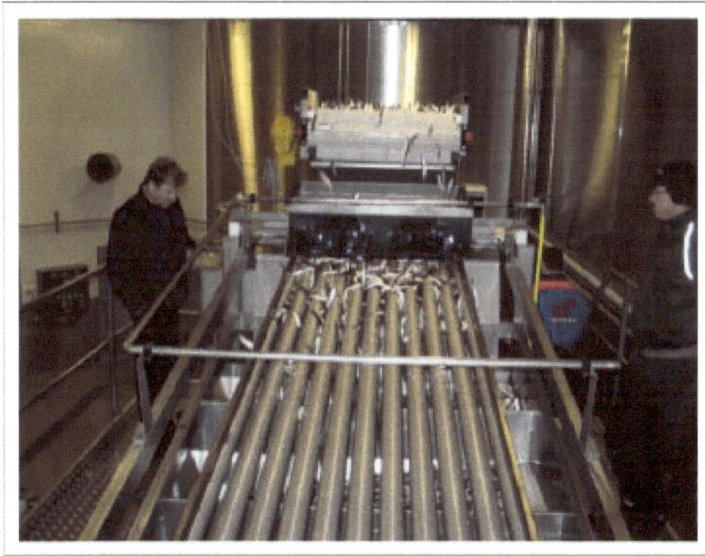

Fig. (2). Roller grader machine performing fish grading (courtesy of Timars manufactor).

The fish fall into the containers or belts positioned below, from where they are separated into storage boxes or transferred to other conveyors for further processing. These equipment's might include water sprays that wash the fish at the same time as it passes down the tracks. In order to clean fish and diminish bacterial contamination, washing is usually performed with potable water in a minimum proportion 1:1 water and fish [5]. The optimal washing is reinforced by the mechanical friction of a rotating drum (with vertical or horizontal axis), depending on rotational speed, fish washer height and angle, angle of water supply and vane angle. This operation step may be performed when the fish enters the processing unit after grading or after head/tail removal and gutting. In small fishes, such as sardines, horse mackerel and herring, the head and guts will be removed in a single operation named nobbing. Head and tail removal and gutting help in bacterial content elimination from fish core but will expose muscles to external conditions and to microorganisms access; the same can be reported in filleting [6].

Nobbing machines are available for heads, tails and viscera removal of sardines or other small pelagic fishes (Fig. **3**).

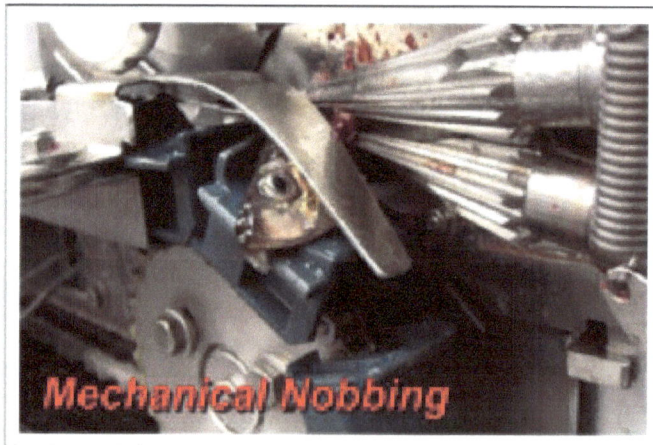

Fig. (3). Nobbing machine for small pelagic fishes (courtesy of Baader® manufactory).

The removal of guts without injuring the throat was developed by Lubeck Werner Wenzel (1974) [7]. The head cut is performed without cutting the guts connected to it; their extraction is favored by traction when removing the head. Fish may be fed to nobbing machines manually or automatically. Nobbing machinery has two rotating knives to cut heads and tails with adjustable position for different sizes of fish. Head cut for large pelagic or demersal species can be performed around the operculum named round cut, or cross or oblique section, or V-shaped, and the main objective is to cut the head with minimal loss of muscle. Fillet can be defined as a cut of dorsal and abdominal muscles, cutted on a plan parallel to the backbone, being the most popular presentation of fish in the market. Filleting is usually performed on large pelagic and demersal fish and can be done manually or automatically. Filleting machinery must have a high performance, sensibility and should be easily cleaned and disinfected. De-skinning is performed after obtaining the fillet. Fish skinning machines are old in the art. In fish skinning machines, the fish tail is delivered first to the nip between a feed roll and the pinch roll which pulls it downwardly toward the cutting edge of a cutter for separating the skin from the flesh. Equipment developments have been made in order to improve the delivery of the fish to the nip between the feed roll and the pinch roll, or using a frozen drum where the skin is glued for a better positioning and operation of the cutter, adjustment of the cutter with respect to the feed roll, clearing the ends of the cutter, prevention of oil loss and ingress of water and attachment of the feed conveyor as a separable unit to facilitate repair and adjustments [8]. For the fillets de-skinning, there is simple equipment, mainly for small units' use, that consists of a space-saving table machine for the smooth skinning of fine fishes without damaging the silver skin. A water spray device ensures that the skin waste is rinsed away with continuous cleaning of the tooth roller. Compared with manual operations, this machine improves the skinning process. Up to 40 fillets can be skinned per minute depending on fillet size and machine type, being required a conveyor to move the fillets when the speed is high [5]. In the last years, a new trend has emerged in fish products processing with the production of minced meat after separation of non-consumable portions, including bones, skin, scales. Fish bone separators offers a new perspective in products innovation that could gain customer's appreciation and be successfully marketed. In filleting operations, a substantial quantity (30-50%) of meat remains attached to the fish's backbone and ribs. Less valuable fish species are also used to produce minced meat, after deheading and body cavities cleaning operations. This minced meat may be used immediately or frozen and stored up to 6 months at -25 °C to -28 °C. Minced meat is used to produce fish burgers, fish sticks, canned fish, vegetable mixes and fish dumplings. Minced meat can also be dried and later added to fish soups. Fig. (**4**) shows a meat separator where fish fillets or carcasses are squeezed through holes

(3-7 mm in diameter) into the cylinder under pressure applied by a conveyer belt partially encircling the cylinder [9]. The cylinder rotates slightly faster than the conveyor. According with the type and size of raw product and with the holes' diameter, the pressure applied by the conveyor to the cylinder can be regulated.

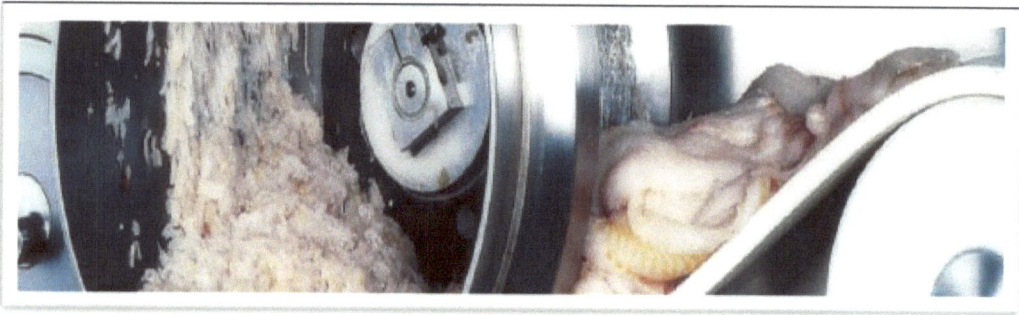

Fig. (4). Fish meat separator machine for fillets or fish carcasses (courtesy of Baader® manufactory).

A shrimp peeling system was originally developed by Gregor Jonsson in 1938. Nowadays, various production machines are available in the shrimp processing industry for processing shrimp. In a conventional production operation, shrimp is transported to various work stations where the shrimp shell is first cut and then the cut shell and vein are removed. Often shrimps are transported to the work stations by being deposited in a plurality of trays located on a conveyor traveling at intermittent speeds through the shrimp processing equipment. A shrimp to be processed is removed from the tray and retained in a clamp assembly which delivers the shrimp to a plurality of work stations to remove the shells and devein the shrimp [10]. There is a continuous improvement in equipment performance and capabilities, regarding a faster processing with less handling in order to avoid raw material exposure to time/temperature abuse and to diminish bacterial contamination contributing to safer and higher quality product. In modern fish products processing, the use of automated machinery in preliminary steps is usual, but considerations must be taken regarding dimensions and performances, to have an efficient and effective production; the equipment sequence should avoid crossing lines processes, transport systems must be available for continuous production lines. A complete automation without manual work should be planed because raw material characteristics (regarding different species, sizes and forms) could give, in some cases, constriction of line processes with quality loss of the final products. Cephalopods and bivalves are used as raw materials in the processing industry. Their preliminary preparation includes washing, de-sanding, blanching and shelling. Cephalopods are usually sterilized in cans. Mussels, cockles, clams and scallops can be pickled or sterilized and presented in cans with salt water excipient. Scallops can be prepared as a very popular ready to cook product breaded or baked in shell.

TRADITIONAL PRESERVATION TECHNOLOGIES
Technologies Based On Water Activity Control
Salting

Salting is the simpler and most used process to preserve fish, based on a long experience and tradition all over the world. Salting might be considered as a method based on osmotic desiccating, with water activity reducing, since salt is included in fish muscles, and during or after this process water may evaporate [11]. The mechanism of salt preservation is inherent to the increase of medium osmotic pressure and consequent water activity reduction, when the salt binds water. This will prevent spoilage microorganisms multiplication thus delaying fish spoilage. With salting, enzyme activity in the fish muscle is reduced, allowing for a longer shelf life, but flavor and textures are kept. Most frequently used raw material is whitefish, either split or filleted. To assure proper processing and curing of salted fish it is important to know which factors influence the variability of raw material such as season, fishing areas and fish age and size [12 - 17]. Also, fishing techniques, raw material handling, and storage technology may have an impact on salted fish production and curing processes' election. Better products are obtained with higher knowledge and fish stock control, *e.g.* fishing at the right time of year. Improved handling practices of raw materials and better control of temperature and relative humidity during processing, curing and storage, as well as packaging should be implemented to obtain higher quality products. The salting process control is essential, being salt concentration and temperature the major features responsible for mass transfer. However, salting is influenced by other factors related to salt purity, particle size and its microbiota.

In fish products, the distribution of water and salt within the fish is influenced by the lipid content, species and freshness, muscle thickness and rigor mortis state; among climate related factors there is temperature and relative humidity to consider [13, 14]. Lipids oxidation may be accelerated if metals like copper (Cu) and iron (Fe) surpass defined salt concentrations. Additional salt metals, including cobalt (Co) and manganese (Mn) might also have pro-oxidant properties; however standard parameters for their presence in salt are not defined.

Lipid hydro-peroxides decomposition will be catalyzed by metals, initiating unsaturated fatty acid derivatives oxidation and production of free radicals. Oxidation might result in fish discoloration, diminishing fish products' quality and value.

The presence of calcium and magnesium in the salt might also affect fish color, taste and texture. These chemicals should be used within certain quantitative limits, to improve fish quality, making it whiter. On proteolytic enzymes calcium has a catalyzing effect and the recommended maximum concentration of use is 0.2% (0.5% $CaCl_2$ or 0.7% $CaSO_4$) [18, 19]. The Ca^{2+} and Mg^{2+} ions also influence the ability of muscle to retain water, by promoting cross-linking of myofibrils and diminishing the area available for water within muscle cells [20].

Elevated ions concentrations are responsible for undesirable alterations in salting rate. When ions bind to proteins of the fish surface layer, they form an obstacle that reduces salt intake and muscle desiccation [21]. For a more proper curing and to eliminate problems, it is recommended to use equal parts of thick and thin salt. Fine salt, with small crystals, has a fast penetration at the beginning of the salting process, however, its penetrating power decreases by the concentration which causes coagulation of proteins in the muscle surface, contributing to poor preservation of the product. Thick salt acts slowly and there is no coagulation of proteins, however, its slow action during the curing process leads to undesirable changes, especially if the curing is processed on hot days. Salt can carry out a contaminant microbiota of halophilic and resistant to salt bacteria such as the Sarcinas halophilic chromogenic bacteria causing red coloring in salted fish products rich in protein. Not all are harmful microorganisms and among halotolerant, there are some species which contribute to the curing of products, such as those from the genus *Halobacterium* and *Micrococcus*. To obtain a salted fish of good quality, good sanitary conditions in processing are of utmost importance. Salted fishes in "rigor mortis" lose less weight than those salted during the "autolysis" state. The rate of salt penetration in fish tissue is inversely proportional to its fat content. In addition to slowing down the salting process, fat content still produces rancidity, giving fish an unpleasant taste. The greater the thickness of fish muscle, the longer will be the curing time, since to get to the fillet center, a longer way will have to be crossed, even with a high speed of salt penetration. The environment temperature in which salting processes occur is of great importance, because of its influence in the salting speed, the higher the temperature, the faster the process will occur. In winter, salting process occurs with greater speed than during the summer, due to high environment relative humidity, thus favoring the rapid formation of brine and, consequently, a rapid penetration of salt in the muscle of fish. Cod, saithe, tusk and ling are the most significant fish stocks used in salted fish production. Over the years, salting methods have progressed from being uniquely dry salting to diverse salting modes. Dry salting consists of the direct application of coarse grain size salt to big sized fishes that have been eviscerated, washed, cut or scaled or filleted. Layers of salt and fish, in the proportion of 1:8 or 1:3 according to the desired amount of salting, are prepared in a pile with the maximum height of 2 meters during several days. For a better salting uniformity fishes must be turned in position daily. During the salting period time, a fluid will flow from the pile with substances in dissolution [6]. The wet salting consists of fish dipping in brines titrated to 18% of NaCl. This option is used particularly on fat fishes such as mackerel and tuna (cutted) since the process counteracts access to oxygen and delays its fat oxidation and subsequent rancidity. Also, dipping in brines avoids the forthcoming of "redness" due to halophytic bacteria present in salt and its multiplication in air. When brine has a salt concentration inferior to 12% fish takes salt from brine, but if that concentration is above 12%, fish will also loose water. A mix salting may be performed and it results by a combination of the two previously described methods. The dry salting of fish is made in a container or barrel with salt. Afterwards, a fluid will flow remaining in contact with the fish and later on saturated brine (36%) can be added. Nowadays, the main steps of white fish salting comprise pre-salting (injection and brining), dry salting and storage until the fish is cured. Machinery was developed to perform fish brine injection. The equipment for fish processing industry was designed with a needle holder prepared with fine needles providing a injection pattern. The injection low pump pressure prevents damages of the meat structure and ensures that the fish is being injected at an accurate and constant level (Fig. **5**).

Fig. (5). Brine injector machine (courtesy of Fomaco).

According to Corominas (2010) [22]; the fluid-injecting machine for fish products comprises one or more injection heads, each of them carrying a plurality of parallel hollow needles that can be retracted by contrast elastics positioned in the opposite end of the needle tip. The injection head moves in up and down double movement, by (generally hydraulic) drive means in relation to the conveyor plane. Each needle has an inlet opening leading into its cannula that can be arranged in a pressurized fluid injection chamber connected to the fluid supplying source.

Generally, salting is an operation combined with other water activity reducing technology, such as drying or smoking. Salted fish is packed and kept in storage. It is important to control both temperature and relative humidity during storage time in order to avoid fish deterioration, particularly its sensorial characteristics. When the product's humidity is too low, changes in water content can take place if relative humidity of the air in the storage room is too high. It is recommended that product's storage occurs at approximately 76% of air humidity in order to preserve all the required properties that buyer's demand. Salted fish is experienced by the consumer in the rehydrated form, regarding taste, smell and texture. The technologists' efforts are directed to the production of salted products with good hygienic and sensorial characteristics. However, health concerns and mandatory directives demand the reduction of sodium chloride level and its replacement with other substances considered less harmful to consumer's health, while maintaining the characteristic salted fish product's taste; new additives that allow a better stabilization and safety during storage of salted products; new material and packages development for salted fish; new salted fish products development with similar taste, resembling that of more valued species (salmon, trout, sturgeon).

SALTING AND RIPENING FISH EXAMPLE: ANCHOVY PROCESS

Anchovy process is a common and traditional method of preparing fish in European countries. It consists of a salted and ripened fish preserved product with specific sensorial characteristics. Chosen species for industrial use include little pelagic fishes including the Atlantic herring (*Clupea harengus*), sprats (*Sprattus sprattus*), sardine (*Sardina pilchardus*) and anchovy (*Engraulis encrasicholus*) .

Fish is wet salted in a container or barrel interposed with layers of salt. The saturated brine draining to the bottom of the container will be reused to sprinkle the fish; this salting process will be finished after one month. The water activity (aw) usually decreases from 0.99 (raw fish) to 0.80-0.84 [23], which controls the eventual pathogenic or deteriorative microorganisms development during subsequent manipulation. The barrels are kept in a different compartment for 3-12 months to promote ripening, according to chamber's room temperature and the desired level of maturation. Temperature is one of the process variables applied for controlling the brining period [14 - 28]. Variations on anchovies chemical composition during harvest is another factor that must be considered for brining stage optimization. This is common in pelagic fish species such as *E. anchoita* or *Sardina pilchardus* with important variability of lipid and water contents, as have been reported, 1.68-10.04 g. 100 g^{-1} and 72.34- 79.56 g. 100 g^{-1}, respectively [12 - 17]. The intrinsic variable - fish lipid content- has a major influence in mass transfer rate during salting, as fatter fishes require more time reach stability

[29, 30]. Also, brining temperature promotes the rise of the mass transfer kinetics.

Still, these factors do not affect water and salt concentrations at the equilibrium stage. The fish will get a pleasant sui generis flavour, a reddish colour and a firm texture which allows to easily remove fish head, guts and skin and filleting. The fish fillets will be packed in a rectangular can extended or rolled in capers. The cans will be filled with olive oil or vegetable oil and sealed. The lower water activity achieved will prevent microbial spoilage. This product is not submitted to a thermal treatment but has a long shelf life being categorized as a semi-preserved fish can (Fig. **6**).

Fig. (6). Semi-preserved anchovy products (original).

DRYING

The aim of drying is to reduce the free water available for microorganism's multiplication in order to avoid their spoilage. Bacteria and yeasts stop multiplying when the water content is less than 25% of the fish total weight, and moulds do not support water contents inferior to 15%. Fish is dried by water evaporation from muscles and skin and by water diffusion from deeper layers to the surface. Drying takes place in two distinct phases. In the beginning, drying rate relies on air velocity and relative humidity around the fish and depends directly on the evaporation rate and water diffusion velocity to surface; when the surface becomes dried, the drying speed is reduced. In this first phase called the 'constant rate period', if the surrounding air conditions remain constant, the rate of drying will remain constant. The second phase of drying called 'falling rate period' begins once all of the surface moisture has been carried away. This phase depends on the rate at which moisture can be carried to the fish surface. The rate of moisture movement to the surface is reduced as the fish moisture concentration falls and the drying rate becomes slower [11]. Temperature, air relative humidity, fish form and orientation to air flow and speed have direct influence on the dehydration rate. So, in the process beginning it is possible to increase its rate with an increase of temperature and air speed, however, if in excess, it will form an impermeable crust. Fish drying methods comprehend a large range of procedures from traditional sun-drying to high technology industrial processes computer controlled.

NATURAL DRYING

Drying fish by deliberately exposing it to the effects of sun, wind or bonfires hanging, are probably the oldest methods to preserve this perishable raw material. The time needed to complete the natural drying process is variable and depends on fish size and thickness, processing unit, geographical and local weather. This last factor is not possible to control by man and the standard quality of the final product might not always be achieved. Natural drying is a slow process and it is weather dependent. Weather instability, with cloudy days and high relative humidity, can cause bacterial and molds multiplication in the fish, while hot days might superheat the fish giving it a "cooked" aspect. Also, sardines and other fatty fish can become rancid during the drying process; in fact, their dehydration is slowed with water diffusion being decreased when fat content increases. It is common to salt fish before drying it. The reason is to increase the osmotic pressure of cellular fluids, having an inferior water activity for the same humidity. The constant rate period of dehydration is shortened and dehydration rate is reduced due to water vapour pressure reduction in the fish surface. Drying rate becomes slower during dehydration at decreased speed, since water diffusion through the fish muscle is lowered by salt addition. This is relevant because dry fish exposed to variations of relative humidity might absorb water and microorganism's multiplication, particularly molds, is favored. With the previous addition of salt to fish, the water activity as well as the microbial multiplication will be better controlled.

ARTIFICIAL DRYING

Fish controlled drying started in 1940 by Torry Research Station (England), by the use of equipment with thermodynamic controlled drying conditions [5]. Artificial drying reduces the product's moisture content to a suitable level for its preservation. Artificial drying or mechanical drying can be done through air drying or lyophilisation. In air drying the fish is suspended or placed horizontally in a controlled environment regarding airspeed, temperature and relative humidity to assure the optimal drying rate. The drying camera is a ventilated space with hot air. Steam or an electric heater with an axial fan is used as a convective heating source, increasing thermal efficiency. The hot air current moves along the baking trays. Cool air enters through an inlet, discharging from a damp outlet, contributing to maintain the relative humidity inside the oven.

Before or during application, it will be possible to adjust parameters (ventilation, temperature or relative humidity), in order to find the best conditions according to the fish state. To have high quality products, the drying temperature should be in the range of 30 to 40 °C, the air velocity between 2 and 3 m/s and relative humidity should be 45 to 55% in the dryer.

FREEZE DRYING

Freeze dried fish products are commercially available. Drying can be performed by freezing and water sublimation by vacuum. This process produces final products with a different quality of that of natural drying. Nevertheless, only costly products, including prawns and shrimps, are freeze dried because of the process high cost. These can be present as components of soups mixtures and ready- to-eat foods. Several by-products are dehydrated, such as fish meal, hydrolysates and protein concentrate [11].

SALTED AND DRIED FISH EXAMPLE: COD FISH

Salting and drying fish are common preservation techniques around the world. The most common dry fish products are dried cod fish without salt, and salted and dried cod fish. *Gadus morhua*, is the most frequently used specie on heavy salted fish production, however further species are used, such as Gadus macrocephalus,

Gadus ogac, saithe (Gadus virens), ling (Molva vulgaris), tusk (Brosme bromse), haddock (Melanogrammus aeglefinus) and blue ling.

Cod is considered a lean (<1% lipid) fish, storing its fat in the liver. The fat content depends on the season, the size and the age of the animal [21, 31]. The initial preparation of cod comprehends the head removal and gutting, with a completed bleeding and washing to eliminate blood clots and foreign materials. The operations of gutting, removal of head, scale (cutting the muscles and reflection in a plane spinal) and salting are done on board. Nowadays, these operations can also be made in land in fresh or in frozen/ thawed fish (in the case scaling is performed because the other operations were made before freezing). Fish salting occurs afterwards filleting or splitting. Salted fishes on board are landed and dried immediately or transferred to a cold room (0-5 °C) to wait for drying. Originally, the fragmented fish was piled along with alternating coats of salt, *i.e.* pile salted. In this dry salting, the fish can be restacked several times to obtain a better curing. Traditionally, cod fish was salted during 12 or more days at a moderate temperature (12 °C). Salting was normally done by dry salting with thick salt on scaled fish, but technological changes occurred fast in recent years, and wet salting is increasingly used, including brining and injection. These procedures promote a robust water retention within the muscles, as they influence muscle proteins and spacial structure [32]. Currently, procedures of heavy salted fishes include diverse stages, including pre-salting, dry salting, and curing in storage.

By the end of last century, manufacturers in Iceland begin to pre-salt fish by pickling and brining, previously to the formation of salting piles. Room temperature control during storage allows reducing salting period to 10 to 12 days and produces lightly cured foods.

Nowadays, the combination of several technological methods are applied, including injection, brining, pickling and dry salting/pile salting/stacking, according to manufacturers and respective countries. Brine formulation may also differ, with salt concentration ranging from 12 to 24%. In salt fish products, the polyphosphates have been used with an antioxidant effect. Consequently, the product stays white, and does not turn to yellow. Nevertheless, European food legislation does not accept the inclusion of phosphate additives in salted cod products. Meanwhile, food legislation requires for manufacturers to use approved antioxidants, to avoid salted fish darkening. Before drying, the fish has to be washed, brushed and pressed. In the case of yellow curing, cod fish is dipped in fresh water for a few hours and dried afterwards. This procedure gives the product a yellow color on the muscle surface, with strong taste. For natural drying, cod fish skin is horizontally stretched with the down. When climatic conditions are adverse, it is necessary to reap and stretch the fish several times during the drying period (ranging from several to about 60 days), resulting in high labor costs. From time to time during the process, fishes are stacked and weighed. After a few days in this position, they will be stretched as before. Artificial drying is performed in air drying tunnels, being the cod fish horizontally stretched with the skin down in trays with a bottom grid. These trays are placed in cars with spaced shelves to allow for air circulation. The drying program (with temperature and humidity control) has drying cycles followed by resting cycles during which occurs the migration of water from cod deep layers to the surface. Usually, for cod drying, 6 hours cycles of are used. To obtain a dehydration of 47%, 24 to 48 hours are necessary, depending on the fish thickness.

Subsequently to salting, food products can be categorized by size and observed aspect, packed and kept under storage. Fish curing continues throughout storage and transport, and additional transformation and curing period vary according to market and consumer's preference. In Portugal, for example, fish is desiccated and cured.

Cod is mainly commercialized with less than 47% of water content and more than 16% of salt. To be consumed, it needs to be desalted [33]. Changes in the salting technological process with curing time reduction, lead to higher water percentage and mass of salted products. These deviations result in a greater salt acquisition, the prerequisite for salt saturation of muscle liquid phase and in longer shelf-life products. Storage temperature and moisture must be precise to decelerate halophilic bacteria's growth and maintain mass increase. However, several microorganisms and enzymes keep their activity in elevated salt concentrations and low storage temperatures [34, 35]. In order to decrease the possibility of halophilic bacteria development and oxidation, the temperature of processing and curing operations must be remain lower than 8 °C . If temperature rises to 20 °C or more, halophilic bacteria, namely *Halobacterium cutirrubrum* or *Pseudomonas salinari* and *Sarcina litoralis*, may multiply producing a "red" defect or "cod rouge". Halophilic bacteria development gives rise to pink discoloration on the fillets surface while oxidation promotes yellow or brownish discoloration. To decrease mass fluctuations through storage, the equilibrium between storage relative humidity (RH) and the aw of salted fish must be maintained [36].

Before consumption, products are rehydrated (by water soaking), resulting in water intake and muscle desalting. Rehydrated products' shelf life at refrigerated conditions is shorter than that of the salted products, but can be prolonged several weeks if the products are frozen. Recent alterations in salting and curing protocols have changed salted products features, which show increased mass and better commercial value [37]. In fact, reduced curing times and inferior curing and storage temperatures result in delicate flavors and whiter aspect of the final product [37, 38]. Another important factor to final product quality is a better quality of raw material due to technical improvement of catching, handling and storage.

As heavy salted fish rehydration is slow, light salted cod fillets (only ~2% salt) were commercialized two decades ago, to answer to consumer's demands of easier and faster food preparation. Light salted fish products process and technology differs considerably from the ones of traditional "bacalhau". They are simply injected and/or brined and afterwards are freezed. Because of that, they lack the characteristic flavor and texture of heavy salted cod. The fame of heavy salted cod results from its distinctive sensorial characteristics (flavor and odor) (Fig. **7**).

These products are popular in Southern Europe, including Spain and Portugal, and in Latin America.

Quality grading is done by trained quality assessors that evaluate product's visual appearance, including color, width and presence of injuries or hemorrhages. Quality grading guidelines differ between countries (Table **1**) [21, 30, 37].

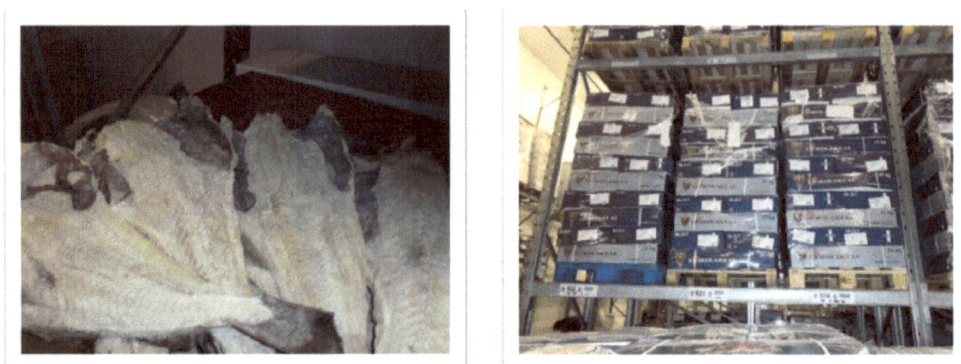

Fig. (7). Salted and dried Cod fish under storage.

Table 1. Salted codfish commercial categories of SPIG and PORT classification and quality description.

Quality Description	
SPIG Categories	**PORT Categories**
SPIG I Fillets should be thick with light appearance but no defects.	**Port A** Fillets that are light in color, thick, without blood stains; only minor gaping is allowed (gaping may appear as opening or ruptures between the myotomes because of the weakening of the connective tissue).
SPIG II Fillets are similar to SPIG I, except that minor defects are allowed.	**PORT B** Fillets that are not of grade A quality because of small defects. The color is darker and the fillets are thinner or with long gaps along the fillets.
SPIG III Fillets are allowed to have more gaping than fish in SPIG II and minor defects due to heading and filleting machines.	**PORT C** Fillets that have quality defects, like gaping or other apparent mechanical defects in the fish flesh. The color of the fillet is too dark. to be graded as B quality. Fillets that have been washed due to slight red discolouration of the fillet (caused by the growth of halophilic bacteria).
	PORT D Fillets of grade D are of lower quality than C due to stronger gaping, discolouration or other visible defects. Then C and D fillets are packaged together and acquire the grade PORT CD.
	PORT AB PORT AB may have small defects and a slightly darker color. Fillets of A and B grade can be packaged together. The maximum ratio of B fillets is 50%in each unit. **PORT E** Fillets of grade E have great defects but are yet suitable for human consumption. The parts of the fillets with visible defects can be cut away.

SMOKING

A wide range of fish and shellfish may be smoked such as herring, salmon, trout, cod, swordfish, eel and oysters. Fish can be smoked with or without head, whole or filleted.

Smoking is an ancient preservation technique and was done originally to increase fish preservation and storage time. Currently, the smoking process is used to confer a pleasant and distinctive flavor to fish. There are two different processes of smoking: cold and hot smoking. Smoking consists in the exposure of a food product to burning wood´s smoke. Hardwood shavings or sawdust are commonly used, like beech (*Fagus silvatica*), oak (*Quercus* sp.), robinia (*Robinia pseudo-acacia*), alder and birch. It is not advisable to use trees of the pine family, grasses and peatmoss for fish smoking. The moisture loss is also uniform in all areas being the total weight loss between 5-20% and mainly reliant on smoking period.

Chilled smoked fish products shelf life may reach 7 days. This shelf life can be extended to 2-3 weeks if there is a good temperature control, below 5 °C. Cold smoking is an intermittent process, since the smoking step is interrupted a few times to allow fresh air to enter into the smoker.

Hot smoking objective is to cook the fish, consequently the smoke temperature is high. It is also possible to apply hot smoking with temperatures higher than 70 °C. Hot smoked fish products are not as stable as cold smoked fish, and therefore they are not adequate for long transport. Still, to stabilize hot smoked fish products during short periods without refrigeration, they should have a sufficiently low water activity. The fish used in smoking process (salmon, swordfish, trout, herring, eel) can be fresh or frozen. Fresh fish will have to be prepared (washed, head and scales removed and gutted). In the case of salmon, trout or herring, they will have to be previously selected (calibrated according to weigh and size) and filleted by removing bones and spine. After removing pin bones, the fish muscles might be dry salted or plunged into brine or injected with brine by multi-injection machinery. The fillets will be hanged and dry and smoked into a smoking camera. The time needed to accomplish this task depends on fish species, size, fat content and type of smoking process applied - cold or hot. Fish final products' flavour and taste depend on the absorption of phenolic compounds by the fish muscles. Afterwards, the fillets will be skinned, sliced and vacuum packaged.

TECHNOLOGIES BASED ON PH CONTROL PICKLING OR MARINATING

Fish preservation using acetic acid brine is a very old procedure applied to little pelagic fishes including herring (*Clupea harengus*), sprats (*Sprattus sprattus*) and sardine (*Sardina pilchardus*) .

The term 'marinated fish' is applied to semi- preserved fish treated with salt and organic acids such as vinegar and kept in brines, sauces or oils. Marinated fish is generally a ready-to-eat product not subjected to high temperatures prior to consumption [39]. The principle of preservation is based in the use of acetic acid and salt, to avoid microorganism's multiplication and to obtain raw fish tenderization or changes in its original taste, textural and structural properties. Fish pickling may be performed cold, cooked or fried. The marinating bath constitution, the fish-liquid ratio, and the exact way of treating the product are of decisive importance for a successful final quality. During cold pickling, beheaded, gutted, washed and filleted fresh fish is immersed in a solution of salt (6-8%) and acetic acid (4-5%) or vinegar during a variable period, according to the considered fish species (lipid content, texture and size among others) from 40 h at 3 °C or 20 °C to 8 days at 18°±3 °C [39 - 44]. The proportion of fish: solution must be at minimum 1.5:1. The fish will acquire particular sensorial characteristics for texture (allowing bones removing), a white rose colour and a sweet and sour taste and flavour. Marinated fish products can be packaged in crystal jars or cans with a shelf life of 60 -120 days at 0 to 7 °C [40, 41]. For marinated herring in a first step of hard cure, the fillets are kept for three weeks in acid brine, being stable up to 6-9 months at 2-7 °C. This step is a buffer stock for posterior processing [45, 46]. In this step, fish is packed in a mild and tastier solution, with salt, acid, spices, sweeteners and aromatic herbs garnish. Different herring commercial presentations are available: Deli, pieces (skin-off), rollmops (skin-on butterfly fillets) and fillets (skin-off) [45, 47]. A third processing step could be required if the final product is to be presented in a sauce. Usually the selection of spices for flavoring is done by the producer (dark pepper, mustard), onion, dill, tomato and curry sauces are common choices. In pickled or marinated cooked fish, the raw fish material in acid brine is placed in a can and covered with a suspension of gelatin salt, acetic acid and seasoning. Thermal treatment aims to prepare a ready-to-eat product. Nowadays, a fish scalding is performed for 10-15 min. in a solution with 3% acetic acid and 6.5% salt, resulting in a product with 0.4 % acetic acid and 1.5% salt in tissues. The pH reduction, when it is higher than 4.5, is done by the solution of gelatin with salt, acetic acid and seasoning, to avoid mold growth and fat oxidation. Pickled cooked fish might have a shelf life of 6 months. Fried pickled fish preserved with salt and acetic acid does not need a curing period, being fried to become a ready-to-eat product. Fish is coated with a batter and fried in oil at 160-180 °C until it gains a superficial golden brown color, as a consequence of Maillard reaction. Frying reduces fish weight by 20-30% water loss. Afterwards, the products will be canned in acid brine. This last will diffuse in fish tissues during 2-3 days and the initial acid concentration of 2.5 % will decrease to 1.3% at the end of the process. Shelf life of these products is larger than cooked pickling fish, reaching 1 year. All marinated products are considered semi-preserved because vinegar and salt are not sufficient for killing microorganisms. Therefore, they should be stored at 0-10 °C. In addition to herring (traditionally used species), eels, hake, cod, mackerel, sea salmon, dogfish and shrimps are also processed as marinades in Europe [46]. In Spain, fresh anchovy, sprat, mussel, and cockle are usually used for marinades [48, 49]. In Great Britain, herring and usually three species of shellfish (cockles, mussels and whelks) are processed as marinades [50].

FERMENTED FISH PRODUCTS

Fermentation by microorganisms can also be used in fish products production and preservation [51 - 54]. Lactic acid bacteria (LAB) are frequently applied for biopreservation, as they are "generally recognized as safe ", being part of fish microbiota. LAB can inhibit the growth of other microorganisms due to nutrients competition and production of several antimicrobial molecules, including lactic acid, acetic acid, hydrogen peroxide and bacteriocins [55]. LAB have been isolated in some processed fish, including those with decreased salt concentration (< 6% w/w NaCl), with preservatives (sorbate, benzoate, NO_2), increased pH (> 5.0), and vacuum package and stored and distributed at low temperature (< 5 °C). These highly valued delicatessen and ready-to-eat products include fish and shellfish that are cold-smoked, pickled, marinated and brined.

The concept of biopreservation can be explored as a hurdle in these foods in order to increase shelf life and ensure safety. However, the application of such cultures in fish production is not frequent, when compared to other foods, including dairy, meat, cereal and vegetables. In the French market a patented *Lactococcus lactis* strain is available [56] for shrimp application. *Carnobacterium* spp., *Lactobacillus* spp., *Lactococcus* spp. or *Leuconostoc* spp. strains could be used in seafood, being commonly isolated at high level in such products [57]. Such strains could be applied in Asian and African countries in traditional fermented products like anchovies, fish sauces, Plaa-som [51], Enam Ne-Setaakye [52, 53], Suan yu and Myeolchi-jeot. In fact, the microbiota of Jeotkal, or Joetgal, the common designation for Korean salted and fermented seafoods [54], used as an ingredient to improve food's taste or alone as a fermented food itself, includes the following microorganisms: *Achromobacter*, *acillus*, *Brevibacterium*, *Flavobacterium*, *Halobacterium*, *Leuconostoc*, *Micrococcus*, *Pediococcus*, *Pseudomonas*, *Staphylococcus*, *Sarcina*, and the yeasts *Saccharomyces* and *Torulopsis* [58, 59]. The same is observed in Plaa-som, a Thai traditional fermented fish with sour-tasting flavor [60], formed by spontaneous fermentation primarily started by LAB present in the ingredients, tools and environment [52, 61]. According to Kopermsub & Yunchalard (2010) [62], the LAB present in the initial fermentation steps include *L. garvieae*, *S. bovis* and *W. cibaria* , followed by *W. cibaria*, *P. pentosaceus* and *L. plantarum* , which is responsible for the conclusion of the fermentation process. A combination of such LAB could constitute a starter culture for Plaa-som fermentation. In fact, the same assumption could be considered for Som fug, a highly nutritious traditional fermented fish cake from Thailand [52, 63], and for Feseekh, a traditional fermented fish from Egypt which is also very nutritive [64, 65].

TECHNOLOGIES BASED ON TEMPERATURE CONTROL
Frozen

Freezing is a technology for fish preservation based on temperature control. This procedure is not used in a large variety of processed fishery products, being more applied in the earlier stages of processing, *i.e.*, for fresh fish immediately after catch, or in industrially processed fish remaining with the fresh fish's characteristics and presented frozen to the final consumer, as in fish steaks and fillets, clean cephalopods, lamellibranches with or without valves, or constituting raw material for other industries. Almost all of the processed fishery products technologies may use frozen raw material which in some cases is thawed before being subjected to other technological processes. For example, cod allocated to salting and drying, sardines or tuna for canning or to be processed as fillets or sticks made from compact blocks of frozen fillets, which are individually cut according to the desired product without thawing. Freezing is used as the final step in processed fishery products: in some products in which the fish is raw and combined with other ingredients that modify fresh fish's characteristics, such as the shells of scallops (scallops) and mussels; stuffed and ready to cook, where the clam is shredded and mixed to other ingredients; and also in some types of fish cakes and seafood (fish cakes), obtained from grinded fish flesh or fish pulp, obtained by mechanical separation from where fish muscle mass even when it is raw shows changed color, texture and flavor, with almost total loss of the characteristics of fresh fish. With features similar to the above products, there is another group that encompasses several products whose composition include fish, but in smaller quantities than other ingredients of vegetable origin such as wheat flour, soybeans, potato starch, breadcrumbs, vegetable fats. Included in this group are the croquettes, fish or shellfish patties and cod pastries. The freezing process applied to these three fish product groups is Individual Quick Freezing (IQF) technique, which fits in the fast freezing concept. This fast freezing can occur in a continuous freezing tunnel with forced air circulation, by mechanical production of cold air, with -30 to -40 °C temperatures, or in a cryogenic tunnel by spraying nitrogen or liquid carbon dioxide, with -70 to - 150 °C temperatures. The average shelf-life of these products, in the usual commercial chain conditions, with storage, distribution and sale temperatures ranging between -18 ° and -15 °C, is of 3 to 9 months, depending on the processed fishery products' fat type and amount. Surimi-derived products is another group of products that is sold frozen and can be considered processed. These products, in most cases presented as shellfish imitations, such as "King Crab" legs and lobster tails or pieces, result from surimi extrusion and coagulation and seafood extracts, binders and colorants addition, to confer the appearance of cooked seafood. These surimi derivatives are normally marketed under the product's designation, frozen and vacuum-packed in sealed plastic films and/or thermally retracted. The storage shelf-life of this group of products may be more than one year, if kept under the temperatures indicated by the manufacturer. Another group of frozen fish-based processed products, consists of fish dishes that contain fish as a portion (fillet or piece) separated from the other ingredients, with or without a cover sauce or other type of ready-to-eat dishes in which the fish is closely linked to other ingredients with no possibility of separation, such as fish rolls (fish loaf) and soufflés. These products will be rapidly frozen, such as the referred above, and the process is usually done in equipment with mechanical production of cold, because of product's thickness the freezing time is increased and the cryogenic fluids freezing is too expensive for the product's commercial value. The product can be frozen by forced air or by contact with horizontal plates, and both systems may be continuous or static. Average temperatures attained in these equipment's can range between -30 °C and -40 °C. These products' shelf-life can reach 3 to 9 months, depending on their composition.

RESTRUCTURED FISH PRODUCTS: FISH STEAKS, STICKS AND BURGERS

Transformed fish products including steaks, sticks and burgers, are highly successfully all around the world, particularly with children and consumers that are not fans of eating fish. These products are ready-to-use but need a thermic treatment in the oven or fryer prior to consumption. The restructured fish products' processing is performed with blocks of fillets or with minced meat. Minced fish meat is used for fish fingers, fish burgers, sausages, restructured fillets and salty pastry. These products need an adequate frozen technology to be preserved. The manufacturing process starts by filleting the fish and removing the skin and the abdominal wall (skin and pin bones off fillets), made on board or on land. The fillets are arranged on trays of rigid walls and constant dimensions, which should be well-filled and then frozen in a horizontal plates' freezer. The top plate of each freezing station is based directly in the surface of the fillets blocks, compressing it. The compression of plates reinforced by the swelling of the fish caused by freezing, gives a high consistency to the blocks. The blocks are demolded and may proceed directly to the fish sticks' production lines, or what is more usual, being wrapped and packaged (fish block) to be sent to other plants, which produce the fish sticks or fingers. The fish blocks can be constituted not only by fillets. It is allowed to include a certain percentage of fish minced from the separation of uneatable portions, including bones, skin and scales, obtained by mechanical separation. This inclusion has the advantage of filling up small spaces between the fillets, originating more compact blocks. Fish fingers, sticks or burgers' process, is easily mechanized, automatic and continuous and, in general, consists of: raw material preparation, cut of raw material, forming, battering, breading, pre-frying, packaging and freezing (Fig. **8**).

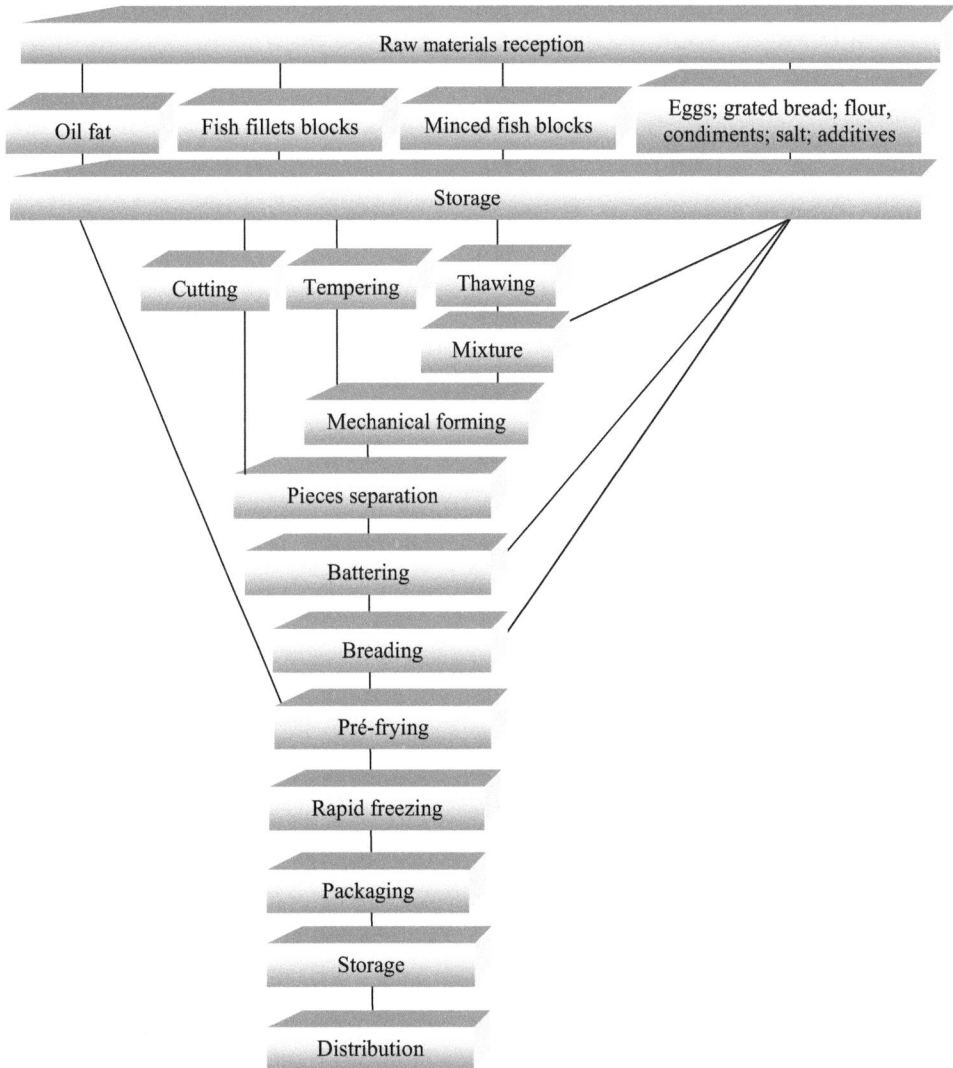

Fig. (8). Flow chart process of coated fish fingers, sticks or burgers.

The frozen fish blocks are cut in portions and then in tablets with band saws, and afterwards the tablets are cut in sticks with multidisc saws. Then, these products are uniformly coated with batter. Several batter equipment's are available in the market. Fish pieces must have all sides well-coated.

Coated fish products may be pre-fried before refreezing or solely refreezed. The fish products are submitted to a frying period of 20-40 seconds at approximately 160 to 195 °C, in order to acquire a golden brown color, taste, and a good batter coat or bread crust fixation. According to the frying oil temperature, coated fish pieces should remain the sufficient time in the frying bath to achieve a satisfying color, flavor and firm structure adherent to the fish pieces, to be preserved frozen. The rapid freezing process can be performed in a spiral freezer equipment (or continuous tunnel with mechanical or cryogenic cooling) and the product's thermal center temperature should reach -18 °C or lower temperature after thermal stabilization. During storage and distribution the product has to be kept deep frozen in order to maintain high quality.

SURIMI

Surimi represents a protein concentrate [66], mainly used in the manufacture of diverse Japanese food products, including itatsuki, kamaboko, chikuwa, hanpen and satsuma-age. The proteins that form this product are mainly myofibrillar, being responsible for its capability to form gels.

More than 60 species can be used to produce surimi, however, Alaska Pollock is 40- 55% used as raw material. Fish's freshness is the most important factor affecting surimi's quality. However, the period of capture is important, since it is related to the state of fish fatness, giving poor quality surimi when the % of humidity is high with low fat and protein content. The humidity, protein content and relationship between sarcoplasmic/myofibrillar proteins are important parameters for gel formation, influenced by fish species. The fish should pass by the preliminary steps of processing in order to be clean and without the presence of fragments such as intestines, membranes, blood coagulates, amongst others, which are hard to eliminate in the following technological steps. To guarantee the fish hygiene, it should be washed two times, firstly after head and guts removal and, secondly, previously to the placement of such products in the meat separator. Soft water use, without dissolved salts and metals, is recommended for fish washing. After obtaining minced fish, its washing is essential to eliminate detrimental components, such as connective tissues and lipids.

The washing process promotes the dilution of blood, pigments and other impurities present in the minced meat, causing product's discoloration or proteins catalyzes [66, 67]. To fish proteins isolation from minced fish, this is firstly mixed with conditioned water in a tank. The washing occurs continuously and the ratio mince:water is automatically monitored and accurately adjusted. This step promotes the leaking from the muscles of water-soluble components, particularly sarcoplasmic protein which is believed to impair fish muscle gel-forming potential. Sarcoplasmic proteins occur in the fluids inside and between muscle fibers, and include many of the fish's metabolic enzymes, which might act to break down the functional proteins of surimi, if not removed. The presence of inorganic salts is also assumed to promote surimi's denaturation during freezing.

According to Nishioka (1984) (cited by [66]), the higher number of washing steps applied to fish muscles, the stronger surimi's gel-forming capability. It is well-known that as much as 50 percent of all the water-soluble components in the fish muscle are removed after the first leaching cycle. At least three cycles of washing are performed (9-10 min. each) with a proportion of water/raw material of 3:1 or 5:1 [6]. The water should be potable and refrigerated (4-5 °C), with a neutral or slightly acidic pH in order to avoid muscle protein denaturation. The last washing is performed with a small proportion of salt (0.1-0.2%), which facilitates the posterior dehydration by centrifugation. To increase yields, eliminating several process stages at the same time, centrifugal separation might be considered as an alternative to the use of rotary screens and presses in the conventional surimi process (Fig. **9**).

In the washing and leaching processes for raw surimi production, the water soluble inorganic constituents as well as the sarcoplasmic proteins in the fish muscle, which are believed to thwart the gel forming capability of the myofibrillar protein, will be removed. The minced and leached fish meat -raw surimi- contains myofibrillar protein as its main ingredient that provides elastic properties to the product when it is processed into kamaboko (surimi-based products). Surimi enters an extruder to be formatted in blocks that will be frozen. However, when frozen, the myofibrillar protein becomes denatured and loses its gel-forming capability. If raw surimi develops a spongy texture, it will be useless as kamaboko's raw material. Consequently, surimi for cold storage is mixed with anti-denaturant additives before being frozen. A surimi that can be stored while retaining its gel-forming capability became a reality after the discovery of additives that can be mixed with raw surimi to protect the myofibrillar protein from freeze denaturation. The dewatered mince from the decanter is pumped into a blending system where it is mixed with sugars (sucrose and sorbitol, 3-5%) and polyphosphates (0.2-0.3%) to prevent freeze denaturation, allowing surimi to keep an elastic texture and water absorbing capability. The mixer has a cooling covering, which maintains the temperature as low as possible and to preserve fish's protein high quality. During mixing, the surimis' moisture content can be adjusted to of 78%. As noticed in Table **2**, fish species might affect surimi's moisture. The final gel has around 16 to 18% of protein, 78 to 80% of moisture and less than 1% fat.

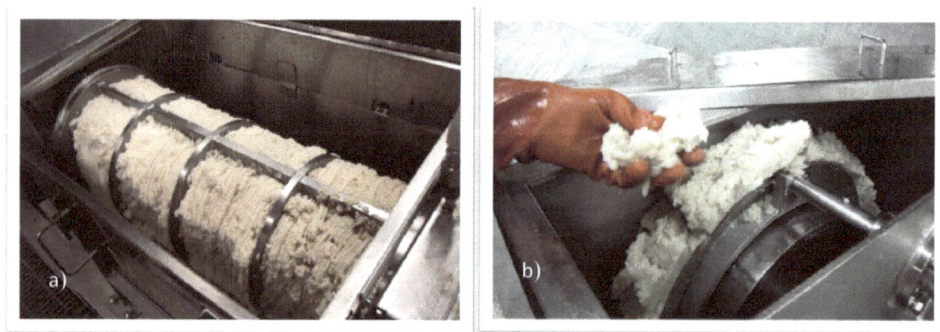

Fig. (9). Surimi dewatering step (**a**) and before adding additives (**b**).

Frozen surimi is a demanded raw material, made of fish muscle. Surimi-based products comprehend a diversity of products made of surimi and different ingredients and flavors. In Japan, kamaboko products are manufactured from frozen surimi using several cooking techniques.

Table 2. Pollack, golden threadfin-bream and hairtail surimi composition (wt%) (from Hsu & Chiang, 2001).

	Pollack surimi	Golden threadfin-bream surimi	Hairtail surimi
Protein	16.9^a	17.6^a	15.0^b
Moisture	73.7^a	75.8^b	78.0^c
Fat	0.83^a	1.05^b	1.02^c
Ash	0.70^a	0.56^b	0.87^c

[abc] Means of at least three determinations, expressed as mean. Values in same row with different letters are significantly different (P < 0.05). One way analysis of variance was conducted, and the comparison of treatment means was based on the multiple ranges test of Duncan.

Presently, frozen surimi production occurs in numerous countries, being acknowledged as an international food product. It is recognized as a healthy food and innovative surimi-based foods with different qualities and providing a wide diversity of tastes are being developed. There are various food products being made based on this raw material and with the addition of other ingredients, particularly with shellfish flavor and aroma extracts, then passing the gel by extrusion dies to obtain filaments that are coagulated under acidic conditions. These filaments, similar to muscle fibers, are aggregated under various formats and dyed, simulating lobster tails and crab legs. Defined quality features for frozen surimi is imperative to assure the efficacious production of surimi-based foods, including kamaboko and crab equivalents, in order to run into consumer's quality expectations. Methods for quality evaluation have been defined and standardized [68]. A commercial assessment of surimi comprises the evaluation of its rheological properties (punch, torsion or fold testing), pH, color, moisture content and observation of flaws (occurrence of skin or connective tissue). Other objective examinations include dynamic analysis, viscosity, and compression, usually applied for research purposes [69]. Pollock surimis' quality is classified according to the Japanese National Surimi Association that formulated a quality-ranking method, embraced by numerous traders. Surimi with the highest quality is graded as SA, followed by FA which is the second highest grade in quality, and by A or AA (Table **3**) which is the third quality grade. The lowest quality grades are graded as KA or K and as RA or B .

Table 3. Example of frozen minced fish paste (surimi) quality standard: physical and chemical standard.

Raw Material	Grade	GS test (Gcm)	Whiteness(%)	Moisture(%)	Impurities
Yellow black sea Breame	SSA	>700	>53	≤76	≤8
Yellow black sea Breame	SA	600-699	>53	≤76	≤8
Yellow black sea Breame	FA	500-599	>53	≤76	≤8
Yellow black sea Breame	AA	300-399	48-50	≤76	8-10
Lizardfishs	A	200-299	48-50	≤77	≤15
Mixed	AB	100-199	45-50	≤77	≤15
Mixed	B	<100	40-50	≤79	≤20

High quality products must be white with gel strength of 1000 g/cm^2.

Surimi's quality relies on fish species [70]. A high quality gel formation depends on ε-(γ- glutamyl)lysine (GL) crosslinks in the proteins, promoted by endogenous transglutaminases [71]. The activity of these enzymes quickly declines after catch, being nearly inactivated by freezing. The application of freezing stabilizers is mandatory for surimi's quality maintenance during long periods of frozen storage, often more than 1 year [72]. Frozen surimi's quality and shelf-life are adequately sustained during storage at temperatures inferior to -20 °C to prevent protein denaturation. The technological potential of surimi is huge, because it allows the preparation of different types of food with high nutritional value with it.

REFRIGERATION AND MODIFIED ATMOSPHERE PACKAGING

Fish is packaged both entire or in small portions, which facilitates its distribution to retailers and the ease of use to consumers. Fish and processed fishery products modified atmosphere packaging (MAP) coupled with refrigeration lead to a shelf-life extension [73]. It requires the elimination of the air and its replacement for a particular gas or gases combination. During storage, MAP atmosphere is constantly modified due to product respiration, biochemical alterations and gases penetration over package constituents.

The use of CO_2-enriched atmospheres prolongs the shelf-life of fishes by impeding the multiplication of psychotropic Gram-negative bacteria and *Pseudomonas* spp. In fact, MAP protects fish, shellfish and by-products against possible contaminants, as showed by several available studies [74 - 82].

This methodology success relies on many factors including gas composition, fish species, fat content, initial microbiota, temperature regulation, packaging characteristics and equipment efficacy. MAP still requires the adoption of good hygiene practices and temperature control during manufacturing and handling of fish and processed fish products, to avoid bacterial development and biochemical changes.

A certain number of microorganisms are denominated specific spoilage organisms (SSOs), which should be identified and evaluated in order to establish shelf-life, including *Pseudomonas*, *Shewanella* and *Photobacterium phosphoreum* [83].

Vacuum pack (VP) can also be considered as MAP since air is not replaced after being removed from the low oxygen permeability package. This system is applied for cold-smoked salmon consumed in Europe.

Despite this economic importance, quality changes limiting cold-smoked salmon shelf-life are not perceived. Under MAP, lactic microbiota led by *C. piscicola* does not probably account for cold-smoked salmon spoilage, being settled that sensory rejection are due to autolytic alterations.

The texture (soft) and flavors (sour, rancid, bitter) development at spoilage are the cause of rejection, irrespective of the total quantification of LAB and Gram-negative bacteria and shelf-life period length [84]. There are so far three types of MAP packaging formats for retail: semi-rigid trays (for meat or fish), sacks or bags of flexible films (for fresh vegetables) and retractable (for ham), and "flow pack" packaging with flexible films (for cheese and bakery products). The equipment used to produce these packages can be divided into two types: a) vacuum machines, such as chamber, sealing and thermoforming machines; b) vertical or horizontal gas flow machines [85]. The vacuum chamber machine, using prefabricated flexible film bags zipped by heat-sealing, can perform only vacuum or, when prepared for this purpose, can introduce modified atmosphere after the vacuum phase. Since it is a discontinuous process in two stages, the operation speed is lower than the prior art gas flow [86]. The single or dual chamber machine has low productivity, reaching 12 to 14 packages per minute. However, they have the advantages of producing different sizes and shapes of packs, require little space to operate, can be easily adapted to production lines and have low cost. Sealing machines are fed with a film roll being the rigid package thermally sealed, after creating vacuum or gas injection. These automated machines allow greater productivity. Thermoforming machines form the package itself from a film roll. After manual placing or automatic packaging of food, a second film is overlaid, followed by the vacuum phase and the gases mixture injection phase (if equipped to do so), completing the cycle with thermal sealing. These machines have high productivity due to the operations automation, allowing rigid and flexible films use. However, they need a larger installation space and have higher costs. Gas flow machines may be horizontal or vertical, and use rolls to form a flexible film packaging bag. The procedure removes the air in contact with the product through a gas flow under moderate pressure. After this, the film is welded. This machine allows different sized food packaging with high productivity and easy automation; however, it can only use flexible films [86]. New developments in MAP systems include active packaging, bio packaging, and bio coating by keeping food products in an atmosphere that has different composition than that of normal air [87].

HEAT TREATMENTS

For the production of safe and tasteful fish products, thermal treatments are used during manufacturing. These treatments induce several alterations, including biological, physical and chemical ones, responsible for changes at the sensory, nutritional and texture levels. Thermal processes in food technology could be defined as the controlled use of heat to modify food's reactions rate.

These treatments can be categorized by the heat treatment strength, into pasteurization (temperature range 65-80 °C); sterilization (110-120 °C); and ultra-high-temperature processing (140 - 160 °C).

These treatments affect enzymes functions, being responsible for microorganisms' inactivation.

Along with the uses of thermal processing during the production of a fish product, heat application is also used by the actual consumer, during domestic cooking, through boiling, frying, steaming, baking, stewing, roasting, using traditional and microwave ovens .

Heating is also applied in other traditional transformation procedures, including drying and smoking processes, canning and pasteurization. For some fishery products such as fish sausages, made with different condiments, it is usual to perform pasteurization. The cooking operation is also used on tuna fish before filleting for canning. Heat transfer occurs in food processing as energy transfers through conduction, convection or radiation. Radiation uses electromagnetic waves to raise a product's energy level, while conduction and convection involve energy transfer between molecules. Different cooking methods differ in their predominant heat transfer mechanism. For foods with high moisture contents, water content is considered as a treating vehicle. As water experiences a phase change, the experienced heat transfer will cook the food product. When boiling a submerged product, a phase alteration from liquid to vapor occurs. The temperature flux dramatically changes among product exterior and hot liquid, depending on the temperature difference [88]. The greater the temperature difference, the faster the heat flow rate. The general mathematical heat transfer equations for convection, conduction, and radiation are: a) $q = hA(Tm - Ts)$ (convection) b) $q = L (TmkA - Ts)$ (conduction) c) $q = \sigma\varepsilon A (Tm - Ts)$ (radiation), where q represents heat flow, h heat transfer coefficient, k thermal conductivity, L heat travel length, A transferred heat cross-sectional area, ε emissivity, σ Stefan-Boltzman constant, Tm environment temperature, and Ts object's surface temperature [89]. The combination of conduction and convection is commonly encountered as heat transfers convectively between a gas or a fluid on one surface, conductively through a solid, and sometimes convective heat transfer again on an opposing surface [90]. This combination occurs in oven baking, where temperature is transmitted by convection from the surrounding air to the exterior of the fish product, and then by conduction from the the exterior to the center of the fish product.

In this atmospheric air-drying, the capacity of water removal from the material's surface depends on drying air characteristics, including humidity, temperature and speed [91]. Simultaneously to heat transfer, mass transfer occurs during the thermal processing of foods, as materials move in fluid systems. Often this material is moisture that is evaporating from the product into the surrounding system. There is little information on the mass transfer of food components aside from water, such as salt, sugar, and flavor compounds.

The main advantage of thermal treatment is the inactivation of undesirable microorganisms and other compounds present in foods. Pasteurization is the minimal heat treatment, encompassing a time- temperature combination which kills pathogenic bacteria including *Mycobacterium tuberculosis, Salmonella* spp. and *Staphylococcus aureus*. Pasteurized fish is usually microbiological safe if packaging is performed without recontamination and stored under refrigeration.

Sterilization is a more intense heat treatment that is able to inactivate sporulated bacteria. If no recontamination occurs, these products are safe and may be stored for a very long period. Also, it inactivates anti-nutritional factors including protease inhibitors and other natural toxins are also beneficial effects of sterilization. Nevertheless, the quality of such food products could decrease in time due to biochemical modifications that might occur. While the thermal processing of foods is used to produce desirable quality changes, its primary benefit is to inactivate harmful microorganisms to an acceptably low level. Different parameters can affect the efficiency of heat treatment in pathogen inactivation, such as pH, salt, temperature and oxygen level [92]. Through numerous trials, thermal death models and inactivation kinetics are developed for specific pathogens and food materials. Target pathogen survivals can then be predicted to better optimize the safety of the food product. Bigelow and Esty (1920) [93] first used the idea of a thermal death point, defined as the length of time to totally abolish a specific amount of microorganisms at different temperatures. The thermal death time is influenced by the growth medium, microbes nature and cell numbers. When a pathogen is subjected to lethal temperature, its viable count decreases logarithmically with time. In modeling the microorganisms thermal destruction, the resistance of a specific microorganism in a specific food product is represented as a "z-value", which corresponds to the thermal death time curve slope.

Another term used in thermal inactivation is the F-value, which is the period necessary to terminate a microorganism at a specific temperature, in minutes [94].

The most common method of measuring microorganisms' inactivation is by decimal reduction time (D-value), which corresponds to the period, in minutes, necessary to attain a 90% reduction (1-log) for a particular microbe at a specific temperature in an explicit growth medium. Since it is nearly difficult to abolish all microorganisms present, a pre-determined amount of surviving organisms is deemed acceptable to reduce to for specific products. For example, processed foods must guarantee a minimum reduction of 12D for *Clostridium botulinum* [95].

FISH PRODUCTS CANNING

The canned fish (Appert) production is the most common among fish products production. The most frequently used fish species are pelagic ones such as tuna, sardine, salmon, anchovies, mackerel, horse mackerel, Ray's bream (*Brama brama*), other deep sea fishes such as cod fish and sole can be also canned. Crustaceans (shrimp, crabs, lobsters) and shell fish mussels, squid and cuttlefish might also be processed by this technology. Ice-covered or frozen fish might be used for this type of production, but should be highly fresh. If fish's autolysis is advanced, the products will have very poor quality. Fish to be canned should be prepared according to the following processing steps that are frequently mechanized: washing, standardization, flaking, eviscerating, drawing the head and tails, bleeding, removing the pimples, salting, removing the skin and fins and cutting in pieces. In small fishes, such as sardines, horse mackerel and herring, head and guts will be removed in a single operation called nobbing. In automatic nobbing, the fish is placed in molded pockets in which it is conveyed under rotating blades for head and tail removal. The fish is then eviscerated by a suction process, with a weight loss of approximately 21% during nobbing. After that, the fish is washed to eliminate exterior blood and mucus. Many types of washing apparatuses, with potable or sea water, may be used obtaining a similar quality level. To obtain a salt concentration of approximately 1-2% of fish weight, the salting operation can be done directly in the fish by plunging it in brine or adding a dry quantity of salt previously weighted into cans or to the liquid added to the can (excipient). Usually, a conveyor is used for the fish transport to a brining machine. Brining process' speed and brine's concentration can be controlled and depend on the fish fatness and size. The fish is loaded into the brining unit being immersed in brine, then are screwed through the unit and eliminated afterwards at the opposed unit termination.

In order to screen particles from the brine, the machine is equipped with a filter. This brining process step allows the fish product to attain the proper salt concentration, having also additional valuable effects, including mucus removal, exterior brightening and skin toughening. Afterwards, the product has to drain previously to packing.

Brine should not include high concentrations of magnesium chloride, which will be responsible for the development of disagreeable bitter flavors. Also, brine must be changed at regular periods, to avoid microbial contamination.

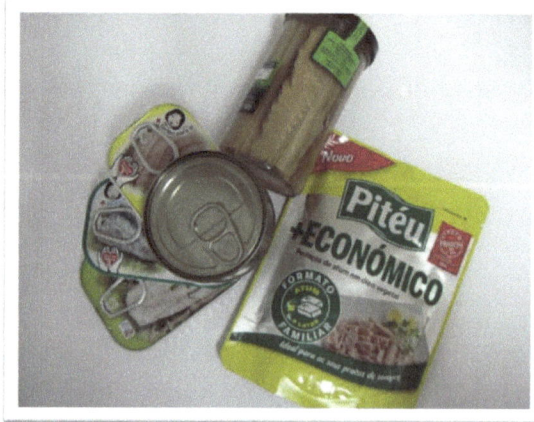

Fig. (10). Canned fish products (original).

Cans size and shape depend on market requirements, processed fish species and final product type (Fig. **10**). Metal cans are coated with an epoxy resin allowed to be in contact with foodstuff that avoids direct contact between metal and food. The canning step is very sensitive for the product's final quality, particularly in small fishes, such as sardines, and is frequently performed manually. After fish preparation, it will be packed in cans with a predefined weight. One of the most classical canned fish products is olive oil canned tuna (Fig. **11**).

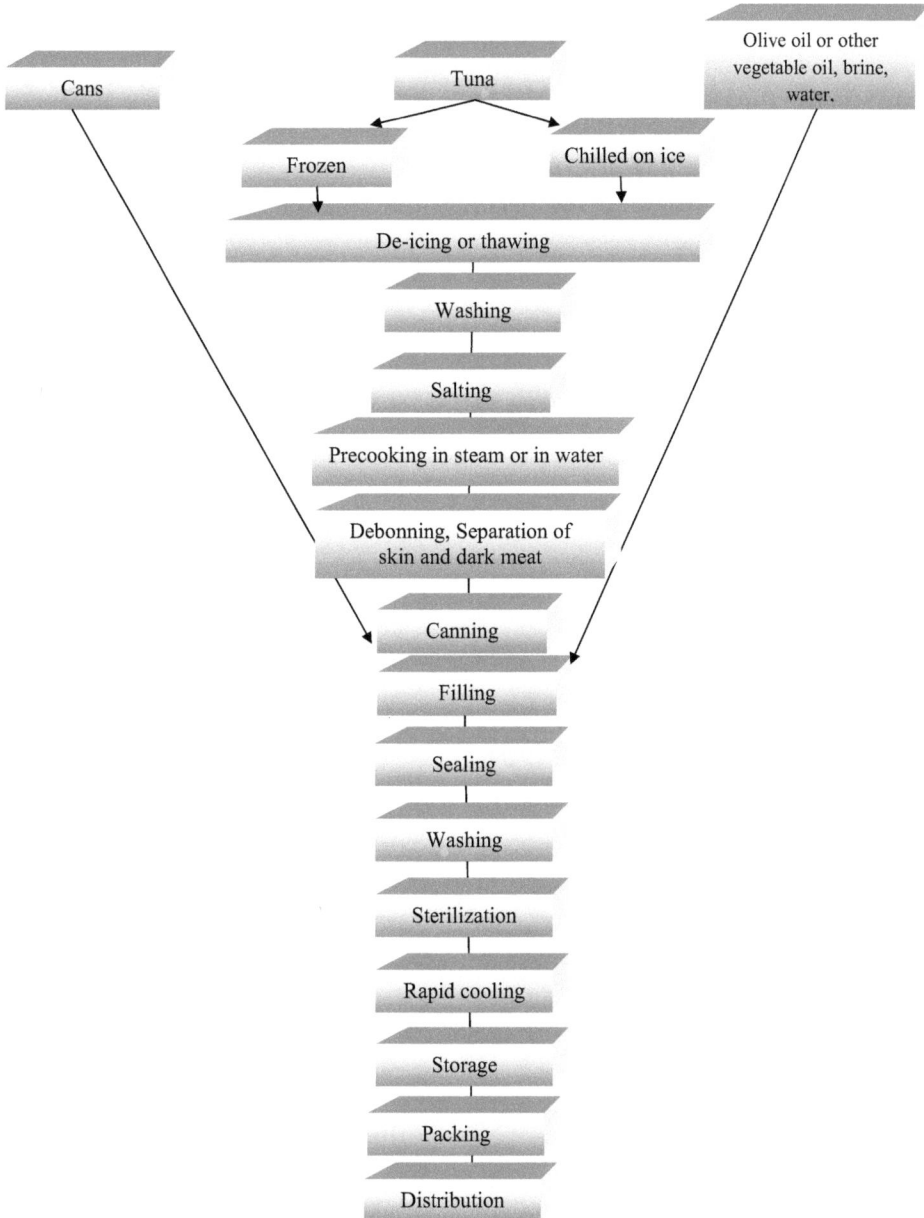

Fig. (11). Flowchart of Tuna cans production.

After preliminary operations, due to its big size, the tuna fish is cooked at 102-103 ºC in steam cameras, until its internal temperature is above 71 ºC. During cooking, tuna fish will lose 20-30% of water and proteins coagulate, changing the product's sensorial and textural characteristics. After cooling, the fish is cleaned from bones, skin, dark muscles, and filleted. Each fish has four loins of clean white muscle that will be used to produce a canned fish product of good quality (Fig. **11**). There are machines that produce tube shaped tuna loins with constant thickness that may be filled to the can and cut can-sized segments of uniform weight are performed (Fig. **12**).

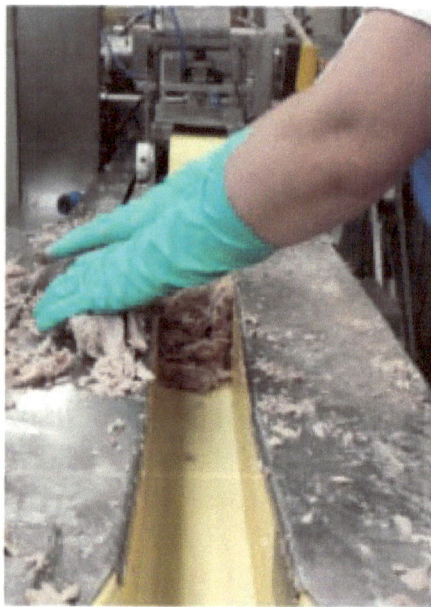

Fig. (12). Tuna loin filling for canned tuna production.

Cans Tuna Frozen Chilled on ice De-icing or thawing Washing Salting Precooking in steam Debonning, Separation of skin and dark meat Canning Filling Sealing Washing Sterilization Rapid cooling Storage Packing Distribution. In sardines, the pre-cooking operation can be performed in full cans, using automatic steam cookers at temperatures around 95 °C (Fig. **13**).

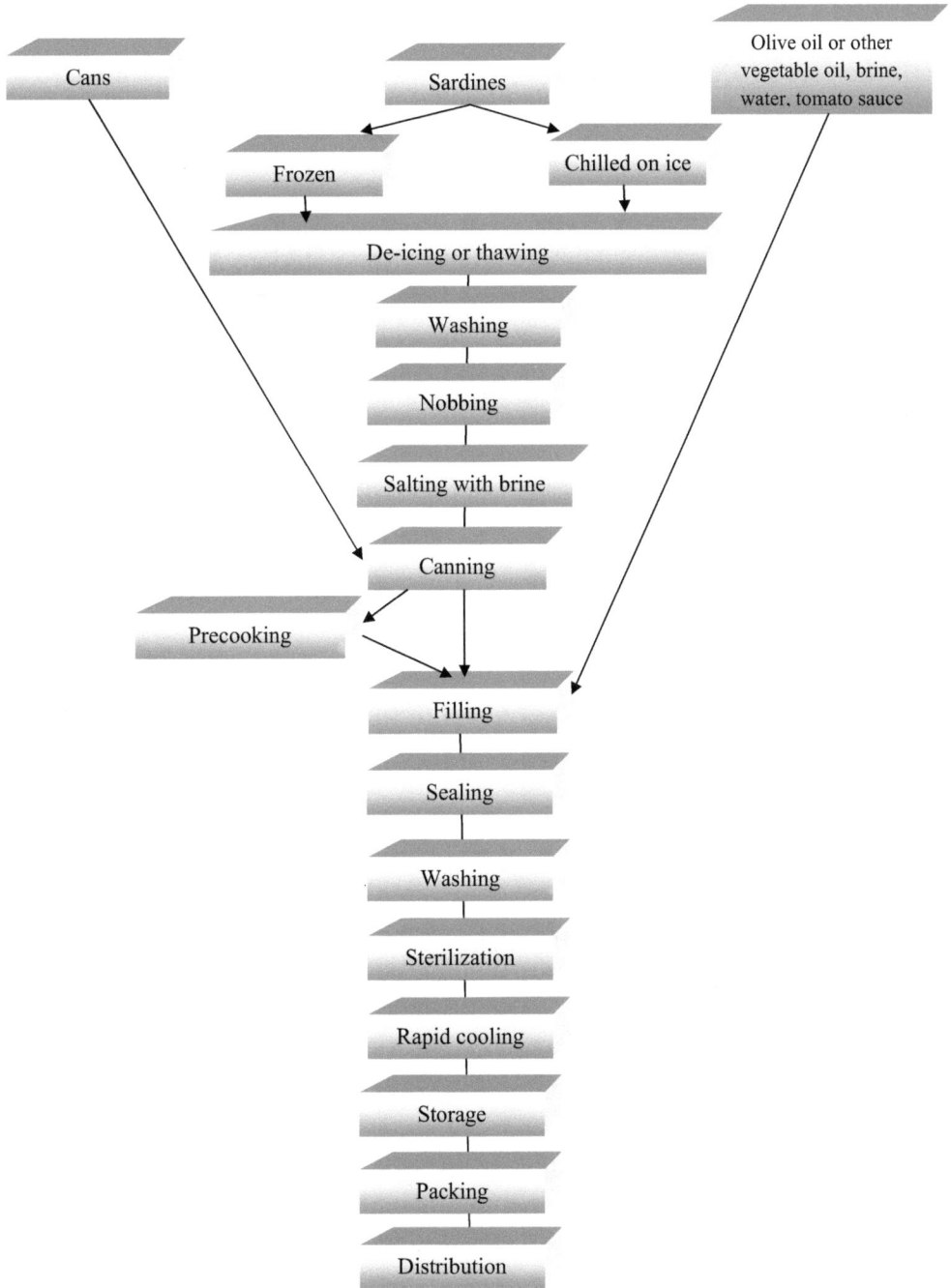

Fig. (13). Flowchart of sardine cans production.

In these steam cookers, cans are upturned on punctured conveyors, to permit for the concurrent entrance of vapor and drainage of steam and fish exudates. Nevertheless, cans' may also be steamed upright in a few pre-cookers, after which they are inverted and drained. Pre-cooking's final stage consists of a desiccating procedure at approximately 130 °C. As a more expensive process, frying can also be performed, especially in sardines.

Pre-cooked fish cans are moved to a liquid filling station. Each can is filled automatically or manually with a liquid or sauce specially selected considering the final product's characteristics (olive oil or other vegetable oil, brine, water, tomato or other kind of sauce, marinade). Afterwards, cans are moved to closing apparatuses (Fig. **14**).

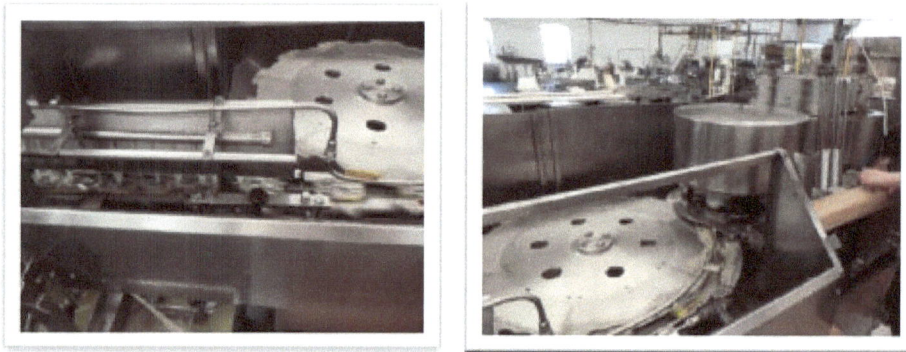

Fig. (14). Sealing step of canned fish production.

In dual seamer machines, these operations can be executed by seaming rolls. The middle and the extremity of cans are secured in a seaming clamp by a weight applied perpendicularly.

This comprehends two operation steps: a) The first operation roll, curl the layers of cover and can's body, in order that they become interlocked as a hook; b) The second operation roll compresses this hook, squeezing in order to complete a hermetic seal. During these operations, the machinery removes the air from the filled cans by vacuum, steam flow and vacuum gas sealing to avoid an increase of the can's internal pressure with edges bursting and convexing, compromising the sealing performance. The success of fish canning industry relates with its ability to hermetically seal metal, glass, plastic and/or foil containers. This is a major step, which failure impairs fish product safety and stability, especially due to post-contamination. Therefore, regular monitorization by specialized personnel is required. If the exterior of the can gets dirty with oil and fish offcuts, washing with water and detergent is required prior being sterilized.

Sterilization's aim to assure the final fish product's safety. Canned fish is a food product that is hermetically sealed in containers and submitted to a heat treatment (commercial sterilization) to destroy or inactivate the product's microbial load including spores, and that also inactivate enzymes, ensuring the product's stability at room temperature for long periods.

Sterilization is generally performed in batch autoclaves (Fig. **15**), using pure saturated steam, hot water or recirculated hot water for heating purposes.

Fig. (15). Canned fish production sterilization step.

Autoclave's time and temperatures are chosen according to the wanted texture characteristics and the process F0 value; for example, 1/4 club and dingley cans should be sterilized at around 115.6 °C for 45 to 60 min.

Melo & Ribeiro (1984) [96] studied the heat penetration in canned sardines in oil and in tomato sauce under ¼ club 30mm cans. The thermal treatment was calculated for three F0 values (3, 5 and 7) at 110 °C, 115 °C, 118 °C and 120 °C with a technological and sensorial appreciation of the final products. The heat penetration in the product is specific for each of them depending on the product itself, can type, form and dimension. All fish cans' should be safe for the consumer, without any potential disease-causing microorganisms or toxins. F0 ought to be enough to assure the absence of *Clostridium botulinum* in the fish product, and Melo & Ribeiro (1984) [96] adopted a F0 =3 equivalent to 15D, considering D the time of decimal reduction of 0.20 min. The thermal treatment of these fish products aims for safety and stability. However, the preservation of the best organoleptic characteristics. Thermal tables calculated to the same lethal value at different temperatures do not have the same effect of cooking, and consequently the product will have different organoleptic characteristics.

After refrigerating, cans are desiccated, labeled and individually packed in cardboards, placed into larger cardboards. When properly processed, canned fish products are high quality foods (Fig. **16**) with a similar nutritional value of those not canned.

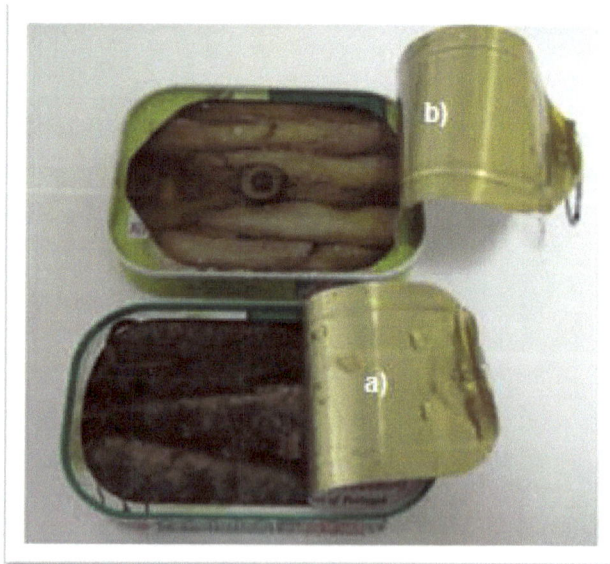

Fig. (16). Skin (**a**) and skinless (**b**) canned sardines (original).

Chemical composition and caloric value vary from one fish product to the other.

When the fish is cooked by heat and canned in its own juice, the caloric value depends on the fish muscle composition, but in the so-called "Specialties", pre-cooking relies on the sort and amount of liquid used (olive oil, vegetable oil or sauce) or on the fish type, so scalding, frying, and smoking can be used.

FISH FINE PASTE OR "PATÉ"

Fish fine paste or "pate" is another example of a pasteurized or sterilized fish product made from a raw material such as sardine, tuna, salmon or other low commercially valued fish. They might be processed from fresh fish, but it is common to use raw materials from other processes such as grinded fish flesh or fish pulp processes, fish filleting, and cutting pieces or parings from fish's smoking or fish used in canned fish products processes. The addition of vegetable oil, potato starch, tomato, salt and spices mixed with fish's raw material can be done in a cutter or in a colloid mill for emulsification. The fish paste will be contained in cans, crystal flasks or tubes and submitted to sterilization. Synthetic casings' filling might be used and a pasteurization step will be performed to finalize the product.

EMERGENT TECHNOLOGIES APPLIED ON PROCESSED FISHERY PRODUCTS HIGH HYDROSTATIC PRESSURE PROCESSING

High hydrostatic pressure (HPP) applied for fish conservation is considered a practical substitute for heat treatment, both financially and technically.

After equipment improvements, HPP use is common in industry using a pressure interval from 100 to 600 MPa, according to the desired objective. As an example, Fig. (**17**) shows available HPP apparatus for food production.

The equipment has an horizontal design being easy to install, operate and maintain. This standard configuration includes two high pressure intensifiers with the possibility of working separately for optimized reliance and uptime, an exclusive characteristic of the Hiperbaric High Pressure Processing range [97]. This process is named isostatic, since pressure is spread homogeneously and immediately, and adiabatic, meaning that regardless product form or dimension, the temperature variation is small considering increasing pressure [98, 99]. This avoids product deforming or heating that may change its organoleptic characteristics. HPP at low or moderate T destroys non-sporulated micro-organisms without changing food odor, flavor and nutritional content.

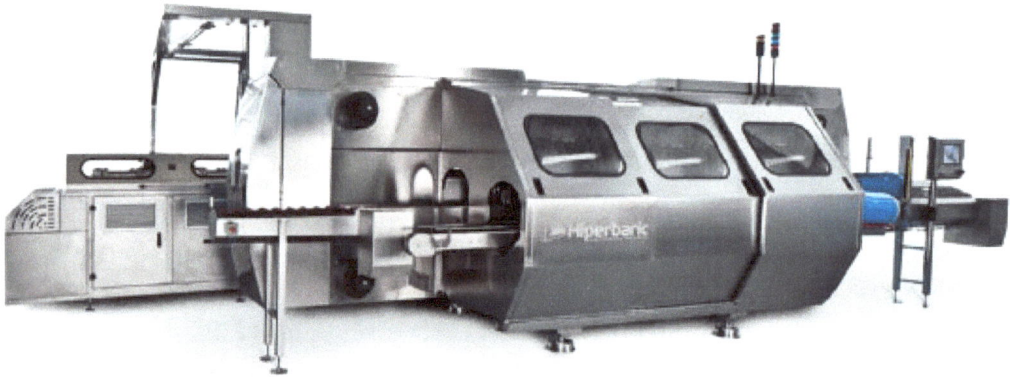

Fig. (17). High hydrostatic pressure equipment for the food industry (Hiperbaric, 2012).

High hydrostatic pressure processing (HPP) has potential use for seafood industry. Studies performed in fish products demonstrated the efficacy of this technology in prolonging shelf-life of fish products' and safety enhancement without decreasing sensorial and nutritional properties, while inactivating microorganisms and parasites [100 - 105]. However, the treatment success or microorganisms resistance are extremely variable and depend on process parameters (P, T and exposure time), on strains, microbial cell morphology and stage of growth and meat matrix [106]. HPP seems to have potential application in surimi and kamaboko's production [107, 108], cold-smoked fish [109], thermal processing [110], and for pressure- assisted freezing [111 - 113] and thawing [114]. This technology allows the inactivation of oxidative endogenous enzymes and the inhibition of lipid hydrolysis prior to products storing and processing [100, 113]. Therefore, these treatments may be responsible for the improvement of fish muscle functional and sensorial properties, allowing them to reach higher market values [115]. This technology is used for the production of ready-to-eat (RTE) fish products. These products are often cut up afterwards being processed and packed, and HPP, alone or coupled with other processes, would avoid post-processing contamination by microbial contaminants [109, 116].

HIGH POWER ULTRASOUND PROCESSING

"High Power ultrasound " (HPU) or "high intensity ultrasound " uses sonication or ultrasonic baths for the application of 20 kHz to 100 kHz frequencies with a sound intensity of 10 to1000 W.cm^{-2}, being advantageous for several processes [117]. It is used in products transformation procedures like emulsion, aggregates dispersion, drying, inactivation of microorganism and enzymes, heating improvement, separation and crystallization adjustment [118 - 120].

Ultrasounds have been tested for its efficacy in poultry and fresh produce surface decontamination, as well as biofilm removal [121, 122]. This technology has the advantage of applying lower temperatures which prevents the breakdown of thermosusceptible elements in high lipid containing foods added to the microbial inactivation [119, 123]. However, disadvantages have been reported related to food physicochemical parameters and components structure modification. Alterations in food color, antioxidant action and viscosity changes seem to the most studied modifications induced by this technology. Literature is scarce regarding HPU application in fish products, although some work has been done in frozen fish thawing. Frozen fish thawing can be a slower process, but for food safety reasons, it is recommended a quick defrosting at low temperatures. HPU application for fish thawing has been reported by Miles *et al.* (1999) [124].

As exterior overheating of frozen fish products may occur at higher intensities, or at higher and lower frequencies, it may be necessary to adjust these factors [125]. Frozen Pacific cod block was thawed rapidly (71% less time than water) without flesh quality changes. Power ultrasound is a promising ecological method, having several uses in food transformation and conservation.

ULTRAVIOLET LIGHT AND PULSE LIGHT PROCESSING

Ultraviolet light (UV) treatment is a major substitute for heat treatments with a positive image to the consumer. The electromagnetic UV radiation spectrum ranges from 100 to 400 nm and can be classified as UV-A (320 to 400 nm), UV-B (280 to 320 nm) and UV- C (200 to 280 nm) .

The latter has a lethal effect in microorganisms when radiation with 200 to 280 nm wavelength is applied [126]. The photons interaction with thymine and cystine in the DNA origins mutations, responsible for the interruption of DNA replication, translation and transcription, and consequently for cell death [127]. UV-C seems efficient for the majority of microbes, which resistance fluctuates considerably according to species or strains, being observed that Gram-negative bacteria are more susceptible than other bacteria, fungi and virus [128, 129]. UV resistance also depends on other factors, including growth phase [130].

UV is applied in the treatment of juices, meats, vegetables, water, air and food contact surfaces, for the control of microbial growth and the extension of shelf-life [131, 132]. However, it is important to refer that UV radiation penetration capability is extremely low, generally limiting industrial application to surfaces [133 - 136]. Several alternative UV sources types have been developed for food application, including continuous UV low-pressure using mercury lamps (254 nm) [131], pulsed ultraviolet light (PUV) and excimer lamp technologies .

PUV is described as a decontamination technique alternative to heat treatment. It uses Xenon lamps for the generation of short intense light intermittent pulses that are able to destroy microorganisms. Pulses wavelengths are near the infrared to ultraviolet spectra, containing UV (around 54%), visible (around 25%) and 20% infrared light (around 20%) radiation [137, 138]. The lethal effect of pulsed light in microorganisms results from photochemical and photothermal effects from a combination sterilizing effect of UV short duration and power of the flash, capability to adjust pulse extent and lamp output frequency [139].

PUV has been shown to be effective against several fish decomposition bacterial species, including *Photobacterium phosphoreum, Serratia liquefaciens, Shewanella putrefaciens, Brochothrix thermosphacta, Pseudomonas* spp., *Listeria monocytogenes, Listeria innocua* and *E. coli* O157:H7 [140, 141]. Pulsed light can also reduce spoilage and pathogenic bacteria in fish products' surface and food contact surfaces, including packaging [142, 143].

PUV is not advised for raw seafood, but it may be applied as an external treatment in cold smoked salmon according to the experiments developed by Shaw (2008)

[143] with cod and smoked salmon. Treatment side effects may include the arising of trimethylamine, staining, sudden temperature increase and humidity decrease [144], along with changes in the structure of unsaturated fatty acids, carbohydrates and proteins [138 - 148].

OSCILLATING MAGNETIC FIELDS PROCESSING

Magnetic fields processing have been evaluated as a technology for microorganisms' inactivation. These magnetic fields can be static (SMF) if their intensity is continuous with time, or oscillating (OMF) if it has a continuous or decaying amplitude sinusoidal waves corresponds to an magnetic field, which can homogeneous (with uniform intensity) or heterogeneous (if intensities decrease with the increasing distance from the center) [149].

OMF can be considered as a substitute to conservative methods, since it inactivates microorganisms present in foods, also being responsible for enzymes denaturation. When OMF is applied with preservation purposes, it is necessary to seal foods in a plastic container, after which it is submitted to several OMF pulses, which number may range from 1 to 100, representing a exposure period of 25 to 100 ms, at 0 to 50 °C. These pulses frequency vary from 5 to 500 kHz, as higher frequencies are less efficient in microorganisms' elimination and have a tendency to increase the food temperature [150].

OMV have several benefits, including allowing the sterilization of many food products enclosed in a malleable plastic pack, promoting an insignificant increase of temperature inside the product, and reducing energy necessities. These treatments can be applied to beef, milk, yogurt, juices and bread roll dough [151]. In fact, Hoffman (1985) [151] showed that a single OMF pulse was enough to reduce bacterial population in foods by 10^2-10^3 cfu/g, using magnetic fields with an intensity from 2 to 25 T and a frequency from 5 to 500 Hz.

To our knowledge no studies have been performed with OMF in fish products, and their mechanism of action is not fully understood, being also necessary to perform safety studies [152]. It is known that sensibility of bacteria to the pulsed magnetic field significantly depends on a variety of microorganisms, and that microbial inactivation depends on microorganisms' growth, magnetic field intensity and electrical resistivity [150].

PULSED ELECTRIC FIELD PROCESSING

Pulsed electric field (PEF) processing can be used for inactivation of microorganisms and food preservation as a non-thermal methodology. PEF requires the administration of high voltage pulses, ranging from 20 to 80 kV/cm to foods positioned flanked by two electrodes . This technology might be applied at ambient temperatures or similar, for 1s or less, minimizing energy dissipation as a result of food heating [153].

The exposure of living cells to electric fields influences membranes permeability, being responsible for structural alterations and membrane break by pore formation, a phenomenon known as electroporation.

These pores can be reversible or permanent, being dependent on the intensity of the electric field, the length and number of the electric pulses.

Studies have been developed to confirm PEF potential for application as a temperature-independent food conservation methodology. PEF technology can produce fresh-like liquid foods, including juice, soups, milk and yogurt and liquid eggs, presenting high quality standards, exceptional taste, nutritious qualities and increased shelf-life. As a non-thermal method, it maintains food products sensorial characteristics such as aroma, taste, and appearance.

Regarding fish muscle structure, alterations in changes in their molecular structure and consistency can result due to the permeabilisation which affects muscles' ability to retain water [154]. PEF applied to salmon demonstrated that it was susceptible to gaping. Klonowski *et al.* 2006 [155], developed work using PEF treatment technique, aiming at optimizing water holding capacity of cod, haddock and Pollock and tenderization of shellfish products such as common whelk and Iceland cyprinid, it was shown that treating the fish muscle with PEF makes the structure porous. Authors concluded that implications and properties of this structure change needs to be addressed. PEF treatment seems not to be appropriate for preserving fish as the lower field voltage distresses its molecular structure and consistency, without successfully decreasing bacteria multiplication.

IRRADIATION PROCESSING (X-RAY, ELECTRON BEAM, Y RAY)

Microbes and bugs present in food products can be destroyed by ionizing or non-ionizing irradiation [156 - 158]. Ionizing irradiation sources used in food treating comprise gamma rays produced by cobalt-60 and cesium-137 radioisotopes, electron rays (e-beam) and X-rays [159]. They act at microorganisms' DNA level, impairment the cell division mechanisms, leading to cell death.

Food and Drug administration (FDA) and World Health Organization (WHO) stated that, in the future, this technology may have a major relevance for food safety. This process controls the development of contaminant and pathogenic bacteria addressing both food quality and safety. Nowadays, in more than 60 countries, irradiation is approved for application to one food product, at least [160].

Gamma irradiation is proved to have the highest application potential, but its broad use has been hindered by unfavorable public opinion [161]. However, these processes have been endorsed by several food and health institutions worldwide, which increased consumer confidence and enhanced food industry interest [158, 162, 163].

Gamma irradiation is efficient in the inactivation of pathogenic microorganisms and in the reductions of contaminants present in foods. Consequent to dosage used in food irradiation, some modifications might occur in their constituents, including in sugars, proteins, triglycerides and vitamins, depending not only on the food components but also on the applied dose, exposition period and conditions [164]. However, it is stated that a 10 kGy dose has no toxicological threat, introducing no particular nutritional or biological alterations [165]. Badr (2012) [166] developed work with cold-smoked salmon in order to controlling the development of *Listeria monocytogenes*, *Vibrio parahaemolyticus* and of biogenic amines (BA) by gamma irradiation. Authors concluded that gamma irradiation using 3 kGy dosages may be used for the successful enhancement of cold smoked salmon safety, not promoting significant changes in the products' chemical and sensorial qualities.

In fact, irradiation technology can have a part in fish products' conservation and delivery with minimal sensorial problems. Common doses used in fish vary from 0.5 to 5 kGy [167]. Methodology mechanisms include decreasing contaminant and pathogenic microorganisms and BA levels, therefore prolonging food products shelf life, without diminishing fish quality, as demonstrated in chub mackerel (*Scomber japonicus*) and in Atlantic horse mackerel (*Trachurus trachurus*) [167 - 170]. Shelf-life extension depends on the irradiation dose, packaging process, storing temperature and fish species. In fact, irradiation of fatty fish, including flounder and sole , as to be performed without the presence of oxygen, to avoid rancidity development . Irradiation of salmon and trout affects their carotenoids and promotes fish decolorizing, limiting the application of this technology.

E-beam is produced by electricity speeding and conversion, therefore originating ionizing radiation. Using this processing technique, food is exposed to electrons which change their chemical and molecular traits, but it does not alter its temperature. Consequently, food quality does not decrease by exposition to high temperatures. Gamma radiation and e-radiation have similar antimicrobial properties; however, e-beam allows the use of higher dosages rates during a shorter period. Conversely, e-beam shows a restricted penetration profundity, contrasting gamma radiation [171]. Studies of the application of electron-beam irradiation in fish products are rather scarce. Herrero et al. (2009) [172] studied the application of this technology to cold-smoked salmon , and demonstrated that e-beam irradiation may be responsible for alterations in protein secondary structures and for the reduction in carotenoid composition. Krizek *et al.* (2011) [173] in a study performed with trout vacuum-packed stored at 3.5 °C concluded that considering biogenic amines content, dosages ranging from 0.75 to 1.0 kGy increased trout's flesh shelf-life by up to 70-98 days, respectively.

The use of irradiation with X-ray for eliminating food pathogens, including *L. monocytogenes*, has been evaluated in different seafoods including oysters [174, 175], ready-to-eat shrimp [176] and IN smoked salmon [177].

According to Farkas & Mohacsi-Farkas (2011) [157], legislation efforts should be performed regarding food irradiation methodologies, especially in the European Union, to improve public opinion. The authors suggest that additional technological advances in devices radiation fonts, particularly for e-beam and X-ray devices, may also contribute for general acceptance.

MICROWAVE PROCESSING

Microwave can be applied for heating foods by several processes, including drying, pasteurization, thawing, and baking [178]. Microwave heating results from materials capacity to absorb the energy from microwaves, and transforming them into high temperatures. When subjected to microwave processing, dipolar molecules of water or moisture present in foods become polarized, and will orient their structures according to the electric field direction. The constant inner friction of these molecules will result in the heating of the food product. However, microwave heating distribution and efficiency is affected by several factors, including food form, dimension and placing, and also by food humidity and lipid contents.

Microwave processing is extremely popular as it allows attaining extraordinary heating ratios, major decrease in cooking period, even warming, harmless usage, and simplicity of handling and low maintenance [179, 180]. Also, this technology does not alter significantly foods' taste and nourishing characteristics, as compared with cooking [181]. For temperature sensitive food materials, microwave heating is a fast procedure for moist elimination [182], and microwave drying technology is an emergent technology used for food dehydration [182, 183]. It can also be applied together with other similar technologies, including air-microwave drying [184, 185].

Microwave processing is efficient for pathogens elimination and enzyme inactivation. Although it is applied in several food processing systems, substantial studies and method optimization are required for specific improvements, related to products sensorial and nutritious qualities.

Regarding fish processing, only a few studies are available. Duan *et al.* (2011) [186] studied the application of microwave processing to fresh tilapia fish fillets.

OHMIC PROCESSING

Ohmic processing is applied in food processing for the rapid conversion of electrical energy into thermal energy, inside a conductor. This procedure is extremely efficient, as the energy is completely dissipated inside the food product without being required external energy sources or heat transfer mechanisms [187]. In fact, Ohmic processing promotes an uniform heating distribution in these products.

Existing uses for Ohmic processing comprise blanching, evaporation, dehydration, fermentation and extraction . Ohmic heating has an ability to rapidly and uniformly heat food products. This thermal-based process has as major critical factors the power of the electric field used, the application period, and the food conductivity, influencing temperature generation and heating efficacy [188].

This technology can be applied in most of foods, as long as salts and acids present allow the passage of the electric current. In the Ohmic process, the factor that controls the heating rate is the electrical conductivity, as opposed for other methods, in which it is the ingredients lowest thermal conductivity that affects the heating efficiency [189].

Ohmic heating has been experimentally applied to surimi by Pongviratchai & Park (2007) [190] and by Tadpitchayangkoon *et al.* (2012) [191]. These studies concluded the efficiency of the passage of electrical current is dependent on the products' humidity and starch content and on the used current frequency and voltage. They also showed that Ohmic processing is responsible for the inactivation of surimi's proteinases, leading to the decrease of free oligopeptides present and to the increase in moisture retaining by myosins.

NEW TRENDS IN PROCESSED FISHERY PRODUCTS

Society has been changing in the last decades. It is a fact that consumers' lifestyle have changed, coupled to the fact that there is an easy accessibility to information, health concerns and longer life expectancy. According to different locations, customs and age of human populations, the preferences to consume fish and fish products are differentiated. These seem to be factors that influence consumer's behavior regarding fish product demands. Fish processing industry knows consumers demands and works to satisfy their exigencies. Fish processing is developing new products so that consumers spend less time cooking at home: a large range of ready-to- cook and -eat fish based products with tasty recipes have been developed also with old recipes updates. Also, the concepts of ease of use and economy for families will give new approaches to fish products presentations; the concept of familiar packages was developed even for those products which were traditionally presented in a can, like tuna packaged in a flexible pouch for familiar use. Meanwhile, emergent technologies became available, thermal or non-thermal processes can be coupled with traditional ones in a hurdle technology concept and applied to fish products in order to supply them with high quality, nutrition and safety. Consumer's demands for fish products free from chemical preservatives could be the issue that will bring to industry the application of these emergent technologies. There is a big potential in biopreservation technologies, exploring and developing safer and original seafood's, with less chemical preservatives. The application of innovative conservation methodologies may include the use of protective microbiota or bacteriocins. However, their inhibitory properties in food environments should be evaluated in fish product's matrix.

The advantages of the intake of oily fish rich in polyunsaturated fatty acids, including mackerel, herring and salmon , have been extensively revised. The main constituents usually related to these beneficial effects are the omega-3 fatty acids . However, fish and related products present substantial concentrations of additional hypothetically protecting elements that should be explored regarding processed fishery products and their health allegations. According to traditional or innovative food processing techniques, sustainability and supply of fish products should meet the quality standard. The integration of production, processing and distribution systems is a trend to accomplish development and sustainability of small fish processing business but also opportunities for large companies.

CONFLICT OF INTEREST

The authors confirm that they have no conflict of interest to declare for this publication.

ACKNOWLEDGEMENTS

This work was supported through the Pluriannual Program (PEst-OE/AGR/U10276/2014), financed by Fundação para a Ciência e a Tecnologia.

REFERENCES

[1]   Leistner L, Gould GW. Hurdle technologies: combination treatments for food stability, safety, and quality. Springer 2002.
[http://dx.doi.org/10.1007/978-1-4615-0743-7]

[2]   Food and Drug Administration (FDA). Fish and Fishery Products Hazards and Controls Guidance. Fourth Edition In: US Department of Health and Human Services, Food and Drug Administration, Center for Food Safety and Applied Nutrition, Eds . 2011; p. 468.

[3]   Ababouch L. World inventory of fisheries Clean technologies Issues Fact Sheets 2005 In: Food and Agriculture Organization Fisheries and Aquaculture Department, Eds , 2005 [2013 October 27]; Available from: http://www.fao.org/fishery/topic/12381/en

[4]   Thrane M, Nielsen EH, Christensen P. Cleaner production in Danish fish processing ?" experiences, status and possible future strategies. J Clean Prod 2009; 17: 380-90.
[http://dx.doi.org/10.1016/j.jclepro.2008.08.006]

[5]   Food and Agriculture Organization. Freshwater Fish Processing and Equipment in Small Plants FAO Fisheries Circular , 1996 [2013 1 April];59. Available from: http://www.fao.org/docrep/W0495E/W0495E00.htm.

[6]   Madrid A, Vicente JM, Madrid R. El Pescado y sus Productos Derivados. AMV Ediciones 1999; p. 411.

[7]   Werner Wenzel L. Fish Nobbing Machine. US Patent 3843998 A, 1974.

[8]   Mitchell RA. Fish skinning machine. US Patent 4250594, 1981.

[9]   Bykowski PJ. Preparación de la pesca para su conservación y comercialización. In: Sikorski ZE, Ed. Tecnología de los productos del mar: recursos, composición nutritiva y conservación Editorial Acribia, SA. Zaragoza, España 1990; pp. 103-24.

[10]   Betts ED. Shrimp peeling machine and method. US Patent 4769871, 1988.

[11]   Brennan JG. Food Processing Handbook. In: Brennan JG, Ed. Wiley-VCH Verlag GmbH & Co KGaA. Weinheim 2006; p. 607.

[12]   Cabrer AI, Casales MR, Yeannes MI. Physical and chemical changes in anchovy (*Engraulis anchoita*) flesh during marination. J Aquat Food Prod Technol 2002; 3: 19-30.
[http://dx.doi.org/10.1300/J030v11n01_03]

[13]   Barat JM, Gallart-Jornet L, Andrés A, Akse L, Carlehög M, Skjerdal OT. Influence of cod freshness on the salting, drying and desalting stages. J Food Eng 2006; 73: 9-19.
[http://dx.doi.org/10.1016/j.jfoodeng.2004.12.023]

[14]   Gallart-Jornet L, Barat JM, Rustad T, Erikson U, Escriche I, Fito P. Influence of brine concentration on Atlantic salmon fillet salting. J Food Eng 2007; 80: 267-75.
[http://dx.doi.org/10.1016/j.jfoodeng.2006.05.018]

[15]   Massa AE, Yeannes MI, Manca EA. Ácidos grasos poliinsaturados de la serie Omega-3 en ejemplares bonaerenses y patagónicos de anchoita argentina Revista A&G 2007; 69: 568-72.

[16]   Czerner M, Yeannes MI. Brining kinetics of different cuts of anchovy (*Engraulis anchoita*). Int J Food

Sci Technol 2010; 45: 2001-7.
[http://dx.doi.org/10.1111/j.1365-2621.2010.02361.x]

[17] Czerner M, TomAs MC, Yeannes MI. Ripening of salted anchovy (*Engraulis anchoita*): development of lipid oxidation, colour and other sensorial characteristics. J Sci Food Agric 2011; 91(4): 609-15.
[http://dx.doi.org/10.1002/jsfa.4221] [PMID: 21302314]

[18] Bjarnason J. Handbók fiskvinnslunar – Saltfiskverkun. The Icelandic Fisheries Laboratories (Matis ohf) 1986.

[19] Akse L, Gundersen B, Lauritzen K, Ofstad R, Solberg T. Saltfisk: saltmodning, utproving av analysemetoder, misfarget saltfisk. Tromsö: Fiskeriforskning 1993.

[20] Hamm R. Biochemistry of meat hydration. Adv Food Res 1960; 10: 355-463.
[http://dx.doi.org/10.1016/S0065-2628(08)60141-X] [PMID: 13711042]

[21] þórarinsdóttir KA, Bjørkevoll I, Arason S. Production of salted fish in the Nordic countries. Variation in quality and characteristics of the salted products. Skýrsla Matís 2010; p. 46.

[22] Corominas NL. Machine for injecting fluids into meat or fish products. US Patent 8336452 B2, 2010.

[23] Filsinger BE. Effect of pressure on the salting and ripening process of anchovies (*Engraulis anchoita*). J Food Sci 1987; 52: 919-21.
[http://dx.doi.org/10.1111/j.1365-2621.1987.tb14242.x]

[24] Bellagha S, Sahli A, Farhat A, Kechaou N, Glenza A. Studies on salting and drying of sardine (*Sardinella aurita*): experimental kinetics and modelling. J Food Eng 2007; 78: 947-52.
[http://dx.doi.org/10.1016/j.jfoodeng.2005.12.008]

[25] Boudhrioua N, Djendoubi N, Bellagha S, Kechaou N. Study of moisture and salt transfers during salting of sardine fillets. J Food Eng 2009; 94: 83-9.
[http://dx.doi.org/10.1016/j.jfoodeng.2009.03.005]

[26] Corzo O, Bracho N. Effects of brine concentration and temperature on equilibrium distribution coefficients during osmotic dehydration of sardine sheets. Food Sci Technol (Campinas) 2004; 37: 475-9.

[27] Corzo O, Bracho N, Rodríguez J, González M. Predicting the moisture and salt contents of sardine sheets during vacuum pulse osmotic dehydration. J Food Eng 2007; 80: 781-90.
[http://dx.doi.org/10.1016/j.jfoodeng.2006.07.007]

[28] Nguyen MV, Arason S, Thorarinsdóttir KA, Thorkelsson G, Gudmundsdóttir A. Influence of salt concentration on the salting kinetics of cod loin (*Gadus morhua*) during brine salting. J Food Eng 2010; 100: 225-31.
[http://dx.doi.org/10.1016/j.jfoodeng.2010.04.003]

[29] Collignan A, Bohuon P, Deumier F, Poligné I. Osmotic treatment of fish and meat products. J Food Eng 2001; 49: 153-62.
[http://dx.doi.org/10.1016/S0260-8774(00)00215-6]

[30] Gallart-Jornet L, Lindkvist KB. The Spanish salt fish market - a challenge to the Norwegian salt fish industry. Tromsø,: Torskefiskkonferanse 2007.

[31] Ingólfsdóttir S, Stefánsson G, Kristbergsson K. Seasonal variations in physicochemical and textural

properties of North Atlantic cod (*Gadus morhua*) mince. J Aquat Food Prod Technol 1998; 7: 39-61.
[http://dx.doi.org/10.1300/J030v07n03_04]

[32]   Thorarinsdottir KA, Arason S, Sigurgisladottir S, Gunnlaugsson VN, Johannsdottir J, Tornberg E. The effects of salt-curing and salting procedures on the microstructure of cod (*Gadus morhua*) muscle. Food Chem 2011; 126: 109-15.
[http://dx.doi.org/10.1016/j.foodchem.2010.10.085]

[33]   Brás A, Costa R. Influence of brine-salting prior to pickle salting in the manufacturing of various salted-dried fish species. J Food Eng 2010; 100: 490-5.
[http://dx.doi.org/10.1016/j.jfoodeng.2010.04.036]

[34]   Pedro SA, Nunes ML, Bernardo FM. Pathogenic bacteria and indicators in salted cod (*Gadus morhua*) and desalted products at low and high temperatures. J Aquat Food Prod Technol 2004; 13: 39-48.
[http://dx.doi.org/10.1300/J030v13n03_04]

[35]   Rodrigues MJ, Ho P, Lopez-Caballero ME, Vaz-Pires P, Nunes ML. Characterization and identification of microflora from soaked cod and respective salted raw materials. Food Microbiol 2003; 20: 471-81.
[http://dx.doi.org/10.1016/S0740-0020(02)00086-2]

[36]   Doe PE, Hashmi R, Poulter RG, Olley J. Isohalic sorption isotherms. I. Determination for dried salted cod (*Gadus morhua*). J Food Technol 1982; 17: 125-34.
[http://dx.doi.org/10.1111/j.1365-2621.1982.tb00167.x]

[37]   Lindkvist KB, Gallart-Jornet L, Stabell MC. The restructuring of the Spanish salted fish market. Can Geogr 2008; 52: 105-20.
[http://dx.doi.org/10.1111/j.1541-0064.2008.00203.x]

[38]   Barat JM, Rodriguez-Barona S, Andres A, Fito P. Cod salting manufacturing analysis. Food Res Int 2003; 36: 447-53.
[http://dx.doi.org/10.1016/S0963-9969(02)00178-3]

[39]   Yeannes MI, Casales MR. Study of processing variables for marinated anchovies (*E. anchoita*). Alimentaria 1995; 262: 87-91.

[40]   Poligne I, Collignan A. (Quick Marination of Anchovies (*Engraulis Enchrasicolus*) using acetic and gluconic acids. Quality and Stability of the end product. Food Sci Technol (Campinas) 2000; 33: 202-9.

[41]   Gökoğlu N, Cengz E, Yerlkaya P. Determination of the shelf life of marinated sardine (*Sardina pilchardus*) stored at 4AC. Food Contr 2004; 15: 1-4.
[http://dx.doi.org/10.1016/S0956-7135(02)00149-4]

[42]   Duyar HA, Eke E. Production and Quality Determination of Marinade from Different Fish Species. J Anim Vet Adv 2009; 8: 270-5.

[43]   Fuentes A, FernAndez-Segovia I, Barat JM, Serra JA. Physicochemical characterization of some smoked and marinated fish products. J Food Process Preserv 2010; 34: 83-103.
[http://dx.doi.org/10.1111/j.1745-4549.2008.00350.x]

[44]   Capaccioni MJ, Casales MR, Yeannes MI. Acid and salt uptake during the marinatig process of *Engraulis anchoita* fillets influence of the solution:fish ratio and agitation. CiAncia e Tecnologia de

Alimentos 2011; 31: 884-90.
[http://dx.doi.org/10.1590/S0101-20612011000400009]

[45]   Ministry of Agriculture. Forestry and Fisheries of Japan A guide to the manufacture of marinated herring products MAFF Commission Technical report n° 1985; 283: 57.

[46]   Shenderyuk VI, Bykowski PJ. Salazón Y Escabechado De Pescado. In: Sikorski ZE, Ed. Tecnologí de los productos del mar: recursos, composición nutritiva y conservación. Editorial Acribia, Zaragoza: España 1994; pp. 211-9.

[47]   Norwegian Industry Standard for Fish. Processed herring products. Norwegian Industry Standard for Fish 1998.

[48]   López Benito M, Sampedro G. Preparación del Marinado de Espadín informes técnicos del Instituto de Investigaciones Pesqueras 1974; 18: 3-14.

[49]   López Benito M, Sampedro G. PreparaciA3n del marinado de mejillA3n y berberecho. Informes TA(c)cnicos del Instituto de Investigaciones Pesqueras 1975; 30: 1-20.

[50]   Mc Lay R. Torry Advisory Note 1972; 56: 3-10.

[51]   Riebroy S, Benjakul S, Visessanguan W, Kijrongrojana K, Tanaka M. Some characteristics of commercial Som-fug produced in Thailand. Food Chem 2004; 88: 527-35.
[http://dx.doi.org/10.1016/j.foodchem.2004.01.067]

[52]   Valyasevi R, Rolle RS. An overview of small-scale food fermentation technologies in developing countries with special reference to Thailand: scope for their improvement. Int J Food Microbiol 2002; 75(3): 231-9.
[http://dx.doi.org/10.1016/S0168-1605(01)00711-5] [PMID: 12036145]

[53]   Asiedu M, Sanni AI. Chemical composition and microbiological changes during spontaneous and starter culture fermentation of Enam Ne–Setaakye, a West African fermented fish-carbohydrate product. Eur Food Res Technol 2002; 215: 8-12.
[http://dx.doi.org/10.1007/s00217-002-0519-9]

[54]   Guan L, Cho KH, Lee J-H. Analysis of the cultivable bacterial community in jeotgal, a Korean salted and fermented seafood, and identification of its dominant bacteria. Food Microbiol 2011; 28(1): 101-13.
[http://dx.doi.org/10.1016/j.fm.2010.09.001] [PMID: 21056781]

[55]   Ghanbari M, Jami M, Domig KJ, Kneifel W. Seafood biopreservation by lactic acid bacteria ?" a review. Food Sci Technol (Campinas) 2013; 54: 315-24.

[56]   Daniel P, Lorre S. Lactic acid bacteria of the genus Lactococcus lactis and use thereof for preserving food products, WO2003/027268. Patent n° PCT/FR02/03180, 2001.

[57]   Françoise L. Occurrence and role of lactic acid bacteria in seafood products. Food Microbiol 2010; 27(6): 698-709. [Review].
[http://dx.doi.org/10.1016/j.fm.2010.05.016] [PMID: 20630312]

[58]   Um M-N, Lee C-H. Isolation and identification of *Staphylococcus* sp. from Korean fermented fish products. J Microbiol Biotechnol 1996; 6: 340-6.

[59]   Mah J-H, Hwang H-J. Inhibition of biogenic amine formation in a salted and fermented anchovy by

Staphylococcus xylosus as a protective culture. Food Contr 2009; 20: 796-801.
[http://dx.doi.org/10.1016/j.foodcont.2008.10.005]

[60]   Thai Industrial Standards Institute. Thai Community Products Standard 26/2546. Bangkok, Thailand: Ministry of Industry 2005.

[61]   Visessanguan W, Benjakul S, Riebroy S, Thepkasikul P. Changes in composition and functional properties of proteins and their contributions to Nham characteristics. Meat Sci 2004; 66(3): 579-88.
[http://dx.doi.org/10.1016/S0309-1740(03)00172-4] [PMID: 22060867]

[62]   Kopermsub P, Yunchalard S. Identification of lactic acid bacteria associated with the production of plaa-som, a traditional fermented fish product of Thailand. Int J Food Microbiol 2010; 138(3): 200-4.
[http://dx.doi.org/10.1016/j.ijfoodmicro.2010.01.024] [PMID: 20167386]

[63]   Saisithi P, Yongmanitchai P, Chimanage P, Wongkhalaung C, Boonyaratanakornkit M, Maleehuan S. Improvement of a Thai traditional fermented fish product: Som Fug Report to the Food and Agricultural Organisation of the United Nations 1986.

[64]   El-Sebaiy LA, Metwalli SM. Changes in some chemical characteristics and lipid composition of salted fermented Bouri fish muscle (*Mugil cephalus*). Food Chem 1989; 31: 41-50.
[http://dx.doi.org/10.1016/0308-8146(89)90149-0]

[65]   Mohamed R, Livia SS, Hassan S, Soher ES, Ahmed-Adel EB. Changes in free amino acids and biogenic amines of Egyptian salted-fermented fish (Feseekh) during ripening and storage. Food Chem 2009; 115: 635-8.
[http://dx.doi.org/10.1016/j.foodchem.2008.12.077]

[66]   Sonu SC. National Oceanic and Atmospheric Administration NOAA Technical Memorandum National Marine Fisheries Service Southwest Region Terminal Island, California; 1986; 122.

[67]   Vilhelmsson O. The state of enzyme biotechnology in the fish processing industry. Trends Food Sci Technol 1997; 8: 266-70. [Review].
[http://dx.doi.org/10.1016/S0924-2244(97)01057-1]

[68]   Koseki S, Ootake R, Katoh N, Konno K. Quality evaluation of frozen surimi by using pH stat for ATPase assay. Fish Sci 2005; 71: 388-96.
[http://dx.doi.org/10.1111/j.1444-2906.2005.00976.x]

[69]   Food and Agriculture Organization. Code of practice for fish and fish products. Codex Alimentarius Commission, World Health Organization, Food and Agriculture Organization of the United Nations. 2012.

[70]   Lanier TC, Carjaval P, Yongsawatdigul J. Surimi gelation chemistry. In: Park JW, Ed. Surimi and Surimi Seafood. Boca Raton, FL, USA: CRC Press 2005; pp. 435-90.

[71]   Lee HG, Lanier TC, Hamann DD, Knopp JA. Transglutaminase effects on low temperature gelation of fish protein sols. J Food Sci 1997; 62: 20-4.
[http://dx.doi.org/10.1111/j.1365-2621.1997.tb04359.x]

[72]   Shaviklo GR. Quality assessment of fish protein isolates using surimi standard methods. UNU - Fisheries Training Programme 2006; p. 34.

[73]   Del Nobile MA, Corbo MR, Speranza B, Sinigaglia M, Conte A, Caroprese M. Combined effect of MAP and active compounds on fresh blue fish burger. Int J Food Microbiol 2009; 135(3): 281-7.

[http://dx.doi.org/10.1016/j.ijfoodmicro.2009.07.024] [PMID: 19755204]

[74] Lopez-Caballero ME, Goncalves A, Nunes ML. Effect of CO2/O2 containing modified atmospheres on packed deepwater pink shrimp (*Parapenaus longirostris*). Eur Food Res Technol 2002; 214: 192-7.
[http://dx.doi.org/10.1007/s00217-001-0472-z]

[75] Sivertsvik M, Jeksrud WK, Rosnes JT. A review of modified atmosphere packaging of fish and fishery products ?" significance of microbial growth, activities and safety. Int J Food Sci Technol 2002; 37: 107-27.
[http://dx.doi.org/10.1046/j.1365-2621.2002.00548.x]

[76] Ozogul F, Polat A, Ozogul Y. The effects of modified atmosphere packaging and vacuum packaging on chemical, sensory and microbiological changes of sardines (*Sardina pilchardus*). Food Chem 2004; 85: 49-57.
[http://dx.doi.org/10.1016/j.foodchem.2003.05.006]

[77] Corbo MR, Altieri C, Bevilacqua A, Campaniello D, D'Amato D, Sinigaglia M. Estimating packaging atmosphere-temperature effects on the shelf life of cod fillets. Eur Food Res Technol 2005; 220: 509-13.
[http://dx.doi.org/10.1007/s00217-004-1090-3]

[78] Corbo MR, Di Giulio S, Conte A, Speranza B, Sinigaglia M, Del Nobile MA. Thymol and modified atmosphere packaging to control microbiological spoilage in packed fresh cod hamburgers. Int J Food Sci Technol 2009; 44: 1553-60.
[http://dx.doi.org/10.1111/j.1365-2621.2008.01822.x]

[79] Goulas AE, Chouliara I, Nessi E, Kontominas MG, Savvaidis IN. Microbiological, biochemical and sensory assessment of mussels (*Mytilus galloprovincialis*) stored under modified atmosphere packaging. J Appl Microbiol 2005; 98(3): 752-60.
[http://dx.doi.org/10.1111/j.1365-2672.2004.02512.x] [PMID: 15715879]

[80] Poli MB, Messini A, Parisi G, Scappini F, Figiani V. Sensory, physical, chemical and microbiological changes in European sea bass (*Dicentrarchus labrax*) fillets packed under modified atmosphere/air or prepared from whole fish stored in ice. Int J Food Sci Technol 2006; 41: 444-54.
[http://dx.doi.org/10.1111/j.1365-2621.2005.01094.x]

[81] Torrieri E, Cavella S, Villani F, Masi P. Influence of modified atmosphere packaging on the chilled shelf life of gutted farmed bass (*Dicentrarchus labrax*). J Food Eng 2006; 77: 1078-86.
[http://dx.doi.org/10.1016/j.jfoodeng.2005.08.038]

[82] Sivertsvik M. The optimized modified atmosphere for packaging of pre-rigor filleted farmed cod (Gadusmorhua) is 63ml/100 ml oxygen and 37ml/100 ml carbon dioxide. Lebenson Wiss Technol 2007; 40: 430-8.
[http://dx.doi.org/10.1016/j.lwt.2005.12.010]

[83] Alfaro B, Hernández I, Balino-Zuazo L, Barranco A. Quality changes of Atlantic horse mackerel fillets (*Trachurus trachurus*) packed in a modified atmosphere at different storage temperatures. J Sci Food Agric 2013; 93(9): 2179-87.
[http://dx.doi.org/10.1002/jsfa.6025] [PMID: 23401147]

[84] Paludan-Müller C, Dalgaard P, Huss HH, Gram L. Evaluation of the role of Carnobacterium piscicola in spoilage of vacuum- and modified-atmosphere-packed cold-smoked salmon stored at 5 degrees C.

Int J Food Microbiol 1998; 39(3): 155-66.
[http://dx.doi.org/10.1016/S0168-1605(97)00133-5] [PMID: 9553794]

[85]   Hastings MJ. MAP machinery. In: Blakistone BA, Ed. Principles and applications of modified atmosphere packaging of foods. Aspen, USA 1999; pp. 39-62.
[http://dx.doi.org/10.1007/978-1-4615-6097-5_3]

[86]   Mondry H. Packaging systems for processed meat. In: Meat quality and meat packaging; 1996; pp. 323-33.

[87]   Lee KT. Quality and safety aspects of meat products as affected by various physical manipulations of packaging materials. Meat Sci 2010; 86(1): 138-50.
[http://dx.doi.org/10.1016/j.meatsci.2010.04.035] [PMID: 20510533]

[88]   Wang T, Macgregor SJ, Anderson JG, Woolsey GA. Pulsed ultra-violet inactivation spectrum of *Escherichia coli*. Water Res 2005; 39(13): 2921-5.
[http://dx.doi.org/10.1016/j.watres.2005.04.067] [PMID: 15993922]

[89]   Jamnia A. Practical guide to the packaging of electronics: thermal and mechanical design and analysis. Boca Raton, FL, USA: Taylor & Francis Group 2009; pp. 5-8.

[90]   Toledo RT. Fundamentals of Food Process Engineering,. New York, USA: NY: Springer Science + Business Media 2007; pp. 223-84.

[91]   Lazarides HL. Dehydration System Design. In: Heldan DR, Ed. Encyclopedia of Agricultural, Food, and Biological Engineering. New York, USA: Marcel Dekker 2003; pp. 180-5.

[92]   Rosnes JT. Identifying and dealing with heat resistant bacteria. In: Richardson P, Ed. Improving the thermal processing of foods. Cambridge, England: Woodhead Publishing Limited 2004; pp. 456-70.
[http://dx.doi.org/10.1533/9781855739079.5.454]

[93]   Bigelow WD, Esty JR. The thermal death point in relation to time of typical thermophilic organisms. J Infect Dis 1920; 27: 602-17.
[http://dx.doi.org/10.1093/infdis/27.6.602]

[94]   Townsend CT, Esty JR, Bselt FC. Heat resistance studies on spores of putrefactive anaerobes in relation to determination of related processes for canned foods. Food Res 1938; 3: 323-46.
[http://dx.doi.org/10.1111/j.1365-2621.1938.tb17065.x]

[95]   Dalgaard P. Microbiology of marine muscle foods. In: Hui YH, Ed. Handbook of Food Science, Technology and Engineering. Boca Raton, FL, USA: Taylor & Francis Group 2006; pp. 1-20.

[96]   Melo RS, Ribeiro JJ. Tabelas de esterilização para conservas de sardinha em óleo e em tomate Indústria alimentar 1984; 5: 15-20.

[97]   Hiperbaric Hiperbaric range High process processing , 2013 [2013 May 1]; Available from: http://www.hiperbaric.com/en

[98]   Smelt JP. Recent advances in the microbiology of high pressure processing. Trends Food Sci Technol 1998; 9: 152-8.
[http://dx.doi.org/10.1016/S0924-2244(98)00030-2]

[99]   Wilson DR, Dabrowski L, Stringer S, Moezelaar R, Brocklehurst TF. High pressure in combination with elevated temperatures as a method for the sterilisation of food. Trends Food Sci Technol 2008;

19: 289-99.
[http://dx.doi.org/10.1016/j.tifs.2008.01.005]

[100] Murchie LW, Cruz-Romero M, Kerry JP, *et al.* High pressure processing of shellfish: A review of microbiological and other quality aspects. Innov Food Sci Emerg Technol 2005; 6: 257-70.
[http://dx.doi.org/10.1016/j.ifset.2005.04.001]

[101] Gómez-Estaca J, López-Caballero ME, Gómez-Guillén MC, López de Lacey A, Montero P. High pressure technology as a tool to obtain high quality carpaccio and carpaccio-like products from fish. Innov Food Sci Emerg Technol 2009; 10: 148-54.
[http://dx.doi.org/10.1016/j.ifset.2008.10.006]

[102] Yagiz Y, Kristinsson HG, Balaban MO, Marshall MR. Effect of high pressure treatment on the quality of rainbow trout (*Oncorhynchus mykiss*) and mahi mahi (*Coryphaena hippurus*). J Food Sci 2007; 72(9): C509-15.
[http://dx.doi.org/10.1111/j.1750-3841.2007.00560.x] [PMID: 18034712]

[103] Brutti A, Rovere P, Cavallero S, D?(tm)Amelio S, Danesi P, Arcangeli G. Inactivation of *Anisakis simplex* larvae in raw fish using high hydrostatic pressure treatments. Food Contr 2010; 21: 331-3.
[http://dx.doi.org/10.1016/j.foodcont.2009.05.013]

[104] Ye M, Huang Y, Gurtler JB, Niemira BA, Sites JE, Chen H. Effects of pre- or post-processing storage conditions on high-hydrostatic pressure inactivation of *Vibrio parahaemolyticus* and *V. vulnificus* in oysters. Int J Food Microbiol 2013; 163(2-3): 146-52.
[http://dx.doi.org/10.1016/j.ijfoodmicro.2013.02.019] [PMID: 23545264]

[105] Vasquez M, Torres JA, Gallardo JM, Saraiva J, Aubourg SP. Lipid hydrolysis and oxidation development in frozen mackerel (*Scomber scombrus*): Effect of a high hydrostatic pressure pre-treatment. Innov Food Sci Emerg Technol 2013; 18: 24-30.
[http://dx.doi.org/10.1016/j.ifset.2012.12.005]

[106] Hugas M, Garriga M, Monfort JM. New mild technologies in meat processing: high pressure as a model technology. Meat Sci 2002; 62(3): 359-71.
[http://dx.doi.org/10.1016/S0309-1740(02)00122-5] [PMID: 22061612]

[107] Uresti RM, Téllez-Luis SJ, Ramírez JA, Vázquez M. Use of dairy proteins and microbial transglutaminase to obtain low-salt fish products from filleting waste from silver carp (*Hypophthalmichthys molitrix*). Food Chem 2004; 86: 257-62.
[http://dx.doi.org/10.1016/j.foodchem.2003.09.033]

[108] Uresti RM, Velazquez G, VAzquez M, Ramrez JA, Torres JA. Effects of combining microbial transglutaminase and high pressure processing treatments on the mechanical properties of heat-induced gels prepared from arrowtooth flounder (*Atheresthes stomias*). Food Chem 2006; 94: 202-9.
[http://dx.doi.org/10.1016/j.foodchem.2004.11.005]

[109] Lakshmanan R, Parkinson JA, Piggott JR. High-pressure processing and water-holding capacity of fresh and cold-smoked salmon (*Salmo salar*). Food Sci Technol (Campinas) 2007; 40: 544-51.

[110] Ramirez R, Saraiva J, Perez Lamela C, Torres JÁ. Reaction kinetics analysis of chemical changes in pressure-assisted thermal processing. Food Eng Rev 2009; 1: 16-30.
[http://dx.doi.org/10.1007/s12393-009-9002-8]

[111] Amanatidou A, Schlüter O, Lemkau K, Gorris LG, Smid EJ, Knorr D. Effect of combined application of high pressure treatment and modified atmospheres on the shelf life of fresh Atlantic salmon. Innov Food Sci Emerg Technol 2000; 1: 87-98.
[http://dx.doi.org/10.1016/S1466-8564(00)00007-2]

[112] Alizadeh E, Chapleau N, de Lamballerie M, Le-Bail A. Effect of different freezing processes on the microstructure of Atlantic salmon (*Salmo salar*) fillets. Innov Food Sci Emerg Technol 2007; 8: 493-9.
[http://dx.doi.org/10.1016/j.ifset.2006.12.003]

[113] VAzquez M, Torres JA, Gallardo JM, Santiago JS, Aubourg P. Lipid hydrolysis and oxidation development in frozen mackerel (*Scomber scombrus*): Effect of a high hydrostatic pressure pre-treatment. Innov Food Sci Emerg Technol 2013; 18: 24-30.
[http://dx.doi.org/10.1016/j.ifset.2012.12.005]

[114] Rouille J, Lebail A, Ramaswamy HS, Leclerc L. High pressure thawing of fish and shellfish. J Food Eng 2002; 53: 83-8.
[http://dx.doi.org/10.1016/S0260-8774(01)00143-1]

[115] Aubourg SP, Torres JA, Saraiva JA, Guerra-Rodríguez E, Vázquez M. Effect of high-pressure treatments applied before freezing and frozen storage on the functional and sensory properties of Atlantic mackerel (*Scomber scombrus*). Food Sci Technol (Campinas) 2013; 53: 1-7.

[116] Picouet PA, Cofan-Carbo S, Vilaseca S, Ballbè HL, Castells P. Stability of sous-vide cooked salmon loins processed by high pressure. Innov Food Sci Emerg Technol 2011; 12: 26-31.
[http://dx.doi.org/10.1016/j.ifset.2010.12.002]

[117] Pingret D, Fabiano-Tixier A-S, Chemat F. Degradation during application of ultrasound in food processing: A review. Food Contr 2013; 31: 593-606.
[http://dx.doi.org/10.1016/j.foodcont.2012.11.039]

[118] Higaki K, Ueno S, Koyano T, Sato K. Effects of ultrasonic irradiation on crystallization behavior of tripalmitoylglycerol and cocoa butter. J Am Oil Chem Soc 2001; 78: 513-8.
[http://dx.doi.org/10.1007/s11746-001-0295-y]

[119] Chemat F, Zill-e-Huma , Khan MK. Applications of ultrasound in food technology: Processing, preservation and extraction. Ultrason Sonochem 2011; 18(4): 813-35.
[http://dx.doi.org/10.1016/j.ultsonch.2010.11.023] [PMID: 21216174]

[120] Awad TS, Moharram HA, Shaltout OE, Asker D, Youssef MM. Applications of ultrasound in analysis, processing and quality control of food: A review. Food Res Int 2012; 48: 410-27.
[http://dx.doi.org/10.1016/j.foodres.2012.05.004]

[121] Lillard HS. Decontamination of poultry skin by sonication. Food Technol 1994; 48: 72-3.

[122] Baumann AR, Martin SE, Feng H. Removal of *Listeria monocytogenes* biofilms from stainless steel by use of ultrasound and ozone. J Food Prot 2009; 72(6): 1306-9.
[PMID: 19610346]

[123] Piyasena P, Mohareb E, McKellar RC. Inactivation of microbes using ultrasound: a review. Int J Food Microbiol 2003; 87(3): 207-16.
[http://dx.doi.org/10.1016/S0168-1605(03)00075-8] [PMID: 14527793]

[124] Miles C, Morley M, Rendell M. High power ultrasonic thawing of frozen foods. J Food Eng 1999; 39: 151-9.
[http://dx.doi.org/10.1016/S0260-8774(98)00155-1]

[125] Kissam AD, Nelson RW, Ngao J, Hunter P. Water-thawing of fish using low frequency acoustics. J Food Sci 1981; 47: 71-5.
[http://dx.doi.org/10.1111/j.1365-2621.1982.tb11029.x]

[126] Sastry SK, Datta K, Worobo RW. Ultraviolet light. J Food Saf 2000; 65: 90-2.
[http://dx.doi.org/10.1111/j.1745-4565.2000.tb00623.x]

[127] Sizer CE, Balasubramaniam VM. New intervention processes for minimally processed juices. Food Technol 1999; 53: 64-7.

[128] López-Malo A, Palau E. Ultraviolet light and food preservation. In: Barbosa-Cánovas GV, Tapia MS, Cano MP, Eds. Novel food processing technologies. Madrid, Spain: CRC Press 2005; pp. 464-84.

[129] Sommer R, Lhotsky M, Haider T, Cabaj A. UV inactivation, liquid-holding recovery, and photoreactivation of *Escherichia coli* O157 and other pathogenic *Escherichia coli* strains in water. J Food Prot 2000; 63(8): 1015-20.
[PMID: 10945573]

[130] Bucheli-Witschel M, Bassin C, Egli T. UV-C inactivation in *Escherichia coli* is affected by growth conditions preceding irradiation, in particular by the specific growth rate. J Appl Microbiol 2010; 109(5): 1733-44.
[PMID: 20629801]

[131] Koutchma T. UV Light for Processing Foods. IUVA News 2008; 10: 25-9.

[132] Gayán E, Serrano MJ, Raso J, Álvarez I, Condón S. Inactivation of *Salmonella enterica* by UV-C light alone and in combination with mild temperatures. Appl Environ Microbiol 2012; 78(23): 8353-61.
[http://dx.doi.org/10.1128/AEM.02010-12] [PMID: 23001665]

[133] Shama G. Ultraviolet-irradiation apparatus for disinfecting liquids of high ultraviolet absorptivities. Lett Appl Microbiol 1992; 15: 69-72.
[http://dx.doi.org/10.1111/j.1472-765X.1992.tb00727.x]

[134] Wallner-Pendleton EA, Sumner SS, Froning GW, Stetson LE. The use of ultraviolet radiation to reduce *Salmonella* and psychrotrophic bacterial contamination on poultry carcasses. Poult Sci 1994; 73(8): 1327-33.
[http://dx.doi.org/10.3382/ps.0731327] [PMID: 7971677]

[135] Wong E, Linton R, Gerrard D. Reduction of *Escherichia coli* and *Salmonella senftenberg* on pork skin and pork muscle using ultraviolet light. Food Microbiol 1998; 15: 415-23.
[http://dx.doi.org/10.1006/fmic.1998.0185]

[136] Kim T, Silva JL, Chen TC. Effects of UV irradiation on selected pathogens in peptone water and on stainless steel and chicken meat. J Food Prot 2002; 65(7): 1142-5.
[PMID: 12117248]

[137] Oms-Oliu G. MartAOn-Belloso O, Soliva-Fortuny R. Pulsed light treatments for food preservation. A review. Food Bioprocess Technolb 2010; 3: 13-23.
[http://dx.doi.org/10.1007/s11947-008-0147-x]

[138] Shriver SK, Yang W. Thermal and nonthermal methods for allergen control. Food Eng Rev 2011; 3: 26-43.
[http://dx.doi.org/10.1007/s12393-011-9033-9]

[139] Elmnasser N, Guillou S, Leroi F, Orange N, Bakhrouf A, Federighi M. Pulsed-light system as a novel food decontamination technology: a review. Can J Microbiol 2007; 53(7): 813-21.
[http://dx.doi.org/10.1139/W07-042] [PMID: 17898836]

[140] Lasagabaster A, de Marañón IM. Sensitivity to pulsed light technology of several spoilage and pathogenic bacteria isolated from fish products. J Food Prot 2012; 75(11): 2039-44.
[http://dx.doi.org/10.4315/0362-028X.JFP-12-071] [PMID: 23127714]

[141] Ozer NP, Demirci A. Inactivation of *Escherichia coli* O157:H7 and *Listeria monocytogenes* inoculated on raw salmon fillets by pulsed UV-light treatment. Int J Food Sci Technol 2006; 41: 354-60.
[http://dx.doi.org/10.1111/j.1365-2621.2005.01071.x]

[142] Gomez-Lopez VM, Ragaert P, Debevere J, Devlieghere F. Pulsed light for food decontamination: a review. Trends Food Sci Technol 2007; 18: 464-73.
[http://dx.doi.org/10.1016/j.tifs.2007.03.010]

[143] Shaw H. Report On Pulsed Light Processing Of Seafood. Ccfra Technology Ltd Sea Fish Industry Authority, Campden and Chorleywood Food Research Association Group. 2008; p. 38.

[144] Yang W, Mwakatage NR, Goodrich-Schneider R, Krishnamurthy K, Rababah TM. Mitigation of major peanut allergens by pulsed ultraviolet light. Food Bioprocess Technol 2011.

[145] Chung SY, Yang W, Krishnamurthy K. Effects of pulsed UV-light on peanut allergens in extracts and liquid peanut butter. J Food Sci 2008; 73(5): C400-4.
[http://dx.doi.org/10.1111/j.1750-3841.2008.00784.x] [PMID: 18576985]

[146] Mwakatage NR. Efficacy of pulsed UV light treatment on removal of peanut allergens MS thesis, Department of Food and Animal Sciences, Alabama A&M University 2008.

[147] Yang W, Chung SY, Ajayi O, Krishnamurthy K, Konan K, Goodrich-Schneider R. Use of pulsed ultraviolet light to reduce the allergenic potency of soybean extracts. Int J Food Eng 2010; 6: 1-2.
[http://dx.doi.org/10.2202/1556-3758.1876]

[148] Nooji J. Reduction of wheat allergen potency by pulsed ultraviolet light, high hydrostatic pressure and nonthermal plasma. MS thesis, Department of Food Science and Human Nutrition, University of Florida 2011.

[149] Pothakamury UR, Barbosa-CAnovas GV, Swanson BG. Magnetic-field inactivation of microorganisms and generation of biological changes. Food Technol 1993; 47: 85-93.

[150] Barbosa-Cánovas GV, Schaffner DW, Pierson MD, Zhang QH. Oscillating magnetic fields. J Food Sci 1998; 65: 86-9.
[http://dx.doi.org/10.1111/j.1750-3841.2000.tb00622.x]

[151] Hofmann GA. Deactivation of microorganisms by an oscillating magnetic field. US Patent 4,524,079., 1985.

[152] Food and Drug Administration. Kinetics of Microbial Inactivation for Alternative Food Processing Technologies - Oscillating Magnetic Fields , [2013 June 1];2000 Available from:

http://www.fda.gov/Food/FoodScienceResearch/SafePracticesforFoodProcesses/ucm103131.htm.

[153] Takhistov P. Pulsed electric field in food processing and preservation. Food Sci Technol (Campinas) 2006; 1: 126-49.

[154] Gudmundsson M, Hafsteinsson H. Effect of electric field pulses on microstructure of muscle foods and roes. Trends Food Sci Technol 2001; 12: 122-8.
[http://dx.doi.org/10.1016/S0924-2244(01)00068-1]

[155] Klonowski I, Heinz V, Toepfl S, Gunnarsson G, Azorkelsson G. Applications of pulsed electric field technology for the food industry Rannsóknastofnun fiskiðnaðarins / Icelandic Fisheries Laboratories , 2006 [2013 May 1]; Available from: http://www.avs.is/media/avs/Skyrsla_06-06.pdf

[156] Farkas J. Irradiation as a method for decontaminating food. A review. Int J Food Microbiol 1998; 44(3): 189-204.
[http://dx.doi.org/10.1016/S0168-1605(98)00132-9] [PMID: 9851599]

[157] Farkas J, Mohacsi-Farkas C. History and future of food irradiation. Trends Food Sci Technol 2011; 22: 121-6.
[http://dx.doi.org/10.1016/j.tifs.2010.04.002]

[158] Lacroix M, Outtara B. Combined industrial processes with irradiation to assure innocuity and preservation of food products: a review. Food Res Int 2000; 33: 719e724.
[http://dx.doi.org/10.1016/S0963-9969(00)00085-5]

[159] Codex general standard for irradiated foods Codex STAN 2003; 106-1983. Rev. 1e2003;

[160] Blackburn C. Irradiated foods for immuno-compromised patients and other potential target groups. Food Environ Protec Newsletter 2011; 14: 4.

[161] Resurreccion AV, Galvez FC, Fletcher SM, Misra SK. Consumer attitudes toward irradiated food: results of a new study. J Food Prot 1995; 58: 193-6.

[162] DeRuiter FE, Dwyer J. Consumer acceptance of irradiated foods: dawn of a new era? Food Serv Technol 2002; 2: 47-58.
[http://dx.doi.org/10.1046/j.1471-5740.2002.00031.x]

[163] Ouattara B, Giroux M, Smoragiewicz W, Saucier L, Lacroix M. Combined effect of gamma irradiation, ascorbic acid, and edible coating on the improvement of microbial and biochemical characteristics of ground beef. J Food Prot 2002; 65(6): 981-7.
[PMID: 12092732]

[164] Kuan Y-H, Bhat R, Patras A, Karim AA. Radiation processing of food proteins: A review on the recent developments. Trends Food Sci Technol 2013; 30: 105-20.
[http://dx.doi.org/10.1016/j.tifs.2012.12.002]

[165] Diehl JF. Food irradiation: past, present and future. Radiat Phys Chem 2002; 63: 211-5.
[http://dx.doi.org/10.1016/S0969-806X(01)00622-3]

[166] Badr HM. Control of the potential health hazards of smoked fish by gamma irradiation. Int J Food Microbiol 2012; 154(3): 177-86.
[http://dx.doi.org/10.1016/j.ijfoodmicro.2011.12.037] [PMID: 22285535]

[167] Ozogul F, Ozden O, Ozogul Y, Erkan N. The effects of gamma-irradiation on the nucleotide

degradation compounds in sea bass (*Dicentrarchus labrax*) stored in ice. Food Chem 2010; 122: 789-94.
[http://dx.doi.org/10.1016/j.foodchem.2010.03.054]

[168] Mbarki R, Ben Miloud N, Selmi S, Dhib S, Sadok S. Effect of vacuum packaging and low-dose irradiation on the microbial, chemical and sensory characteristics of chub mackerel (*Scomber japonicus*). Food Microbiol 2009; 26(8): 821-6.
[http://dx.doi.org/10.1016/j.fm.2009.05.008] [PMID: 19835766]

[169] Park SY, Kim B-Y, Song H-H, Ha S-D. The synergistic effects of combined NaOCl, gamma irradiation and vitamin B1 on populations of *Aeromonas hydrophila* in squid. Food Contr 2012; 27: 194-9.
[http://dx.doi.org/10.1016/j.foodcont.2012.03.015]

[170] Mendes R, Silva HA, Nunes ML, Empis JM. Effect of low-dose irradiation and refrigeration on the microflora, sensory characteristics and biogenic amines of Atlantic horse mackerel (*Trachurus trachurus*). Eur Food Res Technol 2005; 221: 329-35.
[http://dx.doi.org/10.1007/s00217-005-1172-x]

[171] Tahergorabi R, Matak KE, Jaczynski J. Application of electron beam to inactivate *Salmonella* in food: Recent developments. Food Res Int 2012; 45: 685-94.
[http://dx.doi.org/10.1016/j.foodres.2011.02.003]

[172] Herrero AM, Carmona P, Ordonez JA, de la Hoz L, Cambero MI. Raman spectroscopic study of electron beam irradiated cold-smoked salmon. Food Res Int 2009; 42: 216-20.
[http://dx.doi.org/10.1016/j.foodres.2008.10.010]

[173] Krizek M, MatejkovA K, Vacha F, Dadakova E. Effect of low-dose irradiation on biogenic amines formation in vacuum-packed trout flesh (*Oncorhynchus mykiss*). Food Chem 2012; 132(1): 367-72.
[http://dx.doi.org/10.1016/j.foodchem.2011.10.094] [PMID: 26434303]

[174] Mahmoud BS, Burrage DD. Inactivation of *Vibrio parahaemolyticus* in pure culture, whole live and half shell oysters (*Crassostrea virginica*) by X-ray. Lett Appl Microbiol 2009; 48(5): 572-8.
[http://dx.doi.org/10.1111/j.1472-765X.2009.02573.x] [PMID: 19291215]

[175] Mahmoud BS. Reduction of Vibrio vulnificus in pure culture, half shell and whole shell oysters (Crassostrea virginica) by X-ray. Int J Food Microbiol 2009; 130(2): 135-9.
[http://dx.doi.org/10.1016/j.ijfoodmicro.2009.01.023] [PMID: 19217681]

[176] Mahmoud BS. Effect of X-ray treatments on inoculated *Escherichia coli* O157: H7, *Salmonella enterica, Shigella flexneri* and *Vibrio parahaemolyticus* in ready-to-eat shrimp. Food Microbiol 2009; 26(8): 860-4.
[http://dx.doi.org/10.1016/j.fm.2009.05.013] [PMID: 19835772]

[177] Mahmoud BS. Control of Listeria monocytogenes and spoilage bacteria on smoked salmon during storage at 5 AC after X-ray irradiation. Food Microbiol 2012; 32(2): 317-20.
[http://dx.doi.org/10.1016/j.fm.2012.07.007] [PMID: 22986195]

[178] Chandrasekaran S, Ramanathan S, Basak T. Microwave food processing - A review. Food Res Int 2013; 52: 243-61.
[http://dx.doi.org/10.1016/j.foodres.2013.02.033]

[179] Salazar-Gonzalez C, San Martin-Gonzalez MF, Lopez-Malo A, Sosa-Morales ME. Recent studies related to microwave processing of fluid foods. Food Bioprocess Technol 2012; 5: 31-46.
[http://dx.doi.org/10.1007/s11947-011-0639-y]

[180] Zhang J, Davis TA, Matthews MA, Drews MJ, LaBerge M, An YH. Sterilization using high-pressure carbon dioxide. J Supercrit Fluids 2006; 38: 354-72.
[http://dx.doi.org/10.1016/j.supflu.2005.05.005]

[181] Vadivambal R, Jayas DS. Non-uniform temperature distribution during microwave heating of food materials ?" A review. Food Bioprocess Technol 2010; 3: 161-71.
[http://dx.doi.org/10.1007/s11947-008-0136-0]

[182] McLoughlin CM, McMinn WA, Magee TR. Microwave vacuum drying of pharmaceutical powders. Drying Technol 2003; 21: 1719-33.
[http://dx.doi.org/10.1081/DRT-120025505]

[183] Orsat V, Yang W, Changrue V, Raghavan GS. Microwave-assisted drying of biomaterials. Food Bioprod Process 2007; 85: 255-63.
[http://dx.doi.org/10.1205/fbp07019]

[184] Datta AK, Ni H. Infrared and hot-air-assisted microwave heating of foods for control of surface moisture. J Food Eng 2002; 51: 355-64.
[http://dx.doi.org/10.1016/S0260-8774(01)00079-6]

[185] Duan ZH, Zhang M, Hu QG. Characteristics of microwave drying of bighead carp. Dry Technol 2005; 23: 637-43.
[http://dx.doi.org/10.1081/DRT-200054156]

[186] Duan ZH, Jiang L-N, Wang J-L, Yu X-Y, Wang T. Drying and quality characteristics of tilapia fish fillets dried with hot air-microwave heating. Food Bioprod Process 2011; 89: 472-6.
[http://dx.doi.org/10.1016/j.fbp.2010.11.005]

[187] Salengke S. Electrothermal effects of ohmic heating on biomaterials: Temperature monitoring, heating of solid-liquid mixtures, and pretreatment effects on drying rate and oil uptake. PhD Dissertation, The Ohio State University 2000.

[188] Bengston R, Birdsall E, Feilden S, Bhattiprolu S, Bhale S, Lima M. Hui YH. Ohmic and induction heating. Handbook of Food Science, Technology and Engineering. FL, USA: CRC Press, Taylor and Francis Group 2006; p. 120.

[189] Halden K, de Alwis AA, Fryer PJ. Changes in the electrical conductivity of foods during ohmic heating. Int J Food Sci Technol 1990; 25: 9-25.
[http://dx.doi.org/10.1111/j.1365-2621.1990.tb01055.x]

[190] Pongviratchai P, Park JW. Electrical conductivity and physical properties of surimi-potato starch under ohmic heating. J Food Sci 2007; 72(9): E503-7.
[http://dx.doi.org/10.1111/j.1750-3841.2007.00524.x] [PMID: 18034719]

[191] Tadpitchayangkoon P, Park JW, Yongsawatdigulet J. Gelation characteristics of tropical surimi under water bath and ohmic heating. Food Sci Technol (Campinas) 2012; 46: 97-103.

SUBJECT INDEX

www.ingramcontent.com/pod-product-compliance
Lightning Source LLC
Chambersburg PA
CBHW050809220326
41598CB00006B/157